MONTGOMERY COLLEGE
ROCKVILLE CAMPUS LIB
ROCKVILLE, MARYLAND

P9-AOW-137

Network Security Illustrated

Jason Albanese
Wes Sonnenreich

McGraw-Hill

New York   Chicago   San Francisco   Lisbon
London   Madrid   Mexico City   Milan   New Delhi
San Juan   Seoul   Singapore   Sydney   Toronto

SEP 02 2004   292759

The McGraw·Hill Companies

Library of Congress Cataloging-in-Publication Data

Sonnenreich, Wes
    Network security illustrated / Wes Sonnenreich, Jason Albanese.
      p. cm.
    Includes index.
    ISBN 0-07-141504-1
      1. Computer networks—Security measures. I. Albanese, Jason. II. Title.

TK5105.5.S635 2003
005.8—dc22

                         2003059292

Copyright © 2004 by Jason Albanese and Wes Sonnenreich. All rights reserved. Printed in the United States of America. Except as permitted under the United States Copyright Act of 1976, no part of this publication may be reproduced or distributed in any form or by any means, or stored in a data base or retrieval system, without the prior written permission of the publisher.

1 2 3 4 5 6 7 8 9 0   DOC/DOC   0 9 8 7 6 5 4 3

ISBN 0-07-141504-1

The sponsoring editor for this book was Judy Bass and the production supervisor was Sherri Souffrance. It was set in ITC Century Light by MacAllister Publishing Services, LLC.

Printed and bound by RR Donnelley.

McGraw-Hill books are available at special quantity discounts to use as premiums and sales promotions, or foruse in corporate training programs. For more information, please write to the Director of Special Sales, McGraw-Hill Professional, Two Penn Plaza, New York, NY 10121-2298. Or contact your local bookstore.

 This book is printed on recycled, acid-free paper containing a minimum of 50 percent recycled de-inked fiber.

Information contained in this work has been obtained by The McGraw-Hill Companies, Inc. ("McGraw-Hill") from sources believed to be reliable. However, neither McGraw-Hill nor its authors guarantees the accuracy or completeness of any information published herein and neither McGraw-Hill nor its authors shall be responsible for any errors, omissions, or damages arising out of use of this information. This work is published with the understanding that McGraw-Hill and its authors are supplying information but are not attempting to render engineering or other professional services. If such services are required, the assistance of an appropriate professional should be sought.

Dedication

I would like to dedicate this book to my incredible wife Emily, who provided me with the strength and courage to complete this project and to the memory of my grandfather, Irving Monchik, whose spirit and intellect inspires me each and every day.

Jason Albanese

I dedicate this book to my parents, who might actually have been right once or twice, well… maybe just once. Their attention, devotion and unending love have made them the best parents one could hope for, despite Mom's paranoia about whether I've eaten enough or Dad's endless supply of really bad jokes (like the one about the koala and the prostitute).

Wes Sonnenreich

Contents

Introduction

This book is both a reference guide and something that can be read end-to-end. It has been designed to provide decision makers with essential knowledge about information security concepts and technologies. After reading this book, will you be able to run out and implement a specific solution based on instructions contained within these pages? No. Will you know where any one technology fits within the framework of securing your business? Definitely. Will you feel comfortable discussing the technology in a business capacity? Absolutely.

Here is some advice to help you take full advantage of the experience:

Use the Website: Designed as a companion to the book, the website has loads of additional information. The structure of the site mirrors that of the book. There are additional chapters posted on the site that did not make it into the book. For each chapter, in the book and on the site, you'll find links to additional resources including security articles, whitepapers and websites. You'll also find corrections and clarifications. Finally, topic specific forums allow readers and community members to discuss security concepts, best practices and issues in an intuitive, organized manner. See http://www.sagesecure.com/nsi.

Use the Map: The book and the map are designed to be symbiotic. When used together, the reader can gain powerful insight about security technologies in a relatively short period of time. The map contains a large number of icons that have been strategically placed within a simulated business environment. All of the icons refer to technologies covered in the book. Corresponding chapters within the book are labeled with these icons. For example, on the fold out map, a tank icon is used to represent a virtual private network. The tank is located on the title page for the chapter on virtual private networks, and in the upper right margin of subsequent pages. This makes it easy to reference information about concepts on the map by just flipping through the book and looking for the appropriate icon.

Make the Connections: Many topics covered in this book have strong connections with one another. While reading a chapter, be sure to check out the "Making the Connection" section. The references provided will lead directly to concepts that surround the current chapter and give you tremendous insight to related technologies.

Read the Rest of this Introduction: We know; you want to jump right in and get your security-groove on. Resist the temptation for a few more minutes and finish up the introduction. We promise not to let it drag on and on (we ramble often enough later on in the book). You will get a lot more out of the book by understanding how content is organized and why we bothered to write this book in the first place.

Drop us a line: We may not be able to answer every question, but we'll try. Frequent questions and their answers will get posted to the website, so check there first. Our address: nsibook@sagesecure.com.

How This Book is Organized

Security concepts are organized based on business needs, as opposed to technological similarity. We've tried to focus on how these concepts relate in terms of practical business functionality. For example, network monitoring is discussed in Part 1, *"Managing Security"* rather than in a section on intrusion detection. For people with a technical background, this method of organization may seem strange. But one of our goals is to change the way people think about security. As we'll say many times throughout the book, security is not a technological issue; it's a business issue.

As an end-to-end experience, we've organized the chapters into parts based on a managerial view of security. We can best explain our view with an analogy to building and securing a house:

- A house sits on a parcel of land. Securing the land is the first and most critical step to securing the house. An alarm system won't help if the house is demolished by a natural disaster, or if it's located in such a bad neighborhood that the police barely take the time to respond! These are examples of management level issues, and they are dealt with in the first few chapters of the book.
- The foundation of the house represents the network design. A network built with security in mind makes every other aspect of security much easier. A poorly built network can collapse on itself, and is very hard to secure after-the-fact. The house itself represents the information that moves around the network. That's why the bulk of the book deals with information security.
- The security systems on the house are the finishing touches. Alarm systems, automatic lights, and insurance all contribute in the event that everything else falls apart. This is equivalent to ensuring availability and intrusion detection, found at the end of the book.

If it were possible to talk about information security in a straightforward manner, there'd be no need for the map in the back of this book. Actually, there'd be little need for the book, since linear concepts are often easy to understand. Alas, information security is a like a bowl of pasta. Twirl one strand and next thing you know you've got half the bowl wrapped around your fork. You can't talk about one security topic without talking about three or four others... and talking about those means talking about three or four more... you get the picture.

Challenging as it was, we tried to make each chapter stand on its own without relying on the knowledge found in other chapters. As a result, a technology or concept is not mentioned in passing unless it has been previously given a clear explanation. That said, understanding the surrounding concepts always helps, which is why each chapter contains a section that links to related chapters ("Making the Connections"). In a similar vein, each part also starts with a "Connecting the Chapters" section that shows how the part's chapters interrelate.

How the Chapters are Organized

Each part of the book has a title that describes a business-level need. Examples of part titles are "Managing Security," "Accessing Information," and "Storing Information." The chapters within each part discuss technology concepts related to the business need.

Every part begins with a quick reference page. On the page is a "Summary" that describes the business need and how security fits into the picture. The page also highlights some "Key Points" made throughout the part's introduction and shows how the part's chapters interrelate ("Connecting the Chapters"). After the reference page, the introduction explores general security issues faced when servicing the business need.

Within the chapters, we've tried to organize things consistently. The following six sections can be found in almost every chapter, in this order:

Technology Overview: This covers the basics—what the technology or concept is all about and how it functions in a business environment.

How it Works: Without getting too technical, we try to describe the way in which the technology or concept works in practice.

Security Considerations: This is where we talk about the security problems caused by the technology or concept. In chapters that describe networking topics, we focus on security issues inherent in whatever's being covered. In security topic chapters, we look at the limitations of the given security technology/concept and how they can be overcome.

Making the Connection: Here we tie concepts to other chapters in the book. In general, reading the connected chapters will improve your overall understanding of any particular security topic. In a few cases, making the connection is critical to completely understanding the chapter at hand.

Best Practices: We've collected some tips and suggestions based on our experience and the experience of others in the security field. These are techniques that can improve the effectiveness of a security technology or prevent failures.

Final Thoughts: This is where we summarize key issues or mention anything that didn't fit in one of the other sections. If we've got nothing else to say, we might just blab for a few paragraphs to fill up space.

Why We Wrote this Book

This book was written to provide a general business audience with the knowledge they will need to properly integrate security into their company. The concept is based on our vision that in the years to come, business will no longer be able to afford to be reactive about security. We firmly believe that information security will become a fundamental part of all business infrastructures. Organizations of all shapes and sizes will reorganize, plan and spend a lot of money to properly protect and defend the core of their business: information.

We don't believe there's anything like this book in the realm of information security. What does exist tends to fall into a few basic categories:

Trade Media: There are hundreds of magazines and journals that rant and rave over the latest in network and security technologies. These sources are a great way to stay informed. However, many of these articles skirt the line between paid advertisements and devout worship. It's very difficult to get an honest picture of a particular technology from these sources alone.

Books for "Simple" Needs: These books are designed to give people who lack technical backgrounds an understanding of isolated security concepts. They can often provide the average user with simple solutions for their needs, but won't provide managers with enough information to feel confident about their choices.

Hacker Books: On the other end of the spectrum are security books for system administrators and hackers. Frequently written by an infamous hacker or security expert, these titles focus on specific "hands-on" security for Unix and Windows machines. They also discuss methods in which to break into these machines. These books are usually full of riveting inside jokes like:

```
% \(-
(-: Command not found.
```

Technical Documentation: Concerned about wireless security? Why not just read the original specifications for your wireless system and analyze it yourself? Or, grab a whitepaper and a cup of coffee and solve your dataflow problems. This includes the many excellent books on particular technologies, such as TCP/IP Illustrated (a book that we've read cover to cover many times).

After years of looking closely at these options we realized something was missing: a comprehensive reference guide written for intelligent business people. This is a book that provides the reader enough information in a few pages to make business-level decisions. A compilation that relates security concepts and technologies based on the way they're used in real life—not based on technological similarities or ideals. In other words, a practical guide to information security.

So here it is... we hope you find it useful. If you like it (or don't) please let us know how we can make the book better by sending feedback to: nsibook@sagesecure.com

Jay Albanese
Wes Sonnenreich

Acknowledgments

This project began as a dream of ours, to write a book that clearly bridges the gap between the worlds of technology and business. We don't know if we've actually done that, but we couldn't have even tried without the help of many individuals. Thank you to everyone who gave his or her time, thoughts, support and encouragement during the long and grueling process of making our dream a reality.

Neil Burstein for representing us, guiding us, reviewing our material and becoming part of our team.

Robert Altman for encouraging our idea and providing invaluable feedback that gave us the confidence to put the initial proposal together.

Robert O'Brien for reviewing our work and feeling strongly enough about it to attach his name to the cover.

Ola Peterson and Tom Fogerty for their review and input of many critical business sections of this book.

Kenny and Michael Faltischek for reviewing specific portions of this book, and for their ongoing advice, guidance, and friendship.

Tom and Caroline Yates for their thorough review of the technical accuracies of many of the chapters, and more importantly for being great friends over the years.

James Deverell for giving thorough comments and criticisms (and helping us to design the 1-page summary at the beginning of each part). He visited us in NY expecting to have a fun weekend on the town, and instead we handed him 100 pages and a red pen.

A special thank you to the entire XPLANE team including David Grey, Bill Keaggy, and Judd Knight who gave this book character, personality, and a fighting chance on the bookstore shelves. We admire their professionalism and dedication and thank them for doing an outstanding job.

To our editors, Marjorie Spencer for believing in our idea and making it a reality, and especially to Judy Bass for keeping the torch lit during the downpours we endured, guiding us to a successful conclusion and putting up with our constant needs and detail oriented obsessing. Also, a special thank you to Cary Sullivan, for opening the doors to McGraw-Hill and for being a friend during some crazy times.

Beth Brown and her team at MacAllister, for putting in heroic effort at the 11th hour by stomping out copy issues, getting us the layout we envisioned, and enduring many extra rounds of obsessive tweaking.

Linda Orton for all of the feedback, ideas for marketing and sushi dinners. Her friendship and support was a key factor in our success and in our desire to form SageSecure.

Bruce Stout, whose Rainmaker's Forum proved invaluable in connecting us with people that have helped shape our book and our business. These people include Charles Jones, Gary Osland and Jack Gold.

Rochelle and Randy Blaustein once again provided much appreciated support and feedback during the planning and writing process. Without Rochelle's initial faith years ago, this and Wes's other books would never have been written.

Brendan Hammond for his friendship, hospitality in Perth, and highly enjoyable discussions on business, ethics and life. Particularly relevant to this book were the unforgettable experiences at Argyle Diamonds, which has a unique need for both physical and information security. Furthermore, for introducing us to Lionel Louw and Kevin Russell, whose thoughts and ideas about management consulting gave us additional confidence in our own approach.

Jess, Eggy, EK and Mike (161) for listening to Jay's ideas and reviewing portions of the book, all while pretending to enjoy the topic at hand. Daryl Klein for listening to Jay rant about the process of writing a book on more than one occasion. Without great friends like you it would have been impossible to remain sane and focused throughout this project. Likewise, Wes owes a big thanks to his friends who had to endure a year of whining about the pain and agony of the writing process. Lisa Braun and Sabrina Walton deserve a special mention (along with those who have already been recognized), since they not only listened to the endless complaints, but actually read chapters and gave extremely valuable feedback on our approach to writing. The writing process even led to making new friends: Young, Jeff and the crew at Kudo Beans contributed a steady supply of high quality caffeine, a comfortable writing environment, and fun times when the writer's block made work impossible.

Rory, because of his concern for our success and for being there to make us laugh when it was very much needed.

Jason would like to particularly acknowledge James and Terry Albanese for their unending encouragement, love and support from the time when he was focused on building Lego cities instead of companies. It is only with their faith that Jason has been able to chase down his dreams.

And finally, a great big thank you to our beds, which will be supporting us tonight when we pass out from a year's worth of accumulated exhaustion.

I

Managing Security

Summary

Information security is a business issue that needs to be managed effectively. Good security management can provide consistent protection from compromised data and downtime. Although complete security is impossible to achieve, too little security can cost a company dearly. The appropriate amount of security is unique to every organization. The following chapters explore some of the methods and tools used to manage security.

Key Points

- Information security is a business problem, not a technology problem.
- Total security is impossible. A trade-off has always existed between security and usability.
- Some amount of security is possible, but this can only be achieved after an organization identifies its security philosophy and integrates that philosophy into its business processes.
- Security policies are used to integrate a security philosophy with business processes. They should be driven by the needs of the business, not the needs of the technology.

Connecting the Chapters

When developing a security philosophy, a security assessment can provide necessary information on how business processes use network technology. It also identifies critical points of security within the business.

Once a philosophy has been established and security policies have been developed, *systems and network monitoring* tools provide feedback. This feedback can be used to refine policies and the overall philosophy.

- **Chapter 1, "The Security Assessment,"** gauges the risks facing a network and uses the analysis to select and evaluate potential solutions.
- **Chapter 2, "System and Network Monitoring,"** describes tools to enable centralized control and analysis of network systems.

Introduction to Managing Security

Information security has little to do with technology; it's a business problem. If a business needs security, it needs to build security into its very core—its mission and vision. It can be thought of as the fabric an organization's vision is embroidered upon. Ideally, security should be integrated into a business when it is first created or whenever the mission is refined.

Incorporating security into an organization's vision is an executive role that can't be delegated. The core focus and mission of the business must be evaluated in the broadest manner. One way to start this process is to ask, "What makes our customers/investors/partners believe in us?" Look at how confidence is created and then think about how that confidence could be destroyed. What disasters, either natural or man-made, could ruin the business overnight?

Of course, for most of us it's too late to get security in at the start, and significant changes to an organization's vision don't occur all that often. Instead, we're stuck retrofitting security into an already mature business model. That makes things a little more difficult, but by no means impossible.

Visions of Security

The purpose of a vision is to set expectations and goals. Security adds confidence to the vision. Look at one of the biggest companies on the planet: Coca-Cola. In many parts of the world, it's the only beverage a tourist may feel completely safe drinking. It's not even a question in most people's minds; it's just a fact, and it isn't that way by chance. Part of Coca-Cola's core vision is to ensure that their beverage is always safe to drink everywhere. The same goes for McDonald's food. It is unlikely you'll get sick from a McDonald's burger in any part of the world. Finally, have you ever felt in the least bit threatened at a Disney park (phobia for giant helium-voiced rodents aside), even though no visible security is present?

Security and Business Processes

Whether it's to improve efficiency, cut costs, or prepare for future changes, at some point every business process gets reevaluated. This is the best time to factor in security. With just a little more effort, you can apply your company's security philosophy to every aspect of the process.

As you evaluate your business processes, it's important to avoid falling into the trap of treating network security as a "separate" issue. Your network provides information and information services, which are used in larger business processes. What are those processes? Which ones are critical to the business? How can these processes fail? Generally, you'll find the network is just one of many factors that can lead to a business process failure. Your strategy must go beyond the individual factors, protecting each process as a whole.

Breaking down Business Processes

The most important aspect of any business process is people (or robots, if you work for Honda, Sony, or Matsushita). People need access to resources and information. They need to communicate with others. They need tools to help them operate efficiently. They need support when things go wrong. They need to be monitored, but they also need some privacy.

Every one of these needs factors into your business process. These needs also have direct security implications. Here are a few questions to think about as you look at the human resource components of your business process. After each question, the related security concepts, technologies, and the parts that cover them are listed.

- How will you control usage of critical resources without hampering the efficiency of the people who need those resources to do their work? (Part 4, "Determining Identity")

- How can you monitor the productivity and compliance of your employees while protecting their privacy? (Part 1, "Managing Security," and Part 5, "Preserving Privacy")

- How will people access/exchange information? (Part 10, "Accessing Information")

- What happens when the tools needed pose direct threats to the security of the business process? (Part 12, "Detecting Intrusions")

- What happens when people need technical support? (Part 2, "Outsourcing Options")

The Business Needs Should Dictate the Nature of Security

As a rule of thumb, business procedures should never be overhauled to satisfy security needs. A common mistake is to pick a security solution and then force related business processes to adapt. Unfortunately, doing this can seriously disrupt these business processes. Eventually, the processes may fail, or people might circumvent the security in order to get their job done.

If you can't make the existing process secure enough, the problem escalates back to the executive level. Somebody has to decide to reengineer the process. Even then, the new process must be primarily driven by the needs of the business, even if it ultimately means compromising security.

Obviously, the ideal situation is to factor in security from the start. If you happen to be reevaluating a business process that had sensible security considerations built-in from the beginning, you're incredibly lucky. Buy some lottery tickets and send us a few. In the meanwhile, we'll be teaching the rest of the world how to shoehorn some security into their existing processes.

Information is a critical part of any business process. It's an initial ingredient, an intermediate component, and part of, if not the entire, final product. As you look at your processes, look at the flow of information throughout the process and think about the following questions:

- Does critical information reside in a secure environment? (Part 8, "Storing Information")
- Do you need to control and protect the information as it moves throughout your organization? (Part 6, "Connecting Networks," Part 7, "Hardening Networks," and Part 9, "Hiding Information")
- Does the information need to be controlled once it travels outside the company? (Part 3, "Reserving Rights")
- What happens if people can't get the information? (Part 11, "Ensuring Availability")
- How will you know if information has been maliciously altered? (Part 12)
- What will you do if the information is damaged or destroyed? (Part 11)
- Does the information need to be encrypted and/or authenticated? (Part 9)

The process of collecting the data to answer these questions is called a security assessment. The assessment forces you to acknowledge and address all the critical security issues associated with your business. When a security assessment is complete, you will be left with all the information and analysis needed to formulate your security policies. We will go into much more detail in the next section of the chapter, which specifically covers security assessments.

The Harsh Truth

The concept of complete security is an illusion. It's impossible to make something totally secure and usable at the same time. You can build a room with only one door and put all the security in the world around it, but in order to get in the room, the door needs to open. Once that door is open, an intruder has an opportunity to get inside. Every technology thrown at the problem is limited by the reality that some form of access must be granted to legitimate users.

Security technologies and systems attempt to anticipate how an intruder might come through the open door. This sounds like a reasonable approach, but it ultimately fails because the systems themselves are fallible. Machines can only do what they've been programmed to do. People can be tricked and make mistakes. Intruders exploit these facts to get around the best and most elaborate security systems.

Managing Perception

If security is an illusion, managing security is about managing the perceptions of your observers. Some of these observers are the attackers you're protecting yourself against. For many of these attackers, you're not a specific target; you just happen to be in their line of sight. Think of a mugger. He doesn't specifically want *your* money;

anybody's money will do. As callous as it sounds, you want the attackers to look at someone else who appears more vulnerable.

The best way to get passed over is to make the attacker think that you're more trouble than it's worth. This is the tacit principle behind every form of practical self-defense; learning the techniques gives you confidence, which deters potential attackers. After all, even criminals don't want to get hurt. They'll just wait until someone defenseless comes along. Likewise, having a strong-looking security system is often enough to make hackers and other criminals pass over your network in search of easier prey.

But don't kid yourself. The criminals aren't actually afraid of you; they're just doing a perverse cost-benefit analysis. Your training or expensive security system is no match for street smarts in a real brawl. You certainly don't want to brag or otherwise encourage a challenge.

What does it take to encourage attention? Oracle did a good job with their "Unbreakable" campaign, which offered a reward to anyone who could find a hole in their database software. The marketing slogan alone was enough to attract the attention of vigilante hackers; the reward simply pushed it over the top. Let's just say it didn't take long before shattered code littered the ground of Redwood Shores.[1] Other companies, including Microsoft and a number of security vendors, have issued similar challenges with similar results.

If you keep a low profile, you'll improve your chances of being ignored. Although companies like Microsoft and Oracle can get away with baited remarks, you don't want that sort of attention drawn to you or your organization. So don't walk around the Internet with your tae kwan do black belt tied around your waist.

Unfortunately, keeping a low profile isn't enough. Sooner or later, you're going to have a problem. The key is to *manage the risk* ahead of time. Many industries have their own regulatory bodies that offer

---

Perception Versus Reality

Our expectations, formed by our experiences in life, predispose us to certain notions of security. For example, a door guarded by a heavily-armed person is seen as more secure than one that has just a lock. This perception has nothing to do with reality. For example, the heavily-armed person might be easily bribed or led away from his or her post. Or maybe a window is open around the corner where the guard can't see it. In reality, an unguarded but locked door might present a greater challenge to an intruder, yet most observers will say an armed guard makes the door more secure.

---

[1]It could be argued that this was actually an intelligent ploy on Oracle's part. By offering a reward, they got thousands of hackers around the world to discover and notify them about flaws they otherwise would not have found. They used psychological judo, manipulating their adversaries into using their strength against themselves. Within a short time, they were able to clean up the "low-hanging fruit," making their system significantly more secure.

guidelines or dictate requirements for security. Part of the risk-management process involves becoming compliant with the accepted security practices within the industry.

Some industries are behind the eight ball when it comes to security. Their regulations and recommendations are either outdated or flawed in principle. The result is that achieving compliance might actually weaken an organization's "real" security. To the astute security manager, this means that security needs to be dealt with on the political level as a 'business risk management issue, as opposed to a technological one. Ironically, these managers are likely to have the most successful security policies. Why? Because by accepting that the technology battle is a lost cause, they focus on the thing that matters most: perception.

The Security Philosophy

The expectations and goals created by an organization need to be supported by a compatible philosophy toward security. This philosophy is the way an organization approaches the topic and concepts of security. The philosophy establishes and defines a stance on security that dictates the operational parameters throughout the organization. It's easier to create a successful security philosophy if some flexibility exists within the vision.

The ideal time to work on developing a security philosophy is during a business-level reorganization. Is the business moving in the next year? Is the business laying off a percentage of employees, or are many new people joining the firm? Is management unhappy with business flow and thinking about revising the general process? These are drivers of organizational change. Make sure that security is part of the change process.

The Disaster Spectrum Philosophy

Basically, two types of disasters can befall information and information services. The first is the destruction or corruption of data and denial of service. The second is unauthorized access of data and service. As bad as both are, one is often preferable over the other. Which one? It depends on the nature of the organization with the bad luck.

The Security Cycle

A philosophy leads to policies, which lead to procedures, which lead to enforcement, which can be monitored and analyzed to provide feedback that can be used to adjust the original philosophy and policies.

It turns out that service firms are most sensitive to data or service destruction and corruption. These are companies such as medical, legal, and accounting firms. Data destruction means downtime, which directly impacts revenue. Law firms can lose tens of thousands of dollars in potential billable hours if their networks go down

The Disaster Spectrum

Illustration by SageSecure

Security can provide protection from two extreme scenarios, theft of data and destruction of data. Some organizations are not concerned about theft, but worry about the destruction side of the spectrum. Other organizations need to protect their intellectual property and want to avoid data theft at all costs. Many organizations end up somewhere in the middle, defending against both.

Accounting Firms

Law   Firms

IT Products and Services

NEWS

Marketing Media and PR

Intelligenc

Research and Design

Figure I-1

for a short while. Doctors might not be able to access patient records, which could cost a life.

For these organizations, information theft isn't as big of an issue. This doesn't mean that it's irrelevant; it means that if they had to choose, they'd protect against data loss and destruction first. Their security philosophy for information is *keep it available*.

Likewise, companies with a heavy investment in intellectual property (R&D) are often much more concerned with information theft. Plenty of prototypes and blueprints are lying around, so destroying data wouldn't be a major setback (and didn't the fusion group accidentally blow up the server room last week anyhow?) But if the competitors get early wind of the latest breakthrough, the entire product line might become worthless. Here the security philosophy is to focus on *secrecy and privacy*.

Then there's the half and half organizations. Here's where *information technology* (IT) companies (product/service firms) usually end up. These companies are equally miserable whether data is destroyed or stolen. If you're lucky enough to be reading this book at the start of a new company, do whatever you can to avoid this pit of despair. Either end of the spectrum is easier to secure than the middle. The security philosophy for the middle is to avoid getting burnt or cut by the flaming knives you're juggling. If you're already stuck here, at least you're in good company.

The Security Policy

Your security philosophy will have to be interpreted for each business process. These interpretations are known as *security policies*. Some organizations have just one general policy; others have specific policies for each department. Some policies become legal contracts that bind employees, such as compliance and nondisclosure agreements. Others are more technical, prohibiting or authorizing the use of various services such as telephones, computers, and other company resources. Regardless of the presentation, a security policy is just a process-level application of the security philosophy.

Security policies are powerful, because they can both positively and negatively impact your organization's culture and morale. Security policies can even be used as a tool to change organizational behavior, but security should never be the rationale, or driving force, behind business model or culture changes. What happens far too often is that a security approach is chosen and the business model is altered to fit the approach. This is a recipe for disaster.

A successfully implemented security policy will support your business processes while providing necessary protection. It's even possible for a well-planned set of policies to significantly enhance your business culture. However, a policy not in line with your business model can erode morale or prevent work from getting accomplished.

Security policies have organizational and technical components. The organizational part of a policy creates rules for employees to follow, often taking the form of a legal document or a set of procedures. Some examples of organizational policies include confidentiality/nondisclosure agreements and privacy policies. Procedures for handling sensitive information would also be considered organizational policies.

The technical component of a security policy enforces the organizational rules. Hardware and software systems are used to control the flow of information. Many of the technologies described throughout this book can be used to implement technical security policies.

Not all security policies need technical components. Some rely solely on management and legal procedures to ensure compliance, but many policies are best monitored and enforced with the help of technology. For example, the organizational component of a security policy might prohibit general access to the Internet. The corresponding technical policy would be implemented with software that actively blocks access to unauthorized Internet resources.

Many examples of both organizational and technical security policies are available on the Internet. Some are fill-in-the-blank forms; others are samples from real companies. Working with an example policy is a great way to ensure your policy addresses all the critical issues. But be careful—we do not recommend using any of these samples verbatim. Your security policies must be carefully tuned to the needs of your business.

Security policies are generally toothless on their own. They're just rules on paper unless they're enforced. Procedures and oversight roles can be used to enforce legal and organizational policies, but we won't get into the details here. This book deals with technology systems that enforce or manage technical security policies.

Common Types of Security Policies

Network technology is a broad field. Accordingly, many different types of security policies exist within this field, and this section briefly summarizes the following policies that apply to most businesses:

- Acceptable use
- Email
- Local and remote access
- Assessment

Acceptable Use Policy

The Internet can be a helpful research tool as well as an extremely powerful marketing resource for your products or services. But unrestricted access to the Internet can lead to countless hours of total distractions for employees who misuse it.

An acceptable use policy dictates what users can and cannot do on your network. The policy allows and denies various common activities. For example, personal browsing of the Web might be allowed, but certain types of sites might be denied. Activities that harm, or have the potential to harm, the business will also be denied in the policy.

Practicality prevents an acceptable use policy from explicitly handling every possible use of a network. Therefore, a good policy will state up front whether any other forms of network use are allowed or denied. This is truly a question of company culture. Companies that trust their employees often leave the decision in the hands of the user. Those in sensitive industries may prohibit any activity other than those explicitly stated in the policy. This is the most secure stance, but it requires that the policy be very well thought-out.

Many different technologies can be used to enforce the acceptable use policy, including systems that limit outgoing access to designated Web sites. They can prevent access to sites and services such as Web-based email, news, instant messaging, and games.

A good acceptable use policy can have numerous benefits. If your company culture is focused on efficiency, you can remove the temptation to waste time with personal email and aimless Web browsing. *Restricting Internet usage also can minimize exposure to viruses and hackers that target corporate users.*

A downside exists though. Overly restricting access to resources can negatively affect company morale. You do not want to gain productivity or security at the cost of treating your employees like children. If you need prohibitively high levels of security on the network, you might want to consider providing a separate, easily monitored system for unrestricted online access.[2]

Email Policy

Email has become an indispensable form of communication. Day in and day out, most companies rely on email to do business. Unfortunately, email is also extremely insecure. It's an open channel through which any type of information can enter or leave your network. Uncontrolled email undermines all your network security systems.

Although absolutely necessary, email control is always a sensitive topic for a company. People tend to take their email access very personally. It's difficult to take away or restrict email access without seriously harming company morale.

Creating an email policy requires some thought. Figuring out a strategy that satisfies your business model and your security needs while keeping your employees happy isn't easy. The following are some of the questions to consider:

- Should email from outside the company be allowed into your network?
- Do you have the facilities to host your own email, or should you outsource it?
- Do you want to permanently store all email or force deletion after a certain amount of time? What does the law allow?
- How will you prevent email viruses?
- Is secure email a general need, a limited need, or unnecessary?
- What should users do about personal email? Will you monitor their personal email use?
- Do users need off-site access to company email?

Many aspects of an email policy can be transparently enforced using technological solutions, but the best way to make email secure is by properly educating your users. Installing a few basic techniques, such as how to recognize a potentially harmful message, can be far more effective than most email security technologies. Thus, education and technology should be used together to create a successful policy that balances productivity and security.

Local and Remote Access Policies

Local and remote access policies specify how access to the systems on your network will be obtained. Local access policies cover authentication within the company's network, whereas remote policies deal with connections and network-to-network connections.

---

[2]For example, placing a few full-access computers in high-traffic areas will allow employees to check personal email, stocks, and Web sites during short breaks. The location makes it easy to see if people are abusing the service and therefore discourages abuse in the first place.

For most networks, local access is controlled physically, involving simple things like guards and locks on doors, and by using more complicated techniques like passwords. Often a user will need more than one password to access a variety of resources. The more passwords users have to supply, the more likely they'll either choose bad ones or simply write them down somewhere. Neither situation is good for security.

A good password policy minimizes the number of passwords a user needs. It also ensures that users select good passwords that are easy to remember. The policy also must address situations such as lost passwords and changes. Will these passwords be centrally managed? Will biometric devices be used? These and similar questions determine a local access policy.

Remote access is another problem altogether. Allowing a machine that you don't control to access your network is risky. In general, *no* remote access is the best policy, but sometimes this isn't possible. A good remote access policy ensures that the machine accessing your network does so in a secure manner. It limits the services available to remote users or requires the connecting machine to use a secure network connection.

Assessment Policy

A company must be able to see what's happening on its network. Assessing the network allows the company to detect failures, trace intrusions, observe user activity, prevent abuse, and generally nose around for anomalies and red flags. Although a company has the right to look at anything it wants, its employees have a right to know that they're being watched. Setting up an official assessment policy ensures that you're appropriately communicating with your staff.

The trickiest part of designing an assessment policy is deciding how, and to what extent, employees are monitored. If you want to track employee network abuse, unscheduled assessments will have a direct impact. If you are looking to verify data and take inventory, a routine assessment may make life easier on the IT department's busy annual schedule.

An audit policy makes the boundaries of personal privacy clear to everyone in the company. If employees know that someone could be watching at any moment, they won't feel violated by an unexpected audit. A good audit policy will balance privacy with security and peace of mind. It should allow the company to assess and examine usage, while allowing the staff enough freedom to build a positive office culture.

Final Thoughts

Having a security philosophy is critical to a company's long-term health. An organization without a well-defined security philosophy will be severely disrupted when a disaster happens. Perhaps years may go by without a problem, but eventually something will happen. Placing security at the heart of your business is a necessity.

"Planning is everything; plans are nothing." Count von Moltke's thoughts on planning should be kept in mind when considering security. Realize that policies and plans change frequently. The process of creating a security philosophy (or planning) builds the skills needed to rapidly adapt to a volatile environment. Von Moltke also was fond of saying, "No plan survives contact with the enemy." Nothing could be more accurate when it comes to securing a network against hackers and other disasters.

Part of managing security risks is to stay on pace with, or slightly ahead of, the rest of your industry. Don't be afraid to look at competitors or companies in similar industries. How are they addressing their security issues? No organization should want to be too far ahead of or too far behind the pack. Being too far ahead means playing the role of the guinea pig. Being too far behind means the competition has the edge.

Managing security is a skill, just like any other type of management. There really is no magic to the process. It takes time to learn, and a lot of trial and error is involved. Start simple: Incorporate security into the evaluation of a basic business process. Practice by looking at security issues specific to that process.

In the following chapters, we'll look at how security assessments and network monitoring systems can help in designing, implementing, and enforcing security policies.

Chapter 1
Managing Security:
The Security Assessment

A security assessment gauges the risks facing a
network and is used to select potential solutions.

Technology Overview

You can't manage problems if you don't know they exist, and you can't manage successful execution if you don't measure deliverables. A security assessment identifies a company's technical and organizational security fallibilities. The goal of such an assessment is to gather information in order to create or revise security policies.

No "standard" security assessment exists. It's a process that is custom-tailored to each organization. Templates, guides, and software tools are readily available to help conduct a security assessment for any organization, and consultants who specialize in conducting security assessments can also be hired. However it is accomplished, a security assessment will vary depending upon the security goals of the organization being analyzed.

Don't confuse security assessments with security audits. In our opinion, these are two very different concepts. The term *audit* refers to an established compliance procedure used to satisfy legal or regulatory obligations. An *assessment* is an internal initiative used to create a baseline picture of a network's security, usually for the purpose of making improvements. It's pointless for us to discuss audits here, because their requirements change based on the the industry, regulatory, and legal requirements. Recent historical events such as September 11, 2001 and the barrage of corporate accounting scandals have raised the bar significantly in terms of security requirements. Security assessments are something that every organization should periodically perform.

How the Security Assessment Works

In our professional capacity (as opposed to our starving-author capacity), we've been called upon to assess the security of various companies. This section outlines the procedure we use when conducting a security assessment.

Phase One

The first step in this assessment process is a preliminary meeting with the organization's chief information or technology officer. The purpose of the meeting is to gather some general background information on how network technology is being used by the business. This information will help guide the rest of the assessment and highlight specific areas that need closer examination. Engaging the CIO will help to ensure security is being taken seriously within the organization. The following are questions that should be asked during this meeting:

- Who manages the network? What is his or her background?
- Who are the people in charge of development, maintenance, and support?
- What are the primary uses of the computer resources?
- What types of business information are stored on the network?
- Who uses what? Which departments rely on which resources?
- How critical is a functioning network to business operations?
- What disaster recovery plans are in place? Are they tested?
- What are the thresholds for downtime and information loss?
- How much time is available for prevention? Monitoring? Response?
- What are the budget constraints for implementing security solutions?
- Is insurance available for risk protection?
- What kind of change management processes are implemented?
- How effective are your deployment and configuration management procedures?

Phase Two

The next step is to begin gathering information about network resources, such as the hardware configurations and services. Much of this information can be discovered using automated tools. The data collected includes the type of equipment or service, its purpose, its physical location, and any custom configuration information.

Throughout this process, the *physical security* of network devices and connection points should be closely examined and documented. Every detail ascertained by this investigation should be keyed into a *system specification database*. In addition to being necessary for the later phases, this deliverable will become a valuable reference tool for the company.

The combined data can be used to create a visual map of the network, known as a *network topology diagram*. This diagram makes it easy to identify and locate all the critical hardware, servers, and services available on the network. A variation of the topology diagram is a *dataflow map*. This dataflow map provides more detail than the topology diagram, charting the physical pathways and destinations used by critical business information. It thereby turns a simple technical diagram into a powerful business tool. A dataflow map and can be used to gain a global perspective on how a business operates over a network. This empowers executives and managers to make decisions about business process efficiency. See Figure 1-1 Dataflow Map.

The information requirements for phase two are comprehensive. Much like cleaning your attic, it will be a trying and potentially annoying process, but the results will be rewarding. The goal is to be able to address the following key issues by the end of the phase:

- Are there any blatant network topography issues?
- How is the network exposed to the outside world?
- Can somebody easily gain unauthorized access to the network by obtaining physical access to key network devices and connection points?

Completing this phase should produce three potential deliverables: a network topology diagram, one or more dataflow maps, and a system specification database.

Phase Three

This part of the assessment looks at server configuration. Examining all the local servers and the services they run will provide important information about what needs to be protected. Document the specifications and the configuration of the server hardware and software. Also, look closely at usage profiles, the types of applications being used, and who uses them. The level of detail amassed should be significantly greater than in phase two. This adds a complete configuration profile of the servers in the network to the system specification database.

The Value of Deliverables

In general, it is important to conclude each phase of an assessment with a tangible result that can be presented to management. Working toward this outcome helps maintain focus throughout the assessment process. It also keeps everybody on the same page and allows errors and omissions to be caught before they compound into serious problems. Management should get a short document that summarizes the key findings and future steps. It is analogous to having a good roadmap while driving in unfamiliar territory.

Dataflow Map

Illustration by SageSecure

All organizations rely on information flow between departments. Data moves across the desks of workers, into computer systems, through servers, and out through printers. Normal operational procedures expose information and make it insecure. Limiting this exposure is easier to do with a dataflow map.

This diagram represents a highly simplified example of a dataflow map. These maps can help managers decide at a glance critical information access points and bottlenecks. Without a comprehensive understanding of data flow, security initiatives are likely to fail.

Legend

- 📁 - Filesystems
- 📇 - Databases
- 🌐 - Web/Intranet
- 👥 - Customers
- 🏢 - Vendors
- ⚔ - Competition
- ✉ - Email
- 📄 - Inbox/Outbox
- 🧠 - Brainpower
- 📺 - Media
- ⚖ - Regulation

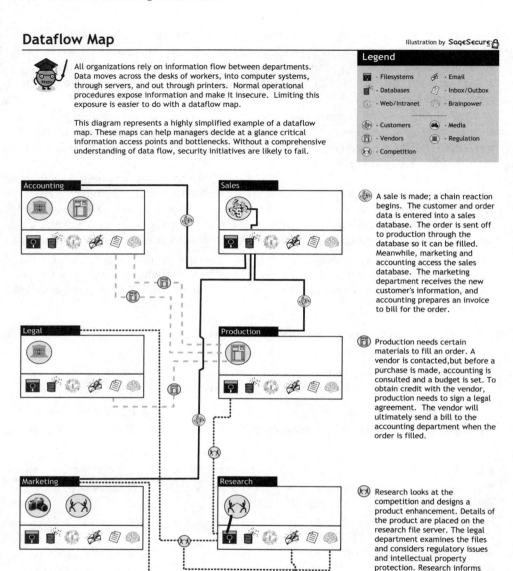

A sale is made; a chain reaction begins. The customer and order data is entered into a sales database. The order is sent off to production through the database so it can be filled. Meanwhile, marketing and accounting access the sales database. The marketing department receives the new customer's information, and accounting prepares an invoice to bill for the order.

Production needs certain materials to fill an order. A vendor is contacted, but before a purchase is made, accounting is consulted and a budget is set. To obtain credit with the vendor, production needs to sign a legal agreement. The vendor will ultimately send a bill to the accounting department when the order is filled.

Research looks at the competition and designs a product enhancement. Details of the product are placed on the research file server. The legal department examines the files and considers regulatory issues and intellectual property protection. Research informs marketing and production of the new product specifications.

■ Figure 1-1

One of the biggest security issues for servers is *trust*. When servers work together, they must be able to trust each other's integrity. A trusted server or user has permission to perform sensitive or potentially damaging actions. Flaws in trust relationships are some of the most common and dangerous security problems.

Part I   Managing Security   **17**

**Chapter 1
Managing
Security:
The Security
Assessment**

The information from this phase will help when making difficult decisions about security policies. By the end of phase three, the following questions should be asked:

- Are the critical servers properly configured?
- Are there any exploitable trust situations?
- Do we really need all these services?
- Is critical data adequately protected from loss and theft by implementing *uninterruptable power supply* (UPS), RAID, or backups?

Phase Four

By now, the assessment process has collected a lot of facts about the technology used by the business. What else is left? Phase four focuses on one of the most overlooked aspects of network security: the workstation. You may wonder why you should be concerned. After all, if only trusted employees have access to company desktops, what can go wrong? A closer look reveals that a lot can go wrong and usually does.

The workstation is the hardest computer to control effectively. Restrict it too much and people can't get work done. Give it too much flexibility and it becomes impossible to maintain. Even basic and necessary daily computing tasks can unintentionally lead to an insecure environment.

Check the workstations to see how close they are to the ideal balance. Take note of each computer's hardware specifications. Then look closely at the software, ensuring that the applications installed are those that have been mandated. The following questions should help identify most of the critical workstation security issues:

- How much access does a hacker have on a compromised workstation?
- Are the workstations sufficiently hardened against hacks?
- Can problems be detected or preemptively fixed?
- Are workstations thin or fat? Are there critical workstations?
- How are applications rolled out to desktops?
- Is workstation access physically secure?
- Are desktop passwords protected and do passwords expire?
- Is wireless network access allowed?
- Are PDAs permitted to connect to the network?
- Can laptops be connected to the network?
- Are consultants permitted to connect to the internal network?
- Do vendors perform demonstrations with their laptops connected to the network?

Phase Five

Now that serious time and resources have been spent collecting this information about the business systems, it should be put to good use. This is the beginning of phase five, the strategic solutions phase. Here all the previous data is analyzed, and potential solutions are presented. It's usually necessary to meet with all key technology executives for a final information review and disclosure. This meeting is where decisions will be made about the organization's policies toward security. The assessor should now be able to do the following:

- Target and identify the firm's security philosophy
- Create a risk management profile
- Make business and technology recommendations for security policies
- Determine how best to protect the network given practical constraints
- Analyze the costs associated with implementing recommended solutions

Best Practices

A security assessment will integrate many of the principles and concepts behind a security strategy. It raises many fundamental questions that must be addressed by an organization's security policies and philosophy. It is an important step in becoming familiar with practical security concerns. Chances are it will expose previously undeveloped issues within the IT infrastructure. It's an extremely valuable process for management to undertake, but it can also be time consuming.

It is possible to purchase technology in the form of a template that generates a security policy, but this does not eliminate the need for experienced IT personnel. Without proper interpretation and management, an IT policy is worthless. A good manager can translate the directives of a security policy into organizational rules. The result is practical benefits from the security policy that can be integrated with business needs. Ultimately, security policy should not change normal operations; instead, normal operations should incorporate an additional security policy.

Final Thoughts

Should you use outside consultants or do the whole assessment with in-house staff? The answer depends on the competency and capacity of the internal staff. It is easier to decide whether or not to outsource a security assessment once the skills of your staff have been determined. In many cases, internal assessments have the potential to be just as effective as those performed by outside consultants, assuming the staff has the capacity to take on additional work. Plenty of tools are available to help with internal assessments, such as ready-made sample audits and various forms of assessment software. However, consultants often have much more experience and may see things that are overlooked by your in-house team. It's up to you to determine which path is the correct one.

Chapter 2
Managing Security: Systems and Network Monitoring

These tools allow for centralized analysis
and control of network systems.

Technology Overview

The Roman Empire had a security management problem. As powerful as it was, it didn't have the resources to monitor and enforce security throughout the entire European continent. Invaders attacked the weak points and slowly worked their way inward. As a result, Rome fell.

Is your network starting to feel like an empire? As your network grows, it also becomes much harder to manage. Problems in some of the least significant machines can eventually turn into network-threatening situations if they're not duly managed. Don't underestimate the seriousness or difficulty of this resource problem; after all, it stumped the most powerful civilization in history.

The key difference between large empires and large networks (as far as this book is concerned) is that modern network systems are designed with remote management in mind.[1] This means that a single central station can control thousands of network devices, workstations, and servers across an entire enterprise. Everything from simple status reports to complex software installations can be done from one place. Technical security policies can be implemented and monitored with ease. Built-in analysis systems can create reports for management.

Centralized network monitoring systems use various techniques to connect to your equipment. For example, the *Simple Network Management Protocol* (SNMP) is a commonly used system that enables network devices to be remotely monitored, controlled, and configured. It's not standard, but most network hardware and operating systems support SNMP.

Unfortunately, SNMP is neither powerful nor secure enough to manage certain complex devices such as routers and firewalls. These systems often have their own remote command interfaces so that control center applications can talk directly to the devices.

Software and services running on your network also have to be monitored and managed. Many control center applications can monitor common software such as Web and email services. However, actively managing these systems is a much more complex task because nonstandard software has already been dominating this environment. It's difficult, verging on impossible, for the creators of command center products to incorporate compatibility with each of the thousands of network applications on the market.

Integrating with modern command center applications takes an unusual mix of knowledge that spans a number of distinctly different fields. That's a major reason to hire consultants to create interfaces for old, nonstandard, and custom systems. Hiring consultants is naturally more common with larger, enterprise-wide networks, due to the prevalence of incompatible hardware and software.

How Systems and Network Monitoring Work

Setting up a network monitoring system is an involved process that will take some time to do correctly. The procedure can be broken down into the following six steps:

1. Strategic design: Before any monitoring happens, someone needs to decide what to monitor and design a system to monitor it. This is a complicated step (touched on in the introduction to this part), requiring a combination of both business and technical analysis.

2. Implementation: At this step, monitoring tools are deployed and integrated with existing hardware and applications..

3. Tuning: One of the first tasks is to tune the amount of information generated by the monitoring tools. These systems can generate a lot of

---

[1]Countries were not built for remote management. Not convinced? Go to London and ask to see the Great British Empire. They keep its remains in a jar somewhere near Piccadilly, so we've been told.

**Chapter 2
Managing
Security:
Systems and
Network
Monitoring**

information, some of which is important, whereas the rest is just clutter. It's easy to monitor too many things or report too many minor problems. You may or may not want to know every time a hacker attacks your site, but you certainly want to know if one has been successful.

4. Monitoring: Once that's done, a lot of information still exists, and someone has to watch for problem indicators. It also would be nice to see if any trends are present. Are hack attempts increasing, and is the increase bigger than the increase in general Internet hacking?

5. Action: Finally, when a problem happens, somebody needs to be notified. This might be trickier than it sounds. Whoever gets notified needs to take action and solve the problem (or officially ignore it).

6. Analysis: At regular intervals, it's useful to dig a bit deeper into the monitored data. Regular monitoring looks at the present, but proper analysis looks at past data as well. Trend and pattern analysis of the complete data picture can help in anticipating future problems. Usage and failure patterns might help pinpoint the root cause of an elusive problem. It can also be useful to keep trend reports on issues that are otherwise too insignificant to watch. For example, a significant increase in the frequency or volume of minor alerts could indicate that a major problem is about to happen. Therefore, even though a system might be tuned to report only major problems, it still shouldn't completely ignore the minor ones.

Security Considerations

It's difficult to enforce security policies. Without a monitoring system, policy breaches and technology failures can go undetected, but a centralized monitoring system can help you ensure policy compliance and make a complex network manageable.

By analyzing the data from the monitoring system, you can figure out if your security policies and enforcement tools are effective. The analysis also gives executive management the feedback necessary to adjust and revise their security philosophy and policies periodically.

Network monitoring systems do have a few caveats. The first is compatibility. If you're using standard equipment and running standard software,[2] your monitoring problems should be minimal. But if you have legacy equipment or custom software that you want monitored, things become much more difficult. You will probably have to get consultants to add support for nonstandard devices.

---

[2]"Standard" means universally accepted and supported. These software and hardware standards include, but are not limited to, the following: Sun, Oracle, Microsoft, IBM, Cisco, HP, Intel.

Another caveat is network design. No central command system can fix a poorly designed network. In fact, if the design is really bad, the system might not be able to properly operate.

Ironically, a central management system can actually weaken security. Think about what happens if a hacker gets into your central management system. He has immediate and total control of your network. In some cases, he doesn't even need access to the central system; he just has to impersonate the system. Although some network devices and servers have secure mechanisms for performing remote administration, others don't. You might want to consider disallowing central administration machines on systems with weak overall security.

These applications will only display what they've been designed to observe. What if the central system is missing something critical or is not working correctly? The bottom line is that security requires the involvement of skilled human beings at some level. You'll need to have some specialists on staff that can tell when something's wrong, even when the system says that all is well.

Some of these systems make it all too easy to define and implement site-wide technical security policies. Security-sensitized *information technology* (IT) staff can be tempted to create arbitrary technical policies that are not mandated by existing organizational policies. This is the classic "solution in search of a problem" situation. It happens because the applications only handle security as a technical process, not as a business or philosophical process.

Faults aside, without central command systems, it would be impossible to build and maintain large networks. Just the monitoring alone is critical. Imagine how difficult it would be if you had to manually check the status of hundreds or thousands of servers and desktops. These tools also automate routines that otherwise take up massive amounts of time, such as deploying updated software. Simplifying basic network management tasks means that less experienced employees can monitor the network effectively. This can save your company money in IT staffing, which always makes management happy.

Having a bird's-eye view of your network is also a big advantage. Many of these systems have visualization tools that can show you a 2-D or 3-D representation of your network and its status. If a picture is worth a thousand words, these pictures are worth about a million pages of status reports.

Making the Connection

Cryptography: Data traveling across the network can be encrypted for added security.

Outsourcing: Network monitoring is frequently outsourced, which has many benefits and issues, enough that it gets its own chapter in the next part.

Disaster prevention: Proper monitoring can help detect minor failures before they become major ones.

Chapter 2
Managing
Security:
Systems and
Network
Monitoring

Proactive security: Data gathered while monitoring can make risk-management techniques and forensics much easier.

Determining identity: Access attempts, successes, and failures can be monitored to detect intrusions.

Preserving privacy: Data can be used to assemble profiles on users. This might be a benefit or a problem, depending on your organizational policies.

Networking hardware: Most network hardware is designed to provide information to monitoring systems, either through SNMP or some other system.

Firewalls/proxies: These devices are the first line of defense from the outside world and can help identify problems and abuse inside and outside the network.

Storage: Full disk drives, exceeded quotas, and file corruption can be detected by monitoring the local file systems of critical machines such as servers. Performance, data access problems, and corruption can be detected and fixed.

Detecting intrusions: Most intrusion-detection systems are built as part of a larger monitoring application or they are integrated with central monitoring applications. The two concepts are closely related.

Expediting recovery: A good monitoring system immediately indicates any problem that requires disaster recovery. The faster a problem is identified, the faster a solution can be executed.

Best Practices

The biggest problem in implementing network monitoring is determining which command center software is right for your network. After all, you won't find this type of software sitting in a shrink-wrapped box at your local computer store.

The current crop of network management systems falls into three classes. In first class, you'll find extraordinarily expensive "solutions"[3], such as IBM's Tivoli, HP's OpenView, and CA's Unicenter. In business class, you'll find smaller vendors and consultants. Back in coach, you'll find a bunch of open-source software developers eating peanuts and hacking into the in-flight entertainment system. They have created free systems called Nagios, Ganglia, and OpenNMS.

Which way you go depends on the size of your company. If your company is huge, with a network consisting of thousands of computers, you'll want to look at the high-end solutions. If you're a business that can free up time but not money, or if you have a savvy IT department with Unix skills, the open-source systems are a great place to start. It can also be helpful to talk to a few consulting firms that offer products or services that fit your needs. Often the right consultants can implement and manage one of the open-source or low-cost solutions provided by niche vendors.

---

[3]By solution, we mean the product + implementation consulting + new servers + new networking hardware + yearly licensing + permanent operational consulting + technical support contract + hardware maintenance contracts + software upgrade contract.

You need to have a clear idea about what you're looking for in order to effectively evaluate these applications. Otherwise, you'll never be able to make sense of the various marketing materials and feature sets. Here are some questions to consider:

- What do you want to know? Do you want to know about hardware and software problems? Do you want to be able to monitor individual users or aggregate usage patterns? Do you want to be able to detect intruders and vulnerabilities?
- What will you do with the information? Do you want the system to automatically fix problems? Do you have procedures for escalating problems that need to be taken into account?
- Do you want to be able to remotely deploy or configure software? Do you want to centralize the management of user accounts?
- Will more than one person access the control system? Will different people accessing the system need different levels of access?
- Is the command system going to be on the network that it's monitoring, or do you need to operate from a remote network?
- Information traveling to monitoring systems is sensitive. Is it properly encrypted to ensure safety?
- What happens if hackers get control of the command center?

Final Thoughts

If you are considering the high-end products, go to their Web sites. Confused by all the different products and solutions? It's intentional. You're supposed to get a sales rep to tell you what you need, but just in case you want to figure it out alone, here are a few pointers:

- Products such as Tivoli, OpenView, or Unicenter are not designed to work right out of the box. What you're buying is a core product and lots of component modules that provide specific types of control and analysis. Components are also provided to connect to the various machines in your network for advanced operations. For example, one component might be specifically designed to interact with an Oracle database.
- These companies have cleverly designed their solutions such that every high-end product in their catalogue is a "necessary component." Put some pressure on the sales rep and watch how quickly things become "optional."
- You might notice that things you thought were related (such as security and network monitoring) are sold as separate products. Many vendors split up their software based on marketing potential, not technical functionality. So what should be a single system is actually five or more separate systems with overlapping functionality.

**Chapter 2
Managing
Security:
Systems and
Network
Monitoring**

- Think twice about the third-party applications on your network. Are they all absolutely necessary? Running a tighter ship will simplify command center integration. With fewer unusual applications running, less interface customization will be required.

If you ask any of these high-end vendors about the smaller solutions (including open source), they may turn their nose up. Ask them to explain why their system is better and they'll present three basic arguments: features, scalability, and support. All three are flawed arguments.

Many of the features the large systems offer will never be used in practice. Think about all the features in Microsoft Office that are hyped up but are unnecessary for conducting normal business in your organization. If the smaller software solutions do what you want them to do, why should you care about features you'll never use?

The big players will tell you their products are much more scalable. They really aren't; they just require really powerful equipment, and that same equipment will generally solve the scalability issues of most software.

Support is also a trick argument. If you use one of the major open-source applications (Ganglia, Nagios, or OpenNMS), you may find cheaper consultants that can support the system. What you may not find is standards. Support might be a fraction of the cost that major vendors charge, but are you getting the same peace of mind and consistent service?

You might find that no single open-source system has all the features you want. The good news is you can try them all indefinitely, because they are free. That said, some aspects of these systems could create extra network traffic. Make sure you're not congesting your network with overlapping functionality.

Open-source systems often require a degree of tinkering that is equivalent to giving the space shuttle a tune-up. Sometimes what is saved in licensing fees is lost in time. Be sure to have the right people working on the project who have specific experience with open-source network monitoring systems.

II

Outsourcing Options

Summary

Some businesses find it easier and more cost effective to outsource security needs. In the physical security realm, this means hiring a security agency. In the world of network security, companies provide the digital equivalent. This part looks at the types of security needs that can be outsourced, and the issues involved.

Key Points

- Outsourcing appears simple, but it's not.
- When you outsource something, you're extending trust. How well do you know the people you're trusting?
- It's crucial to ensure that your organization's security philosophies and policies are understood and embraced by all of your outsource providers.

Connecting the Chapters

This part examines how outsourcing in general can impact information security. The three chapters here focus on outsourcing security technologies and services:

- **Chapter 3, "Outsourcing Systems and Network Monitoring,"** concerns hiring an outside agency to monitor your system's health, performance, and security.
- **Chapter 4, "Outsourcing Disaster Prevention,"** details leveraging external resources to help prevent machine, network, data, software, and human failures.
- **Chapter 5, "Outsourcing Proactive Security,"** looks at using third-party expertise to take preventative steps to mitigate future failures.

Introduction to Reserving Rights

Nothing is simple anymore. Network security is no exception. Before addressing the most basic security issues, your business needs to develop a comprehensive security philosophy. Developing this philosophy is a massive undertaking that requires the discipline and wisdom of a Zen master. But in today's hectic economy, who has time to achieve enlightenment?

If anything has a chance of countering this complexity, it is outsourcing. Outsourcing is the process of hiring an outside agent to perform a business task. Your business is probably already outsourcing services. Does a group come into your office in the evening and perform janitorial services? Is there a company to call when the voicemail system fails? Does your office have an alarm system that is connected to an emergency response service? These are all jobs that have been outsourced to third-party service providers.

The Illusion of Outsourcing

Some people see a magical aura of simplicity around outsourcing. When a process is outsourced, all the associated problems seem to vanish. Outsourcing can look so easy you'll start to wonder why you don't just outsource everything. While you're at it, you can even outsource your own job and run the company from the golf course.

As you have probably experienced, the simplicity of outsourcing is just an illusion. Many of the visible time and cost savings are counterbalanced by equal or greater hidden costs. Outsourcing a process might free up some in-house staff and management resources, but someone needs to manage the outsource provider. This can take a lot of time to do correctly. In many organizations, the high-level managers that can handle this are already busy enough. Adding extra management capacity is expensive and may neutralize the benefits of outsourcing in the first place.

Outsourcing is an alternative way of doing business. In some situations, it might offer significant benefits, whereas in others it could lead to disaster. It's not a guaranteed win

Why Bad Outsourcing Feels So Good

People pay more attention to the simplicity they perceive than the complexity they know. This is why get-rich-quick scams find so many victims. The victims are seduced by the simplicity of the scam, ignoring the obvious practical (and legal) complexities that guarantee failure.

As stated, outsourcing also has a seductive simplicity. Managers can take advantage of the simplicity illusion to hide serious outsourcing problems. Costs can get shuffled into other budgets or buried in general consulting fees. This makes it difficult to identify and acknowledge failed outsourcing initiatives within an organization.

for any given situation. In fact, a number of perils, illustrated in Figure II-1, need to be overcome before you have a chance of success. But if you can get past the hurdles, outsourcing can be a powerful business tool in the right situation.

Much like the board game Othello[1], outsourcing takes "a Minute to Learn, a Lifetime to Master"®. The process of calling and hiring another company to do your bidding is the easy part. The largest difficulty lies in deciding which processes to outsource. The choice may be obvious when you are talking about taking out garbage, but it gets more complicated the closer you get to your core business.

Outsourcing and Data Security

Imagine a group of criminals physically walking into your office and walking out with permanent access to the entire corporate network. You might think this is impossible. You'd be wrong.

Professional criminals do their homework. They will extensively research your company before they ever set foot on your property. They'll observe people coming and going for a while, looking for opportunities. By calling secretaries and human resources, they can learn key employees' names and office locations. They might even sift through the trash for useful information.

In most large companies, unfamiliar faces are always roaming the halls. Perhaps it's a group of repairmen or maybe it's the management consultants. As long as somebody is dressed appropriately and seems to know what he or she doing, nobody asks any questions.

A criminal can usually obtain all the access he needs just by successfully impersonating a temporary worker. For example, suppose a group of electricians has been working in your office for a week. The next week, a new guy comes in to do a final checkover. He's got the right uniform, knows the name of all the people involved, and knows exactly where to look, but he's not really an electrician. As long as he gets

The Client's Responsibility

Most networks evolve over time, at the hands of many different people. It's crucial to have accurate and detailed records of a network's history. This includes notes on present and past configurations, along with the names of the people involved. Understanding a network is a challenging task without this information.

New consultants brought in to work on aspects of a network will spend less time on discovering and more time solving problems if they are provided with adequate history. This could significantly reduce the billable hours of a project, improving the bottom line.

---

[1]Also known as Reversi, which is a simplified version of Go.

#1: Over-Estimating Outsourcing

People think that outsourcing a project will save some of their time.

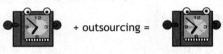

Outsourcing can save time, but it also creates new timesinks (managing the outsource provider, communicating needs, etc.). Whether outsourcing saves time depends on the project.

One thing is certain. If security needs are not adequately communicated, outsourcing can result in serious security problems.

#2: Over-Looking Outsourcing

Some projects and tasks make more sense to outsource than to handle in-house. By not outsourcing, you risk:

- Over-Worked Staff
- Low Morale

- Breakdown of Business Processes
- Poor End Result
- Potential Security Vulnerabilities

The Four Perils of Outsourcing

#3: Communication

Communicating needs accurately is one of the most difficult aspects of outsourcing. Common stumbling points are:

- Incompatible Philosophies
- Different Preconceptions
- Inadequate Interaction

#4: Control

Keeping an eye on security is easier when information and processes are controlled by in-house staff.

Outsourcing puts control in the hands of another organization. Their faults and failures will directly affect the security of the outsourced information and processes.

Concept by **SageSecure** (www.sagesecure.com) | ©2003 XPLANE.com®

Figure II-1   The risks of outsourcing

through the front door, he can get at your wiring closets, which means he can install a tap onto your network.

Of course, you usually assume your outsource providers are trustworthy. But what if they're not? What if the intruder in this example really did work for the electricians and really was part of the project? Let's face it: Few companies can afford to run a background check on every janitor, electrician, and telephone repairperson. Yet when you think about it, who's in a better position to obtain sensitive information without looking suspicious?

The truth is that anyone working within your office environment has the capacity to compromise your security. We'll look at those problems next, as we explore the different issues with outsourcing systems, services, and processes.

Outsourcing Business Technology

A lot can go wrong with an outsourced technology project. Project managers often find it hard to focus on anything other than keeping the project from falling apart. Rarely are the project's long-term effects on the company's security infrastructure a consideration.

Picture a clumsy man standing next to the Hoover Dam on a rainy afternoon. Water levels are rising, and a tiny insignificant crack forms in the dam wall. Thinking he knows better than the dam engineers, he walks over to the crack and sticks some chewing gum in the hole. As he turns to leave, the water pressure shoots the gum out and into his hair. Now the dam is leaking worse than before, he has to get a haircut, and all he's got to say for himself is, "Darn it."

Much like the dam tourist, consultants who have been engaged to solve one set of problems often introduce a new set of potentially worse ones. Network security seems to be particularly sensitive to these added problems or *splash damage*. The sensitivity is present because most security strategies rely on balancing a large set of factors. These factors get thrown out of whack when consultants make changes to the network. Dangerous conflicts and security vulnerabilities can get introduced whenever the design of a network changes, new systems are introduced, or new software is added to workstations.

Some of the worst security problems happen when consultants use a security model that is incompatible with their client's overall security philosophy. Their final product appears to be secure, but a hacker can exploit the incompatibility to bypass the security system. For example, a Web-based application developed by a consulting firm might use a shared internal database that is strictly guarded against all mainstream attacks. The consultants decide to give their application full access to the database, trusting that the built-in security on their application will provide adequate protection to the database. However, a hacker that breaks through their web application will have direct access to a database he might never have seen otherwise.

Even more dangerous is the intentional vendor-initiated security breach. For example, a rogue member of a consulting team might install a remote backdoor on your network. Later on, that backdoor can be accessed for nefarious purposes, compromising your entire network's security.

Selection/Design/Implementation

Over the last 10 years, computer consulting has become a major industry. The emergence of complex business systems has fostered legions of technology experts who possess three areas of expertise: product selection, design, and implementation. Of course a fourth area of expertise not previously mentioned is their ability to mail rather large invoices.

Product selection is the ability to evaluate business needs and choose the right combination of software and hardware based on those needs. A consulting firm that performs this service is aware of all the industry trends. They know the ins and outs of the latest and greatest, and more importantly what to avoid.

Design and implementation go hand in hand but do not need to be performed by the same consulting firm. Software design is often necessary when it is not possible to satisfy unique needs with off-the-shelf products. Network design, however, is a more fundamental necessity. They don't sell off-the-shelf networks, just like they don't sell off-the-shelf businesses (well, they do; they're called franchises, but we digress).

Successfully implementing a custom design requires impeccable project management. The project manager must constantly review the design with both the client and the implementation team. Often the business needs that are driving a project will change over the course of the implementation. A good project manager will work with the development team to figure out how the changes affect the project. He or she will then ensure that the client understands the cost and time implications of changing the scope.

Issues

The first hurdle is getting everyone on the same page in terms of needs, abilities, and budget. Businesses usually want a lot for their money, and consultants are often pressed to exaggerate their abilities. This can lead to unrealistic project specifications. The result? Halfway through the project the consultants realize that fulfilling the contract will take twice as long, cost four times as much, and the final milestone calls for technology that somebody saw on the *X-Files*.

Not all computer systems in an organization are part of the main network. When isolated systems exist, it is often tempting to ignore security issues pertaining to them. One commonly made argument against securing isolated systems is they will never interact with other critical systems. That assumption may be proven wrong over time. Eventually, someone may decide to bring the systems into the main network due to changing business needs. Furthermore, information stored on the isolated systems may be able to help in gaining access to the main network. For example, administrative passwords to the main network may be insecurely stored on the isolated system.

The other side of the spectrum is also a problem. Some systems are heavily integrated into the existing computing infrastructure. Preexisting issues might make securing the new system difficult. Conversely, a security flaw in the new system might undermine the entire existing security infrastructure. Database systems serve as a perfect example. Usually, who can access data is tightly controlled. Applications that access the database often have fewer restrictions. It's up to the application to protect the data it retrieves from the database. A flawed application can give extensive access to a user who would otherwise have tightly restricted access.

Tips

It's crucial to set and emphasize security requirements when hiring consultants to select, design, or build systems. In most scenarios, computer consulting firms need to be efficient to be profitable. On the road to achieving optimum efficiency, corners get cut and details get left out. Don't allow security to become an extra detail that is worked in only if time permits.

Software development requires many different talents, and skill constraints are a problem for software development houses. At any given time, a group of gifted programmers and project managers may have only 85 percent of the knowledge they need to perfect your project. Ensure that knowledge about security is not part of the missing 15 percent.

Installation/Integration/Maintenance

You've acquired the systems and software you need. The outsourcing process went well, and security was intelligently integrated into the project. Now you need the software to be installed on your network. In addition, you will need ongoing support and maintenance.

Some companies outsource all computer-related problems to an external maintenance company. This could include anything from fixing jammed printers and flaky computers to upgrading software and hardware. The guys that show up to fix the broken printer or flaky computer are rarely seen as a serious security threat.

Issues

Do not be fooled into thinking that simple tasks such as network troubleshooting will not affect security. The basic information used by a troubleshooter to get on your local network and fix your printer is often enough to give away the farm. Even if the passwords are changed, support consultants may notice or create hidden vulnerabilities that can be used to compromise the network.

Tips

Changes made to any system should be carefully logged and documented so it can be returned to its previous state if something goes wrong. Here the use of defensive forensics is also quite helpful. A snapshot taken of your system before changes, upgrades, or patches are applied will allow for quick debugging if any problems occur.

Outsourcing Business Services

Signatures of the suburbs include the moist odor of fresh cut grass and the less pleasant din of lawn mowers and leaf blowers. Equally pervasive are the vans of local repairmen and tradesmen. On a particular day, an electrician is paying a visit to the Petersons. They need a more powerful outlet in the kitchen to support their new refrigerator. While making adjustments to their main circuit breaker, the electrician moves a few cables that are in the way. Without realizing it, he's pulled a critical alarm system cable loose. The alarm system still appears to work properly, except it no longer tells the police to dispatch a car when a burglar trips the system. The electrician doesn't bother checking the alarm before he leaves because, after all, he's not the alarm guy. The Petersons are therefore completely unaware of the problem and continue to count on their alarm system as usual.

Whenever you outsource services, you run the risk of creating a security problem. In fact, the less the outsourcing has to do with security, the more likely it is that a problem will occur. Service agents such as electricians, janitors, lawyers, and accountants have little interest in your security policies. It's not that they want to violate your security; they're just blissfully unaware that it's an issue. They do only what they need to get the job done.

Transient Staff

Short-term workers such as tradesmen, janitors, facility managers, and administrative assistants pose a serious threat to network integrity. You might be surprised at the amount of information these often temporary workers have access to around the office. The problem is, companies tend to ignore these individuals. You do not really know who they are or where they came from. This leaves you vulnerable to anonymous attacks of information theft.

Sensitive information is frequently left on memos and notepads around people's desks. Maintenance workers might have physical access to private offices and other critical areas of your business. Someone has to dust the floor and change the air filters in the server room. Is that someone you trust? It would be painfully ironic for a company that spends millions on network security to have its data compromised by a hacker posing as a janitor.

It's possible to train your in-house staff on security awareness. You can also manage the procedures to ensure that the daily routines are fundamentally secure. Although the same can be done with outsourced staff, it's much more complicated. The outsourced staff will need to be trained and their procedures and routines will need to be managed.

This can be especially annoying with administrative staff. Although many companies have internal administrative staff that has been with them for years, many still hire employees for these positions through temp agencies. Unfortunately, even a temporary secretary or office manager may need to view sensitive documents or information. Use caution when delegating work to transient or temporary staff.

Internal Management

When management consultants are brought in to help improve business efficiency, security is rarely a consideration. Remember back in Part 1, "Managing Security," when we reviewed the importance of security policy? Management consultants that work closely with your business will have first-hand knowledge of many policies, including security. Make sure this is information you trust your consultants with, because it could easily be used against you. Once knowledge of a security policy is obtained, it becomes easy to socially engineer a way around your security.

Customer Service

Many companies outsource customer service or technical support. Much like payroll, customer service relies on an infrastructure that is often too costly to build internally. If you are outsourcing customer service, you are giving up access to your most prized possession, your customers. The outsource provider needs access to your customer database and other critical information. Can they provide adequate protection for your information? Are they capable of following and enforcing your security and privacy policies?

Professional Services: Legal/Financial/Sales/Marketing/HR

Although a strong case can be made to be cautious with low-level outsourced services, the same applies to higher-level outsourcing. Specifically, remember to treat your professional services with the same degree of caution. For example, many important digital communications are sent back and forth between management and their accountants and lawyers. Are your professional service firms sending and receiving emails and attached information in an encrypted format? This only scratches the surface of possible precautions.

Later we will take a close look at e-risk management, an area of security directly correlated to outsourcing. E-risk management examines your ability to control or "put a leash on" your active information. Deliberate exchanges of this active information are made with professional service providers. Can you guarantee that the professional service firms you work with will exercise tight control over this information?

Production

Outsourcing assembly and production comes with the same security headaches as other services. Here a third party is being trusted with your products and intellectual property. Without looking very hard you can always find black market items for sale on most city streets. Items such as DVDs and albums that have not been released to the public abound. Often these black market items come from copies made off the assembly line at a production facility.

It is also important to ensure that a production provider is complying with all licensing laws when producing your product. You do not want a graphic designer to use an unlicensed copy of Photoshop to design your marketing campaign.

Outsourcing Security Services

Our livelihood is based on our belief that security is this decade's most important business issue. Obviously, we all think security is incredibly important. Now we're saying you might benefit by perilously leaving your security in the hands of strangers. Sounds a little crazy, doesn't it?

Any consulting company hired to maintain or improve your network security will have complete access to critical network equipment. You now have to protect yourself against threats from their employees or hackers who obtain critical information from the consultants' computers. How do you protect yourself from the consultants that are installing your security systems?

At some level, you're going to have to trust both the skills and ethics of your security service providers. Although it's important that the executives of the security company you hired be beyond reproach, you should be most concerned about the people actually doing the work. Do they have a vested interest in the company's integrity or are they just disposable labor? Highly skilled security implementers will know the ins and outs of whatever system they're installing, including backdoors and vulnerabilities. They will also have an extensive understanding of what they're protecting. If they are not going to break in themselves, they still might sell a few key details to someone who's a bit more enterprising.

Ultimately, this is an unsolvable problem. Although we'll give you some pointers, any problems experienced in this area would most likely be the result of bad luck. The majority of people and companies that you'll encounter are ethical. The security needs of your business will dictate the appropriate amount of confidence you should have in your vendors.

The easiest way to gain confidence in a vendor is through due diligence. As commonsensical as it sounds, few people do enough research when choosing a security vendor. Most businesses will choose by balancing price, quality, and obligations (such as nepotism, kickbacks, and so on), but a further level of due diligence needs to be performed: background checking. Who are the individuals assigned to the project? Ask to look at resumes, and get references from previous projects. It's a lot of extra work, but it's the only way to be sure of what you're getting. It also shows the vendor that you're serious about security.

A more advanced method is to use multiple vendors, where each vendor only sees a portion of the system and nobody knows how the entire system works, but this is incredibly difficult to pull off. Vendors are experts in their area and can usually guess what the rest of a system looks like from a small number of clues. Someone also needs to design the entire system. How will you certify his or her trustworthiness? The overwhelming odds are that a company attempting to build security in this manner will actually end up with weaker security.

In general, decisions on what (and when) to outsource are driven by a cost-benefit analysis. When it comes to outsourcing security, the question is, do you really know what benefits you need and how to evaluate them? If you decide you need to monitor security, does it need to be 24 hours a day? If you decide you need a security audit, how thorough should it be? Do you just want to find potential vulnerabilities, or do you want to test your defenses as well?

Trends in the marketplace show a tendency to overdo security, especially once a decision to outsource is made. After reading this chapter, we hope you will be able to better evaluate your needs. This book and the resources found on our companion Web site (openlysecure.org) contain much of the knowledge you need to perform a fair cost-benefit analysis on outsourcing security. With any luck, you won't under- or overdo security outsourcing. If you do it right, it will cost less, and you'll get more security.

Final Thoughts

Once you can control it, outsourcing will provide your business with great advantages. Outsourcing lets you be more creative and tailor solutions that specifically fit your business processes. Every business wants to save money and time while increasing productivity. Imagine how far you will go if you can achieve these goals while maximizing security.

Chapter 3
Outsourcing Options: Outsourcing Network Monitoring

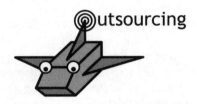

The process of having your system's performance and security
evaluated by an outside monitoring agency.

Overview

Most businesses monitor their systems with an inexpensive, in-house monitoring device: their users. No complaints equal no worries (that's about as mathematical as this chapter will get). For these companies, detecting minor problems is simply not a priority, but like tooth cavities, many of the worst problems start small. By the time the pain hits, you're looking at serious drilling or possibly a root canal.

Chapter 2, *"Systems and Network Monitoring,"* discusses monitoring the systems and services on your network. Criteria are provided for figuring out what you should be monitoring, and the chapter covers the tools that can help monitor systems and services on your network. The chapter also notes that these tools are rather expensive and not always appropriate, which leaves you in a jam if you need to monitor but don't want to spend a fortune.

Luckily, many computer systems and network devices can be monitored remotely, and plenty of companies provide outsourced monitoring services. These companies have assumed the cost of building and staffing multimillion-dollar monitoring systems in the hopes of capitalizing on economies of scale.

A monitoring company can send critical diagnostic information securely from your network to their external network. They see everything your own *information technology* (IT) staff would see, assuming you implemented your own high-end monitoring system. Often this can be done transparently by plugging a "black box" into one or more strategic locations, although some network topologies will require reconfiguration and additional hardware and software.

How Outsourced Monitoring Works

In Chapter 2, we split the process of building and operating a monitoring system into six critical components: strategic design, implementation, tuning, monitoring, action, and analysis. An outsourced monitoring company will be involved in most, if not all, of these phases. The following is what you can expect and what you should avoid.

Strategic Design

Figuring out what to monitor is tough. It requires a unique combination of technical and business knowledge. Consultants with this background can help you identify processes in your business that will benefit from monitoring. They can also recommend the best tools and techniques for implementing a monitoring solution. Whether you monitor in house or outsource, having consultants involved in the strategic design phase is incredibly useful.

You'll also need to watch out for any conflicts of interest if the consultants you use for strategic design will also implement a monitoring solution. You want the consultants to recommend the most efficient monitoring strategy, not the one that makes them the most money down the line.

Implementation

In order to obtain status information from your systems and services, you'll need to reconfigure some machines or install additional devices throughout the network. This can be as simple as plugging in a few specialized monitoring devices, or it might require a complete overhaul of your network topology.

Many monitoring service providers will give you some customized computers to place in strategic locations throughout your network. These "black box" monitoring devices are simple to install, but what do they really do? They gather information, but how much? They send this information to a remote location, but how secure is your information in transit? Can hackers get into your network through the black box? Some simplicity up front can generate a host of difficulties later.

**Chapter 3
Outsourcing
Options:
Outsourcing
Network
Monitoring**

For advanced monitoring, you may have to implement extensive changes throughout your network. This can be time consuming and easily cause functionality, stability, performance, and security problems. If you have the technical competence in house, put it to use, but consider getting some highly skilled consultants to check your network afterward.

Tuning

Wolf! Okay, maybe it no longer inspires the fear of yore. After all, wolves aren't usually roving the city streets, but false alarms are still annoying and potentially deadly if constantly ignored. The average network has plenty of minor problems that are generally ignorable yet can trip the alarms of monitoring systems. It takes time to fine-tune the system to eliminate false alarms while ensuring that real alarms aren't ignored. A good monitoring company will find the perfect middle ground for their clients.

Monitoring

Once the monitoring system has been successfully designed, implemented, and tuned, somebody has to use it. This is the part that makes the most sense to outsource. Economies of scale work in your favor: A fully-staffed monitoring center can watch five networks just as easily as it can watch one, that is, as long as too many problems don't happen simultaneously.

Outsourcing monitoring is a great concept, but certain responsibilities can go drastically wrong. In a lousy economy, your provider might not have enough clients to properly staff their operations center. A late-night critical alarm might get overlooked, or the general competency of the staff might become subpar. Keeping tabs on your provider's active client list is a good start. If your network is mission critical, you might want to think about having a second monitoring company working in parallel as backup.

Action

Sometimes the monitoring company can fix problems remotely, if they have the authority to do so. This, of course, creates a major conflict of interest. It becomes incredibly convenient for the monitoring company to find problems if they get to charge you more money for a solution. In such a situation, you have to worry about monitoring the monitoring company, which begs the question: Have you saved yourself any time? It's often better to have a separate set of consultants available to solve difficult problems.

Analysis

Is your monitoring system properly addressing your business needs? How about your network as a whole? How effective is it? Do the failures throughout your network have a pattern? These are the sorts of questions that an analysis can answer.

With the answers, you'll be able to make better decisions as your business and network change over time.

Basic reporting should be part of any outsourced monitoring service. The basic reports should contain useful information that your own IT team can analyze. Good monitoring providers will offer in-depth analysis, and their expert staff will examine the raw data for you, looking for trends and problems so that you can be proactive.

Which Monitoring Tasks Can Be Outsourced?

Almost any type of monitoring task imaginable can be outsourced and handled by a third party, but certain monitoring tasks are particularly well suited for outsourcing. The following are monitoring tasks that are commonly outsourced.

Internal Services

Basic internal monitoring will inform you if your services are up and running. More advanced monitoring can gauge performance and track usage patterns. For example, you can monitor the load on your database and file servers or scan internal emails and track Web usage.

External Services

Web servers and other external services can go down unexpectedly. Sometimes the server is fine, but the connection to the Internet is blocked. It's often hard to tell when this happens from your own network, especially if the external services are hosted within your network. An outside monitoring company is better suited to watch your external services.

Virus Detection

Most virus-scanning software comes with a system for automatically updating the virus list via the Internet. This often requires a subscription and is essentially the equivalent of outsourcing. These systems are great if everything is configured properly. The only concern is if hackers somehow compromise the virus list distribution system. Even if the system is totally automated, somebody should periodically check to make sure everything is working properly.

Intrusion Detection

This is perhaps the most difficult monitoring task. Intruders don't always leave many trails, and it's incredibly hard to differentiate the tracks left by intruders from those of legitimate users. It's an ideal task to outsource, but it also gives the most information and control to a third party. It also might require significant changes to your network. You may decide not to look for intrusions at all, an option that sounds crazy but might make sense from a cost-benefit point of view.

General Network Threat Information

A couple of organizations with close ties to the hacker community have attempted to gauge the overall level of threatening activity on the Internet. They provide an

**Chapter 3
Outsourcing
Options:
Outsourcing
Network
Monitoring**

early warning service that ostensibly gives a heads-up when major coordinated assaults on the Net are under way. These alerts are sent via email or a thorough custom warning system.

Frankly, we have no idea how these services add value in practice, because most major assaults happen rapidly. The minute one occurs, anyone in the line of fire is toast and therefore probably won't receive the Internet-based warning. Those that do get the warning probably aren't likely to suffer from the attack. A certain level of hacking is constantly happening on the Net, so these services don't tell us anything new there.

The only upside we can see is the notification of a new vulnerability as it comes out. This is useful, but you can get the same information for free. Hundreds of email lists also provide real-time alerts about new vulnerabilities and exploits. It's easy to stay informed by just subscribing to a handful of email lists. See OpenlySecure.org for pointers to these lists.

Crisis Management

Anybody can post anything to the Internet. At some point in time, every company, product, and key executive gets roasted and sacrificed on a silicon altar. Most of the time it's just harmless dissent, but sometimes it gets serious enough to transcend the digital domain and become real world activism. Many a company has been caught off guard by the "sudden" appearance of highly coordinated opposition organized entirely over the Internet.

Your crisis management team probably already uses a news clipping service to track exposure in the major media. A number of services on the Internet perform similar functions. These services scan the Internet for information pertaining to your company, its key executives, and its products. Often such services can be used by crisis management to head off activism, as well as legal and competitive threats. They can also be used to find unauthorized statements by employees or mischief-makers impersonating employees.

It's far more productive to use an information-monitoring service than to look for this information on your own. These services often have access to private data sources, including password-protected and subscription Web sites, and they have customized software for effective scanning. Many of these services also scan the news wires, effectively replacing the traditional print-media clipping service.

Intellectual Property Theft

Although certain types of intellectual property theft are incredibly difficult to control (mp3s and bootlegs of movies are still easy to obtain on the Internet), other types are easy. The illegal commercial usage of copyrighted music, stock artwork, and writings can be detected by searching for digital watermarks and other ownership identifiers.

Using Internet-scanning technology similar to that used for crisis management, some companies will search the Internet and other media for evidence of copyright or trademark infringement. This is particularly valuable for companies that derive revenue from media sales or have strong trademark brands.

Security Considerations

Just on its own, sensitive information is difficult to protect, and things only get worse when a remote monitoring company has the ability to peer into your network. The more the monitoring company does for you, the more they get to see. Monitoring tasks such as remote intrusion detection tend to require full access to your entire network. Unless you use some advanced data hiding system (described in Part 9, "Hiding Information"), the monitoring company will see a lot more than you probably want them to.

This creates a twofold problem. First, you need to trust that the monitoring company is not going to take advantage of their ability to collect and peruse your sensitive data. Although you might have faith in the executives of the company, what will prevent one of their employees from turning to the dark side?

Even if you're confident that the monitoring company is trustworthy, what happens if hackers compromise the monitoring company's systems? The answer is that hackers will have access to your sensitive information. Any two-way data-gathering system used by the monitoring company can become a backdoor into your system. The "black box" systems used by some monitoring companies might become trojans within your network if they get compromised by hackers.

What are the odds a hacker will compromise your network via a monitoring company? It's hard to tell. One thing that works in the hacker's favor is the shoemaker's daughter syndrome. A monitoring company could be so busy trying to protect their client's security that they get a bit lax on their own.

Best Practices

What do you need to monitor? The first step in a successful monitoring strategy is coming up with a realistic needs assessment. The how-to in Chapter 2 will help you decide what you can and can't successfully monitor.

You may not find a single outsourcing company that can monitor everything. If you need to work with more than one company, make sure they communicate with each other.

Also start simple. The easiest thing to monitor is server availability. All you need to know is if a server is up or down. You can also monitor the network by testing server availability from key network segments. For example, can workstations reach the file server? Your monitoring company should be able to do this with ease.

You should have your servers grouped by severity. If a critical server goes down, some sort of immediate notification should take place. A few minutes should be given for less critical servers. This prevents false alarm calls due to reboots and other self-correcting outages. A noncritical service outage notification might be scheduled, perhaps only during business hours.

**Chapter 3
Outsourcing
Options:
Outsourcing
Network
Monitoring**

See if the monitoring guys are paying attention by running the following test during the first few weeks of service. Shut the servers down at various times during the day. See how long it takes them to respond. You'll want to choose strategic times to catch things like late night sloppiness, shift changes, lunch breaks, and so forth. The shutdowns could occur, for example, at 2 AM, 4 AM, 6 AM, 8:30 AM, 12:15 PM, 4 PM, 5:45 PM, 8:30 PM, and 11 PM. Run your tests using both critical and noncritical servers. You might want to set up a fake critical server just for this test. This way you don't have to down any of your truly critical servers. The easiest way to do this is to lie to the monitoring company, telling them that one of your noncritical servers is extremely important.

Once you're happy with your monitoring company's performance, it's time to start monitoring services. Web servers, file servers, and databases are all services that should be independently monitored for failure and accessibility. A monitoring company should also be able to notify you when services are overloaded or otherwise performing poorly. You'll need to set performance thresholds so peak periods of activity don't cause unnecessary alarm.

The next step is system-level monitoring. Here each critical server is closely monitored for signs of failures and potential intrusions. The monitoring company does this by collecting logging information from your servers. Most servers can automatically email these logs, making the collection process relatively simple. However, some monitoring might require direct access to the server or custom software that transfers the information back to the central monitoring facility.

The most complex type of monitoring is network intrusion detection. This requires network and server activity to be constantly inspected. Additional hardware will need to be installed throughout your network, and the monitoring company will need direct access to many of your critical servers. This type of monitoring is the most difficult to outsource and should only be done if absolutely necessary.

Final Thoughts

Don't let the security issues scare you away from outsourced monitoring. It can be a big win for many companies, but certainly ask questions when evaluating potential providers. Beware of any company that says it's got everything covered. No company has ever figured out how to hire only trustworthy people. Conversely, don't write off companies that say they can't solve these problems; they're being honest. You can also protect your company's sensitive information in other ways. These include hiring a second monitoring company, implementing comprehensive encryption (described in Chapter 26, "*Cryptography*"), and purchasing insurance. Ultimately, you'll have to make a judgment call based on your security needs and the quality and honesty of the answers you receive from the monitoring company.

Chapter 4
Outsourcing Options: Outsourcing Disaster Prevention

Leveraging external resources can help lessen or prevent the failure of machines, the network, data, software, and staff.

Overview

Preventing a specific disaster is easy. Preventing disasters in general is impossible. The best you can hope to do is prevent the most common types of disasters. Even then, a lot of things can commonly go wrong. This means a lot of isolated prevention systems and a general management headache. Why not let the pros handle it?

Preventing Machine Failure

Instead of shelling out the cash to have spare equipment on hand (and then risk further problems with installation and configuration), why not get a major manufacturer to bring you a new machine? Some companies, like Dell, can get a replacement machine to you in a matter of hours. Then you just need to do a complete recovery from backup. This is a godsend for critical servers and workstations.

Issues

If you've added a lot of custom, after-market hardware, you may not be eligible for this sort of replacement policy. It also decreases the value of this type of service because you'll have to spend time moving the hardware from one machine to the other. This type of service is best when used with stock equipment.

You might get a replacement machine that is a more recent version of your broken machine (with a faster processor, a bigger hard drive, more memory, and so on). This could be a problem, however. Some systems are finicky when the hardware changes and won't recover from a backup easily. It's better to get an identical machine. Make sure your vendor can do this.

Tips

If the original system uses removable storage (such as a *Redundant Array of Inexpensive Disks* [RAID], as described in Chapter 32, "RAID"), you can just move the drives from one machine to another. Make sure you know the correct procedure for this and test it ahead of time. Some storage systems will think you've installed new drives and will wipe all of your data.

At some point, a vendor may no longer stock identical replacements. One option is to ask the vendor to notify you when your machine is nearing the end of its cycle and either make a trade-in for a newer system or buy an identical spare system. Although it's cheaper to buy a spare, you'll need to periodically test the spare. If it fails before the primary system, you'll just have to upgrade anyway. It's cheapest to buy replacement components, but this means extensive downtime for diagnostics and part replacements. Furthermore, your staff might not be trained to properly perform the repairs. Upgrading to a newer system is generally the best business decision.

Preventing Network Failure

Network failure is usually very simple or very ugly. In the simple case, a core piece of networking equipment fails and replacing it solves the problem. In the ugly case, a design flaw causes the network to incrementally degrade over time. Accurately tracking down this type of network problem requires a lot of information. If consultants are called in to solve a problem cold, they're going to have to gather this information, which may cost a lot of time and money.

A better idea is to get the consultants in when things are working correctly. They'll have a baseline to work from and might even spot some potential problems. Getting your network inspected by consultants can accomplish two goals: 1) It ensures that any significant topological issues will be discovered, and 2) it leaves you

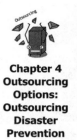

with the necessary equipment and instructions to fix basic network problems on your own.

**Chapter 4
Outsourcing
Options:
Outsourcing
Disaster
Prevention**

Issues

A network is a lot like a golf swing. You can have perfect form, yet the average golf pro will still find something wrong. Likewise, the average consultant will spot many problems within your network, even if everything is working perfectly well.

It's important to differentiate between problems that must be fixed immediately and those that can be fixed when they happen. Immediate problems are those that will either create massive downtime when they happen or cost significantly more to fix later.

The consultants will also gather information that could give intruders more than enough means for invading your network. How well will they protect this information? Are their own systems and processes secure? Who will handle the data besides the consultants directly assigned to your project? Don't forget the story about the shoemaker's daughter—it applies especially well in the IT consulting world.

Tips

Trust and communication are critical. Find a consultant whose opinion you trust, and don't hesitate to get a second opinion.

If you find it tough to trust consultants, we understand. One alternative is to make it clear to the consultants from the beginning that any problems will be fixed by a different consulting firm. This minimizes the consultant's incentive for blowing minor problems out of proportion.

It's always good to stock some basic equipment, even if you're relying on consultants for network repair. Here are a few things worth having on hand: at least two hubs (not switches—hubs are more useful for network diagnostics), a laptop configured like a standard user desktop (to test network services from various places on the network), a rolling cart with a monitor and keyboard on it, electric screwdrivers, spare pre-made network cables, extra spooled cable, a cable crimp tool and crimp ends, wirecutters, cable ties, and a cable tester.

Preventing Data Loss

If done correctly, backup can be a time-consuming process. It's not just about switching tapes. A lot of important questions need to be answered before any backup happens. For example, what are you backing up? What happens when new software and services are added? Who updates the backup system? How do you avoid backing up too much? Do you really want to pay for the expensive infrastructure when then entire process can be handled remotely?

A number of companies provide remote backup services over the Internet. They will back up critical files via a direct connection to your network. Some of these backup companies will provide consulting services, telling you what you need to back up.

Ensure that the company you chose has a secure infrastructure for transferring and storing your data.

Issues

Some remote backup providers automatically choose the files to back up, whereas others will make recommendations. Note the potential conflict of interest, as many of these companies charge by the amount backed up.

How secure is the backup process? Are files transmitted over a secure connection, or is the connection unencrypted? If hackers compromise the remote backup system, your files are at risk. This risk is minimized if the files are encrypted on your network before transmission to the remote storage facility. Consider this backup process a potential backdoor to your establishment.

Tips

When checking out remote storage solutions, be sure to look at the restoration process. Some of these services make restoration quick and easy. All you have to do is browse through the backup list on a Web site, click on the machine or data that needs to be restored, and you're good to go. Other companies make the process expensive and time consuming. Caveat emptor.

Preventing Software Failure

Years of brainwashing by major software vendors has led people to believe that software has become so complicated that failure is to be expected. In addition, the vendors have the further gall to suggest that most failures are actually the user's fault. Because of this, many people think that preventing software failures is impossible. We disagree. In fact, almost all software failures happen for one of three avoidable reasons:

- The software never worked in the first place.
- Conflicting software was added to the system.
- The software relies on another piece of software that is broken.

What about user errors? They all fall under reason number 1 and are clearly the fault of the vendor. Good software doesn't break when something unexpected happens. Years ago, in the days before Windows, vendors would properly test their software before releasing it to the public. User errors rarely, if ever, happened. Today, major vendors don't believe in quality; it's not cost effective. Instead, they hide behind end-user licensing agreements that absolve them of any responsibility toward their software.

**Chapter 4
Outsourcing
Options:
Outsourcing
Disaster
Prevention**

So how do you avoid broken software? One answer is to ask any custom or niche vendor to have their system certified by a professional testing facility. You should use their recommended configuration or ask them to certify and support your own configuration. They should stand by their certification. Any problems that happen in a certified environment are bugs and must be repaired promptly by the vendor at their own cost.

In the case of software from major vendors, you're not going to get this kind of support. Instead, you'll have to work within their parameters. Often a reseller will offer their own support on a particular configuration they've had professionally tested. They're not going to certify something unless it works pretty well; otherwise, they'd lose money on support calls. Any problems with major vendor software will be minimized by sticking with a configuration certified by a supporting third party.

The next most common cause of problems is conflicting software. Any time new software is added to a system, conflict is possible. Preventing users from adding any new software is a good starting point. Any new software should be thoroughly researched for its compatibility with existing software. Using standard software minimizes the chance of a conflict and maximizes the chance that somebody else has experienced a problem, if one exists. If you need to add custom software, make sure the system is fully tested for compatibility before deploying it. A number of labs out there do evaluation and burn-in testing for software. Having your configuration tested by one of these companies greatly minimizes the chance that some conflict will pop up later on.

The final source of problems is closely related to the previous one. In this case, software programs conflict yet are dependent on each other. We all know that most current popular operating systems have bugs. Therefore, all software has the possibility of being crashed due to a bug in the operating system. With most standard software, this won't happen unless the user is doing something extremely unusual. The conflicts that do exist are going to be well documented, and a way to prevent the conflict from occurring is almost always available, such as disabling a feature or installing an upgrade. Consultants can research these conflicts and help your support staff minimize most problems and prepare for unavoidable setbacks.

Issues

Custom-built applications tend to have many bugs and are often susceptible to instabilities in underlying software. Minimum quality levels should be negotiated into any development contract. Often a vendor will refuse to fix problems that stem from conflicts with other software unless the vendor is contractually obliged.

Tips

Do not be in a rush to update the latest version of anything. This includes software upgrades, patches, and driver updates. Often you may find yourself a guinea pig for a mislabeled beta test.

Wherever possible, use system configurations certified by vendors. If you need to use a custom configuration, see if a vendor will certify and support your configuration. If you can't get vendor certification, have a test lab verify configurations before you deploy them. This will help identify potential disasters-in-waiting.

Preventing People Failure

What do you do when critical members of your *information technology* (IT) staff fall ill, go on vacation, or leave the company? One place to turn is a consulting company that provides backup IT support staff. They'll send people to learn about your system, take careful notes on all the basic procedures. If you have a staffing crisis, they'll send in someone who can do a decent job of keeping things running.

The more information the outsourcing company has, the better prepared their staff will be. In a real crisis, they should have enough information and skill to keep things moving while you look for a replacement.

Issues

How much do you want an outside company to know? Are you going to give them master passwords? How secure will they keep this critical information? The information they collect can give hackers and intruders a major edge.

Tips

If you know a critical IT staff member is leaving for a month, have the replacement staff come in a week early to receive specific training. See if the temporary consultants are available for long-term contracts or temp-to-hire work. This can be an ideal way to evaluate and train future employees without having to make a long-term commitment.

Preventing Repeat Disasters

Let's say the worst happens: A hacker gets in and wipes a critical computer. No problem, you can recover with a backup. But now comes the hard part: closing the door the hacker used to get in. In order to do this, you need to analyze the breach. This is a job that requires a highly specialized set of skills. Your staff may not be able to figure it out on their own. You may need to outsource this task to a professional computer forensics company to figure out how the break-in occurred. More information on forensics will be covered in Chapter 5, "Outsourcing Proactive Security."

Issues

A monitoring system exposes sensitive information, but in a limited and indirect manner. A forensics company, however, needs to go through all your systems with a fine-toothed comb. They have direct access to everything, so keep this in mind when working with a forensics team.

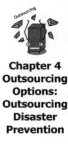

**Chapter 4
Outsourcing
Options:
Outsourcing
Disaster
Prevention**

Forensics skills are nearly identical to those in hacking. Not surprisingly, a number of forensics experts were once (and still might be) hackers. You might be trading one problem for another as a forensics investigator can easily add an undetectable backdoor to your network while "searching" for damage from the original penetration.

Tips

It's best to set up an arrangement with a forensics company before anything happens. It makes a response team's job easier if they have snapshots of all your systems before the penetration occurs. They also might spot vulnerabilities up front, helping you to prevent the break-ins before they occur.

Stopping certain types of prolonged attacks, such as denial of service attacks, requires the coordination of a number of different organizations. Good outsource providers will be able to get these groups on the phone quickly, taking the brunt of the attack off your servers.

Final Thoughts

Many disasters can strike an organization, and it's impossible to prevent every potential disaster. Even if you follow all of our advice, strange things can happen. Being adequately prepared for a disaster is nearly as effective as prevention. Even if you're not prepared, understanding the nature of a disaster can help you minimize the damage. Consultants can help you estimate the potential risks face by your organization.

Chapter 5
Outsourcing Options: Outsourcing Proactive Security

Outsourcing

The expertise of an outside party can help when taking preventative steps against future failures.

Overview

Hackers are infinitely resourceful and are constantly coming up with new ways to compromise "secure" systems. It's flat-out impossible to defend all potential hacker attacks, but it is possible to defend yourself against known vulnerabilities that are commonly exploited. It's also possible to mitigate the risks associated with insecure situations. This is the concept behind proactive security, also known as e-risk management.

The following are the five primary components of a successful e-risk management strategy:

Policy: Your security policies are your first line of defense. A fundamentally secure business process makes a hacker's job much harder.

Auditing: Are your security policies being enforced effectively? An audit can help identify gaps in your defenses.

Defensive forensics: How do you know an intruder has broken in? Because something on your system has changed. How do you know something has changed? You compare your current system with your system when it was considered "secure." Such a comparison is part of defensive forensics. It can be used to rapidly detect intrusions and can serve as a record of evidence in legal proceedings.

Protection: You'll probably deploy security systems to prevent the exploitation of your network. You can also use insurance and other risk-transfer techniques for processes too difficult to secure.

Testing: As your business evolves, periodically testing your defenses can ensure that the proactive security measures you've taken remain effective.

Strategizing and implementing e-risk management requires a highly specialized set of skills. Few companies have the in-house skills necessary to effectively create a proactive security strategy. Although outsourcing is an optional choice in most situations, here it is essentially mandatory.

Being proactive in business is always a challenge. If you plan ahead and use the concepts found in this chapter, you can severely reduce or eliminate any damage to your company from a malicious attack. This alone makes e-risk management a highly worthwhile investment. The rest of this chapter will help you maximize the value of your proactive security investment through effective outsourcing.

Policy

Part I, "Managing Security" explains that a security philosophy is an essential part of a business vision. Developing that philosophy is the sort of task best left to executives and high-level management consultants. Taking that philosophy and using it to create specific security policies is another matter. For this, you're going to want the help of expert security consultants with strong business *and* technical backgrounds. Their job is to figure out appropriate security policies based on your mission, vision, security philosophy, and business processes.

Issues

You're looking for business-level security policies, not technical ones. For example, a business policy is something that says, "All communications between branch offices must be secure." A corresponding technical policy would state, "Branch offices must be connected with a virtual private network." The business policy identifies aspects of a business process that need security; the technical policy determines how that security will happen.

Outsourcing

**Chapter 5
Outsourcing
Options:
Outsourcing
Proactive
Security**

Many consultants will provide this business policy service as part of a security audit. In our opinion, this is a bad idea. The results of the audit will create a predisposition toward *specific* technical solutions, which may influence the focus of the security policies. It's better to create the security policies without any specific knowledge of the existing technology base. This keeps the focus on vision, philosophy, and business processes.

Tips

Get your legal department involved. You'll be dealing with issues such as the acceptable use of systems and services, privacy, and data retention. All these policies will have legal implications and will require legal activity to enforce.

Auditing

You've settled on a selection of policies appropriate for your business. Now it's time to figure out where your business actually is, relative to your security goals. An extensive security audit examines the security of your technology infrastructure. It also looks carefully at how technology is used throughout your organization. The result will include a baseline snapshot of your network that will be invaluable when implementing protection systems and defensive forensic strategies. For details on the security audit process, refer to Chapter 1, "The Security Assessment."

Issues

What you want from a security audit is a concise document that looks at your actual security compared to your intentions. It should highlight critical business risks, and be supported by a separate report containing details that can be validated by technical staff.

What you don't want is what most auditing companies will give you: hundreds of pages that describe the various technical vulnerabilities found throughout your network. Instead of focusing on core business issues specific to your company, many consultants simply go through a checklist of common technical problems and security risks. Then they compile a massive document on each specific problem. If you asked one of these companies to describe a forest, they'd respond with a list of the number of leaves and branches on every tree. They'd also probably give you a list of technologies that would increase the average number of leaves per tree, without ever explaining why such an increase would be needed.

Tips

If you find a trustworthy consulting company that does both policy analysis and auditing, don't be afraid to use them for both tasks. Just make sure the analysis is done before the technology infrastructure is examined.

Defensive Forensics

One of the most significant components of e-risk management is defensive forensics. You may have come across the term forensics the last time you read a good police novel. It is a scientific term used to describe several methodologies that reveal detailed clues to someone trying to unravel a mystery. Forensic scientists may analyze and match a hair found at a crime scene with a hair on a suspect's head. Sometimes they try to match paint from a stolen car to a paint chip found at a hit-and-run accident. Using high-powered microscopes and ultracentrifuges, police can break down minute clues into critical evidence.

Similar to police work, forensics is used in the world of computer science. An entire history of activity can be pieced together by gathering detailed information from users and keeping close track of system and file usage. The important difference between computer forensics and police forensics is that computer forensics can start before the problem happens. This is called defensive forensics.

The idea behind defensive forensics is to create a baseline by collecting as much information as possible when things are working. This is known as taking a snapshot of the system. The more snapshots that are taken, the easier it becomes to track the activities of an intruder. When something bad happens, the same information is collected again and differences are sought out. Snapshots can be used to find backdoor programs, detect data deletions and modifications, and build intruder activity trails.

Defensive forensics is also used to investigate the actions and behaviors of legitimate system users, because many of the biggest security risks come from within. The misuse and abuse of resources by users can result in a significant loss of time and money. User activity trails can be obtained by comparing the snapshots of a system over time and are often invaluable in legal proceedings.

Issues

Forensics experts are often brought in once a problem has already occurred. This is incredibly problematic for computer forensics, because without a baseline and audit trail it becomes impossible to accurately see what happened. The best approach is to bring the experts in when things are working. They'll suggest logging and snapshot techniques that will make their job much easier when a problem happens.

True experts in the field of computer forensics are difficult to find. The average network or security consultant will run a few diagnostics and scan through your logs, but few truly know how to properly examine a violated system. A true expert can retrieve information off a computer hard drive even if it's been erased and physically damaged. They can find backdoors and other hacking evidence that no automated software can detect. They have connections with various law enforcement agencies and Internet providers. These experts should be able to trace the activity of hackers back to the originating computer.

Tips

Finding a good computer forensics expert is tough. We recommend calling state and federal law enforcement agencies, including the FBI, to get the names of companies

**Chapter 5
Outsourcing
Options:
Outsourcing
Proactive
Security**

they've worked with in the past. Another good source is the state police, who will probably be able to give you the name of a local company. Large law firms that handle high-tech cases will also have contact information on forensics experts, as their analysis is often crucial testimony.

Protection

Protection is the most complicated component of a proactive security strategy. Most of this book covers the technologies and techniques for protecting networks and information. Your protection is a result of the technical implementation of your security policies through a combination of hardware, software, and policy enforcement. Some aspects of this will be outsourced, whereas in-house staff will handle everything else.

Issues

In theory, a properly designed set of security policies provides the correct level of protection for your business. In practice, it's difficult to get technology to provide the complete protection dictated by your security policies. Practical necessity dictates that you'll always have gaps in your digital armor, but as long as you're aware of it, you can use other techniques to shield the gaps from exploitation. The hard part is getting consultants to tell you the truth about the limitations of the "perfect" technology they implemented for you.

Tips

Use as little technology as possible to achieve the protection requirements of your security policy. Try to use business techniques to obtain the same results. For example, firewalls and blocking devices are often used to prevent unauthorized Internet usage, but the same result can be achieved through proactive management. By using minimal technology, you lessen the chance that a technology failure will open a gap in your defenses.

Vulnerabilities

Albert lives on a diet of Cheetos™ and Coke™. One day he woke up to find that his giant pile of Cheetos had disappeared. He ran to his local bookstore and grabbed a stack of books from the self-help section. While perusing his selections, he found the answer: When somebody moves your Cheetos, you need go out into the world and find a new pile.

Leaving the bookstore, he spotted a steady stream of orange-fingered people coming from a road he'd never been down before. With the crispy-yet-vaguely-toxic taste of Cheetos in his mouth, he raced along the unfamiliar road. Turning a corner, he suddenly found himself facing a brick wall. He could smell the Cheetos on the other side but saw no easy way around.

Unwilling to be thwarted so close to nirvana, he decided he was going to find a way through the wall. He first took a good long look at the wall to see if it had any holes he could squeeze through. In the security world, this process is called *vulnerability scanning*.

Vulnerability scanning is the *automated* process of proactively identifying well-known network system vulnerabilities. Vulnerability-scanning tools test systems against a database of common security flaws. Often, these are the same programs that hackers use to identify easy targets. In fact, a good vulnerability analysis toolkit includes all the popular hacking programs. In Part 12, "Detecting Intrusions," we explore these programs in detail.

Consultants who specialize in this type of analysis are familiar with all the latest programs. Written by and for hackers, these programs don't usually come with easy-to-use graphical interfaces and beginner documentation. The consultants have taken the time to figure out the most effective way to use each tool. They also know how to interpret the results.

Issues

When a consultant runs a vulnerability scan on your network, you can expect one of two outcomes: a list of problems or a clean bill of health. If you have problems, they should be addressed as soon as possible. These problems can be automatically detected, and even novice hackers have little difficulty compromising easily accessible systems with known vulnerabilities.

A clean bill of health doesn't necessarily let you off the hook. When a hacker's automated programs are challenged by a difficult security system, they might decide to get personally involved. Some systems can't be automatically exploited yet are vulnerable to simple techniques such as social engineering. Also, certain advanced techniques used by skilled hackers require direct human interaction.

Tips

When shopping for a vendor to perform a vulnerability check for you, ask them a few specific questions:

- What tools do they use?
- Are the tools commercial, custom designed, or public domain?
- Which vulnerabilities can they detect?
- New system flaws are discovered and exploited on a daily basis. It is a good idea to ask your consultants how they keep up-to-date.

If you are getting fuzzy answers in response to these questions, look elsewhere for a vulnerability assessment team.

Part II Outsourcing Options **61**

Outsourcing

Chapter 5
Outsourcing
Options:
Outsourcing
Proactive
Security

Penetration Testing

When we last saw our cheesy crusader, Albert was in mortal peril. He could find no obvious ways through or around the wall, and the forces of hunger were marching relentlessly across his stomach lining. In desperation, our Havarti hero passed his hand across each and every brick, looking for loose mortar. Finding a weak spot, he subjected it to a vigorous assault of pudgy fingers. A few minutes later, a fleshy tug liberated the first brick from the wall.

Vulnerability scanning is useful, but it comes nowhere near the resourcefulness of a dedicated (or desperate) hacker. The only way to test your system's effectiveness is to hire real hackers. Hiring hacker consultants to find the loose bricks in your network is called *penetration testing*. In our story, the mysterious being that moved the Cheetos is using Albert to penetration test the security wall.

Penetration testing requires a specific type of skill you *hope* you won't have in-house. People who are good at this have spent a *lot* of time hacking. Any employee who can hack your network probably already has, for his or her own benefit or enjoyment. On the flip side, any security consultant you hire certainly was a hacker at one point in time. Hiring one is like asking a wolf to come in and tell you how he'd get at your sheep. What's to stop him from withholding a few tricks and having his midnight snack at your expense?

Issues

If the hacker you hire is good enough, he or she will successfully hack your network, but this is nothing to panic about—yet. The important question is, how was it done? Did he use social engineering to get at critical passwords? Did he use commonly known vulnerabilities? Did he use hard-to-stop techniques such as email-borne backdoor programs? Or did he use some sort of obscure hacking technique known to only the uber-hackers?

From a business point of view, you'll need to figure out if it's worth defending yourself. Be sure to ask the following questions:

- What will prevent a similar real-life attack?
- How much time and money will it take to implement that?
- Will the attack will be duplicated? What motives would lie behind such an attack?

Armed with this information, you can make a risk management decision as to whether or not the vulnerability should be addressed.

Tips

Worried that the wolves are not telling you everything? Hire two different organizations to do a penetration test. The odds that they'll both hold out on the same vulnerabilities are low (unless they collude). Furthermore, as this is more of an art than a science, one group might be more effective than another. Pay particular attention to any differences in information between the two companies regarding a particular issue.

Unless security is critical to your business model, don't worry about the complicated vulnerabilities. Be much more concerned about damage that can be accomplished by moderately skilled hackers.

Most vulnerabilities can be countered with policy and education. Employees who are aware of security problems are less likely to make basic mistakes that hackers can exploit through social engineering techniques. Hiring penetration testers is a great way to test the effectiveness of your security education.

Final Thoughts

Whatever happened to our hungry brick basher? Employing massive hand strength, built from years of extreme video gaming, Albert ripped out a 20-foot wide hole just large enough for his svelte frame to fit through. Victory and gluttony were his for the taking!

III
Reserving Rights

Summary

How do you secure something that you've already given to someone else? How do you prevent digital information from being duplicated? How do you know when something has been duplicated illegally? Maintaining control of digital rights has become one of the most controversial and complicated aspects of security. A rapidly growing industry is addressing many of these core issues. This part will explore the various problems facing digital rights control.

Key Points

- Businesses that create and profit from digital content need practical ways to enforce their legal intellectual property rights as part of their overall security strategy.
- Rights can be enforced through technology, economics, and the legal system.
- Rights can also be circumvented through technology, economics, and the legal system.
- Businesses often use a combination of these three methods in order to successfully protect their intellectual property.
- Maintaining the relevancy of intellectual property laws has been a struggle since the 1960s because of new technologies and a globalized economy.

Connecting the Chapters

The technologies currently being used to manage digital rights are those that support existing legislation. Digital watermarking enables companies to collect evidence of misuse, which can be used to support legal enforcement. Copy protection techniques attempt to prevent unauthorized use in the first place. This part contains the following chapters:

- **Chapter 6, "Digital Rights Management,"** discusses the business strategies and related technologies for controlling intellectual property rights across digital and traditional media.
- **Chapter 7, "Copy Protection,"** examines technology used to prevent the unauthorized duplication of digital media.

Introduction to Reserving Rights

What's the point of securing the data inside your network if you can't control how it's used when it gets outside? Businesses that create and profit from digital content need practical ways to enforce their legal intellectual property rights as part of their overall security strategy. But what exactly is intellectual property, what rights do businesses have, and how can a business protect these rights?

Intellectual property refers to any original creative work you or your organization might produce. It's the virtual concept behind the physical object. The physical copy of this book that you're reading is not intellectual property; it's your physical property. But the *concepts* in this book are *intellectual property*. If you were to make a reproduction of part of this book, you'd be using our intellectual property. If you were to use a portion of this book in your own writing, you'd be creating a *derivative work* of our intellectual property.

What does that mean? It depends. Different countries have different laws regarding intellectual property. In many cases, a person or organization automatically has rights when they create intellectual property. These rights often extend to derivative works. It's generally illegal to make unauthorized copies or derivative works of someone's intellectual property without his or her permission, but the rule has plenty of exceptions. In special cases, copying may be legal because of the intended use or the age of the original work. Identifying derivative works can also be tricky. If a rapper samples a few beats of a pop song, some legal systems might not consider the resulting hip-hop song a derivative work of the pop song. Whether the sample is recognizable is also a factor. Many popular songs can be distinctly identified from a sample that only lasts a second. Other songs are harder to identify; for example, a long sample of a jazz drum solo might be nearly impossible to identify without extreme familiarity with the sampled recording.

It's important for businesses to understand how local, national, and international laws affect their intellectual property rights. In many cases, the intellectual property laws provide protection for certain types of business information as well as more traditional creative works, such as music, artwork, or literature. Defensively, businesses also want to avoid infringing on the rights of other content owners. In situations where the law provides little or no protection, businesses need to have alternative ways of protecting their creative works and business data.

Protecting Your Digital Rights

Rights can be enforced through technology, economics, and the legal system. Technology is fast and flexible; you can quickly create and test systems for protecting your digital content. The downside is that most technology systems can be circumvented with enough effort. Legal enforcement is often more effective, but can take forever and the applicable laws may not provide the level of protection you need. Economics falls somewhere in between. It takes a while to gauge the effectiveness of a pricing model, but the right model can make unauthorized duplication cost-ineffective. Let's look at each method of enforcement in depth.

Technology

The rapid pace of technology has caused intellectual property protection to become complicated. Therefore, it's only natural to look toward technology for a solution. Most early attempts at protecting digital rights involved copy protection technologies. The idea was to somehow lock the data up so that duplication became impossible. Copy protection is still popular today, and techniques have become much more sophisticated. Modern copy protection solutions attempt to control the environment in which the content is used and are known as *digital rights management solutions*. The chapters cover digital rights management and copy protection technologies in detail.

Economics

What if making a copy is simply not worth it? Downloading a movie from the Internet without paying for it is, of course, illegal. In theory, you can find "free" versions of any movie you'd want to see on the Internet. In practice, it's hard to find most movies. The reason? The economics of supply and demand.

Supply

In order to find a movie on the Internet, three things need to happen:

1. *At some point, somebody has to convert the movie—either from DVD or VHS—into a version that can be stored on a computer.* This is known as a *rip* and is a time-consuming process. Few people know how to do it properly. A rushed or unskilled rip results in a low-quality movie. A rip from anything other than a DVD (such as VHS) also results in a low-quality movie. Many popular movies have been ripped, but a significant number of these rips are unwatchable due to their low quality. Also, thousands of older movies have not been ripped because they're not available on DVD. The economic factors here are access to skilled labor (rippers), raw materials (movies on DVD), and time.

2. *The movie you want needs to be available on somebody's computer system.* Ripped movie files are large, often nearly a gigabyte. Most people have no more than 10 to 20 gigabytes of free space on their systems. A person might keep a handful of movies on their hard drive, but most are going to be stored offline on CDs or DVDs. The economic factor here is the cost of storage resources.

3. *The computer with the movie you want needs to be accessible over the Internet.* Some people keep movies on their system but don't make the files accessible to others over the Internet. The movies might only be available for use over a home or corporate *local area network* (LAN). Sometimes the files are accessible over the Internet, but access is controlled by requiring passwords. This is often done to control bandwidth usage. A few people downloading a movie simultaneously can use up much of the available bandwidth, slowing down the overall network performance. The economic factor here is the cost of bandwidth resources.

Demand

Even if you do find the movie you want, it may take hours to download, even over a high-speed connection. Once you've got the movie, it's going to take up nearly a gigabyte of space on your hard drive. You can burn this to a CD, assuming the movie has been properly designed to fit on one. You can also burn it to a DVD, but it won't play on a regular DVD player (unless you have the right DVD burner and some special software). The result? You've spent hours downloading something you can only watch on your computer screen. Eventually, you need to delete it from your hard drive; otherwise, you're going to run out of space.

Frankly, you would be better off renting the movie. Assume a movie rental costs $5 at your local BigHits video store. Compare that cost to the hours of time spent getting the movie off the Internet (and don't forget to factor in the frustration when you get 95 percent of the file and then the computer you're downloading it from shuts down for the evening). Certainly, most of us would value the time lost in getting the movie as being worth more than the $5 rental cost.

Legal Enforcement

Driving a car too fast is illegal in many parts of the world, yet people continue to speed in flagrant disregard of the law. Many speeders get away with it, however, because there aren't enough cops to pull *everyone* over and issue tickets. As a result, there is safety in numbers. As long as several cars are on the road breaking the speed limit (which is apparently the sound barrier in some European countries), the chances of getting pulled over by the police are greatly diminished . . . unless there's a speeding camera. A speeding camera reduces the sense of safety in numbers by catching violators efficiently and indiscriminately.

In the same manner, publishers cannot stop each individual from getting illegal copies of their software. They are generally only concerned about illegal content distributors. Nobody really cares if you copy a CD or a video game. The publisher's legal fees would most likely far exceed the settlement or award. The negative publicity might also hurt sales if the publisher is perceived as "brutalizing the little guy." However, distributing thousands of copies of a CD or video game can directly hurt the revenue of the publisher. It's worth his or her while to go after anyone who distributes illegal materials on a large scale.

Identifying big-time illegal distributors used to be relatively easy to do until file-sharing technologies went mainstream. Anybody with a few hundred MP3s on his or her computer and a file-sharing program could technically distribute thousands of dollars' worth of content to potentially millions of users. This means the average file-sharing user now fits the profile of a big-time illegal distributor.

As a result, publishers have redefined the concept of a distributor to apply to users participating in file-sharing networks (we call these people "peerheads" for reasons explained in Chapter 31). In April of 2003, the recording industry sued a handful of university students whose file-sharing computers contained hundreds of thousands of songs. They are viewed by the industry as large-scale distributors of illegal materials and are being prosecuted as major civil criminals.

Interestingly enough, an argument can be made that primarily targeting distributors is not effective. As long as users are not prosecuted, they will continue to create a demand. Not surprisingly, within a few months of the initial lawsuits the recording industry started suing small-time peerheads—individuals who have shared only a handful of files. This is the digital equivalent of rolling out speeding cameras. If the lawsuits hold up in court, it may scare off many recreational peerheads.

The Other Side of the Coin

Every coin has two sides; every sword can be used against the one who wields it. Likewise, rights can be circumvented through the same technology, economics, and laws that offer protection. Let's look at these methods from the perspective of the "other side."

Evil Technology

You've got a safe and it's the best in the world. Nobody can crack the lock, but you have to take things out of the safe in order to use them, and once they're out of the safe, they're at risk. That's the fundamental problem behind copy protection technologies: Eventually, the digital content has to be unlocked in order to be used. Once it's unlocked, a copy can be made. It might not be easy, but it's possible. As a result, every type of technological copy protection system can be broken.

Evil Economics

What if your creation is a piece of software that is complex and targeted at a specific market? What if it's a document describing a proprietary and unique business method or technology? In these cases, the individual content is highly valuable. An unauthorized copy of the content might be worth a great deal, even if it's sold for significantly less than the original. It's also worth somebody's time to seek out an unauthorized copy, either via the Internet or traditional gray market channels. In such situations, economics encourages theft and piracy.

Evil Legal Enforcement

Laws can be used to manipulate as well as protect. In many recent cases, large companies have used intellectual property laws to stifle the competition. These companies stretch the interpretation of existing laws, applying them in ways that were not intended by their creators.

For example, recently approved anti-circumvention laws were designed to prevent pirates from cracking and distributing software. This seems reasonable, but these laws have already been used in absurd ways. The following are some examples taken from the Electronic Freedom Frontiers' web site (www.eff.org/IP/DMCA/20030102_dcma_unintended_consequences.html):

- "Lexmark, the second-largest printer vendor in the U.S., has long tried to eliminate aftermarket laser printer toner vendors that offer consumers toner cartridges at prices below Lexmark's. In December 2002, Lexmark . . . [sued] Static Control Components for 'circumvention' of certain 'authentication routines' between Lexmark toner cartridges and printers . . . Lexmark added these authentication routines explicitly to hinder aftermarket toner vendors. Static Control reverse-engineered these measures and now sells 'Smartek' chips that enable aftermarket cartridges to work in Lexmark printers. Lexmark claims that these chips are 'circumvention devices' . . . Whatever the merits of Lexmark's position, it is fair to say that eliminating the laser printer toner aftermarket was not what Congress had in mind . . ."

- "Apple's iDVD authoring software was designed to work on newer Macs that shipped with internal DVD recorders manufactured by Apple. OWC [Other World Computing] discovered that a minor software modification would allow iDVD to work with external DVD recorders, giving owners of older Macs an upgrade path. Apple claimed that this constituted a violation of the [circumvention laws] and requested that OWC stop this practice immediately. OWC obliged . . . Rather than prevent copyright infringement, the [laws] empowered Apple to force consumers to buy new Mac computers instead of simply upgrading their older machines with an external DVD recorder."

Putting It All Together

Businesses often use a combination of legal, technological, and economic techniques in order to successfully protect their intellectual property. The benefits of one approach can be used to counteract the vulnerabilities of another. The following are a few examples of the mix-and-match approach to protecting rights.

Technology + Legal = Watermarking

An advanced technology called watermarking can be used in certain situations to trace copies back to the person who made the first unauthorized copy. Legal enforcement can then be used to recover damages and set an example that may discourage future piracy. This technology is explored in depth later in this part.

Technology + Economics = On Demand

The fundamental problem with digital content is control. Once the content is in the control of an end user, the publisher can't easily protect its usage. The most promising approaches to digital rights management keep control in the hands of the publisher. This implies a completely different economic model, where access to the content is more valuable than the content itself. Consumers subscribe to the publisher's service and receive the content they need when they need it, at the highest level of quality. Often called on-demand services, the value is in the convenience.

On-demand services combine new content distribution technologies with a service-based economic model. Similar services have existed in niche markets for years. For example, nearly every law firm has a subscription to an online legal database. Even though almost all the information is publicly available, the convenience of having quick access is worth the subscription price. The fact that the same material is freely available is mostly irrelevant, because finding the free stuff takes too long. The same model is now being applied to both music and video. Publishers are hoping that consumers will prefer to pay a small fee for guaranteed and instant access to the content they want, as opposed to spending time searching for the content on the Internet.

In some cases, the nature of the content directly lends itself to a subscription model. Traditional subscription content, such as news and periodicals, can naturally transition into a digital environment. Other types of content are emerging that specifically leverage the potential of these new distribution models. For example, the subscription-based video game EverQuest has had remarkable success. Players pay to purchase the game and then pay a monthly fee to play. You can't play without paying because the game employs a central server that checks to see if you've paid your monthly fee before letting you access the game. Even if someone manages to pirate the initial purchase, the monthly fee can't be avoided. Not surprisingly, many new games have been designed to follow this model.

Technology + Economics + Legal = The Future?

It's possible to combine all three techniques in certain circumstances. For example, on-demand consumers may receive content specially watermarked with a unique identifier. If the consumer subsequently copies and distributes the content, the illegal copies can be traced back to the original consumer, and a few high-profile lawsuits might convince other consumers to avoid distributing pirated on-demand materials. As a result, it might become harder to find freely available illegal content, and therefore people would find it more convenient to subscribe to on-demand services.

Why Today's Intellectual Property Laws Are Confusing

Intellectual property laws have been struggling since the 1960s to maintain relevance in the face of new technologies and a globalized economy. Current laws are rather confusing. It's not obvious whom they're designed to protect. Some aspects of today's laws do as much to hamper the creators of digital content as help them. By looking at the historical evolution of copyright law, we can begin to understand why we're in this situation.

Six hundred years ago if you wanted to copy a book, you had to do it by hand. Nobody was really concerned about copyrights. Authors had enough trouble getting a single copy of their books made, let alone an unauthorized one. The demand for important books far exceeded the supply. Then, midway through the fifteenth century, the printing press was invented. Suddenly, books became mass communication tools. It has been estimated that 20 million books were in print by 1500.

Governments and religious organizations worried that the widespread availability of information would encourage subversive thought. Events that leveraged printed media, such as the Protestant Reformation, confirmed the fears. As a result, the earliest legislation pertaining to published media, such as the British Licensing Act of 1622,[1] was focused on censorship.[2] This act essentially set up a publishing monopoly that could be easily controlled by the government.

By the end of the seventeenth century, the public was tired of censorship and publishing monopolies. The Licensing Act was repealed in 1695. Publishing became a completely unregulated business, although the government could halt the publication of works deemed libelous. The result? Enterprising Scottish and British publishers started to sell cheap copies of popular books without paying the author or original publisher.

This free-for-all situation became untenable. Publishers were claiming destitution due to rampant piracy. The result was the Statute of Queen Anne[3] (1710), which laid out the framework of the modern copyright system. It gave authors ownership of their work and protection against unauthorized duplication for a limited period of time. By selling or transferring the copyright, a creator forfeited all rights to the work. At the end of the time limit, the work became public property. Most important, the statute gave publishers legal leverage against pirates.

For nearly 300 years, copyright law would essentially remain the same. However, a few significant changes occurred. The amount of protection time increased over the years. The types of works that could be copyrighted expanded to include maps, charts, audio, video, and software. The concept of "fair use" allowed the limited use of copyrighted material in certain situations. Trade agreements included provisions for enforcing copyrights on an international level.

Since the Licensing Act, publishers have controlled the distribution channels. Authors have always needed to sell their rights to publishers in order to get their works distributed. Once the rights are sold, they often lose control over their works. It's up to the author and publisher to agree upon a compensation package.

With this system, both authors and publishers have the potential to profit. Large publishers give the best authors global exposure, while smaller or academic publishers ensure publication for less prominent writers. Music and video have followed the same pattern once the availability of distribution formats (records and videotapes) made mass publishing possible. For more information, go to www.culturaleconomics .atfreeweb.com/cpu.htm.

---

[1]"An act for preventing the frequent Abuses in printing seditious treasonable and unlicensed Bookes and Pamphlets and for regulating of Printing and Printing . . ."

[2]Interestingly enough, it wasn't until over 200 years after the invention of the printing press that authors' and publishers' rights became a serious issue. Part of this was due to the low levels of literacy throughout the world. Only the wealthy knew how to read, and few people were wealthy. It took many generations of education before the public was literate and wealthy enough to represent a meaningful commercial market.

[3]"An Act for the Encouragement of Learning, by Vesting the Copies of Printed Books in the Authors or Purchasers of such Copies, during the Times therein mentioned."

The Duplication Disaster

For over 250 years, the principles set forth in the Statute of Anne governed the world of publishing. Until about 40 years ago, it was practically impossible to reproduce multiple copies of printed text without a printing press. Likewise, no simple way existed for reproducing vinyl records. Only commercial publishers had the capacity to violate copyrights. Then, in a span of just *three* years, a pair of inventions rocked the foundation of copyright law: the automatic photocopier, commercialized in 1959, and the audiocassette in 1962.

For the first time, end users could duplicate published material with relatively inexpensive equipment. Books could be produced or pirated by anyone who owned a photocopy machine. The same went for music recordings and cassette recorders. This meant that anybody had the potential to be a publisher, or a pirate.

These technologies could have proved disastrous for the publishing industry. But there were two factors that partially cooled the industry's heels: cost and quality.

For photocopying, cost was the main issue. Copying an entire book generally cost more in machine consumables (such as paper and toner) than buying the book. But copying select parts of books proved useful in universities, where the cost of textbooks is rather high. Some photocopy centers went as far as producing and selling "course packs" of photocopied excerpts to students.[4]

Quality was the second issue. Generation loss, as it is called, is when the copy doesn't sound or look as good as the original. With each subsequent generation, an exponential decrease in sound quality occurs. Eventually, you are left with a garbled mess that was formerly your favorite album. For videotapes, generation loss is even more extreme. As a result, most people prefer to buy tapes rather than copy them, because the quality of the original is superior.

In 1976, copyright law in America underwent a significant change. The new developments in duplication technology were a major reason for the revisions. Another factor was pressure to comply with international legislation.

The 1976 Act officially broadened copyright law to cover audio and video recordings. It also extended the term of protection to the life of the author plus 50 years (works for hire were protected for 75 years). Furthermore, the Act officially formalized the concept of *fair use*, which allows the unauthorized copying of copyrighted works for certain purposes, including academics, research, reporting, and criticism. It also codified the concept of *first sale*, which is the right to sell an individual copy of a work over and over again, as long as it is never duplicated in the process. Used bookstores could not exist without the concept of first sale.

These changes were brought about to adjust the protections afforded to businesses that survive on intellectual property. Adjustments are necessary because the world is constantly evolving. Most recently it has evolved into the digital age.

---

[4]This led to numerous legal battles on the exact definition of fair use. The resulting verdicts protected the textbook publishers against abuse by photocopy businesses.

Copyright in the Digital Age

For quite a while, it seemed as if copyright legislation had accomplished its goals. Then computers and digital media came creeping around the corner. A copy of a digital file was identical to the original, an actual clone, be it a document, song, video, or some software. This meant that computers could be used to eliminate generation loss. Digital piracy became a viable, inexpensive alternative to legitimate ownership.

Software publishers responded by building anti-duplication devices into their programs. Known as *copy protection*, these technologies prevented the average user from making copies of commercial programs. The media companies also came up with a copy protection technology, known as *MacroVision*, to prevent the duplication of videocassettes.

From the beginning, copy protection was an exercise in futility. Hackers would immediately release tools that could disable or bypass each new type of copy protection. Some of these tools were sold commercially, such as a small device that eliminated the MacroVision copy protection. The hackers felt that copy protection violated a user's fair use right to back up his or her software or media, and the hackers' efforts enabled legitimate users to exercise their legal rights.

The one saving grace for publishers was distribution. Getting a pirated movie meant going to a big city and finding someone selling videos out of a briefcase. The vast majority of the general public did not have access to pirated material, so the actual impact on the bottom line was minimal. The exception to this was in Asia (see "Piracy in Asia" below).

The options for distribution changed dramatically with the arrival of the Internet. For the first

Piracy in Asia

Many Asian countries pay little heed to U.S. and European intellectual property laws. They see the laws as unfair and oppressive to their third-world economies. As a result, piracy is rampant throughout Asia, with many major stores openly selling pirated software, music, and movies at prices low enough to be affordable to the general population.

Publishers have long claimed they've lost billions of dollars in revenue due to piracy in Asia. They are upset that piracy hinders their ability to successfully market their products to the largest populations in the world. Even if a CD can only be sold for $.50 in Asia, the publishers want that $.50 for themselves. As long as piracy is rampant, publishers will never be able to control the price of their product and realize revenue.[5]

---

[5]One of the biggest negotiating issues for China's entry into the World Trade Organization was international intellectual property rights. But the Chinese weren't about to start paying for hundreds of millions of pirated software programs. China had a parallel initiative to switch its computing infrastructure to Linux, a freely available alternative to Microsoft Windows.

time, an author could globally distribute a work without a publisher. It also meant that pirates could just as easily distribute their *warez*. This term is used to describe illegal versions of commercial software that are packaged and distributed for free by hacker groups all over the world.

This was the publishing industry's worst nightmare: a new distribution channel they had no control over that was more advanced than any of the channels they could control. Most publishing houses did not have the foresight to understand what the Internet might mean to their industry. If they had, they may have begun to change their business model in time to alleviate rampant piracy. Instead, they were left trying to play catch-up.

It wasn't long before most popular audio recordings were freely available over the Internet, with the quality nearly as good as the original. Furthermore, no degradation occurred as copies were made and passed around. Software, which had become an equally massive market by this time, was just as easy to copy. The video game industry, which is now one of the largest segments of the entertainment industry, was the most vulnerable to extensive piracy.[6]

Putting the Genie Back in the Broken Bottle

In the digital age, the new distribution channels blurred the lines between fair use and outright piracy. The publishers found themselves throwing sandbags in front of a tsunami. They were being robbed, but who was really responsible? They couldn't sue the Internet. After all, no official "Internet Corporation" existed, so they did the next best thing.

In 1993, *Playboy* changed the game by successfully suing an *Internet service provider* (ISP) for copyright violations. What made the case interesting was that the ISP didn't actively do anything wrong. A customer of the ISP had publicly posted some of *Playboy*'s pictures on a server maintained by the provider without obtaining permission from *Playboy*. This occurred without the ISP's knowledge. Nonetheless, the service provider was held accountable for the user's actions, even though they were completely unaware of the violation.

The *Playboy* case made it obvious that certain aspects of copyright law were not flexible enough to handle the new technologies at hand. How could any ISP operate if they were to be held accountable for the actions of their customers? The same concept could apply to every major communications infrastructure company liable for any piracy that occurs over the Internet.

Even the *Playboy* ruling wasn't enough to stop Internet piracy. ISPs rapidly found ways to legally protect themselves from future abuses. Then the dot-com

---

[6]Ironically, the one thing everybody thought the Internet would change was completely unaffected: books. For years, pundits claimed that the Web would be the end of books. Then people realized that reading large amounts of text on a computer is about as pleasurable as open-eye surgery. The e-book concept, although much hyped, has gone nowhere. The most successful e-books sell thousands of copies, versus millions for their printed counterparts.

boom happened, and the problems with Internet piracy increased a thousand-fold. Owners of intellectual property, mainly the media moguls and their giant corporations, cried out for new legislation to protect their already over-swelled wallets. Politicians with freshly-lined pockets responded.

In 1998, the *Digital Millennium Copyright Act* (DMCA) became law in America. Its purpose was to respond to the cries for legislation to control copyright in the new digital world. It was also designed in response to the World Intellectual Property Organization's demand that America comply with international legislation in the area of copyright law. A number of welcome changes were mandated in the DMCA. ISPs finally obtained limited liability protection in situations such as the *Playboy* case. Computer repair companies could legally access and make temporary copies of copyrighted materials during the repair process.[7]

The DMCA was immediately controversial and the subject of many heated debates. The majority of the act was uncontroversial, but one section caused an incredible amount of grief. Section 1201 broadly stated that it is illegal to attempt to circumvent any technology that controls access to copyrighted material. This means that breaking a copy protection scheme, regardless of the reason, is illegal. "DMCA Versus Fair Use Versus Privacy" explores one of the major ramifications of this clause.

The DMCA in Practice

For publishers, the DMCA is not a great solution. They have only two viable legal strategies against piracy. The first is to focus on the makers and vendors of copyright circumvention devices. These devices include cable TV cheater boxes, file-sharing applications, and tools for viewing protected media. By prosecuting the vendors under the DMCA, publishers hope to remove

DMCA Versus Fair Use Versus Privacy

The DMCA prioritizes anti-circumvention over fair use. This means, if you need to circumvent in order to exercise fair use, that's tough. This is not good, but what's the alternative? In the digital world, fair use makes piracy possible and relatively easy. How do you know when fair becomes too far? Well, you need to track the usage of all digital media, and that starts to eat at privacy.

So, the content world is sitting on the horns of a triceratops. On the one horn, you have the DMCA, which can eliminate fair use in certain situations. On the other, you have Big Brother, which eliminates privacy but can enforce fair use (and define what's fair). On the third, you have freedom, which relies upon the honor system for rights enforcement. It will be interesting to see how the legislative world solves this complex problem.

---

[7]Unbelievably, it was previously illegal for a repairperson to activate copyrighted software on a computer he or she was repairing. Essentially, this meant that the computer couldn't legally be turned on. Yes, this makes no sense, and yes, a repair company was successfully sued under this principle.

the devices from the marketplace. The other approach is to make public examples out of the people who use circumvention devices.

Prosecuting circumvention device vendors might work when it comes to physical hardware, but it's much more difficult for software, which can easily spread over the Internet. Sometimes the software is free; other times the vendors are based in countries immune from international copyright law. How do you stop a multinational consortium of hackers who are difficult to trace and who release circumvention tools from countries where their work is not illegal?

Making examples of cheaters is a potentially effective tactic, but it

---

DMCA and International Law: DeCSS Goes to Norway

In 1998, no software-based DVD players were available for Linux, yet many people had DVD drives in their Linux computers. If you wanted to view a DVD on your computer, you needed to be running Windows. Then a young Norwegian boy created software that could play his legally bought DVDs on Linux by reverse-engineering the existing DVD player software. The Motion Picture Association of America (MPAA) argued that this software was a circumvention device under the DMCA.

---

can also backfire. If too few cheaters are caught, people will feel that the odds are with them. If population of cheaters is extremely large (such as with MP3s over the Internet), too many "examples made" might be seen as a persecution. If the penalty isn't severe, cheaters will keep cheating until they find a way to avoid getting caught. If it is severe, the publishing company may lose customers. Millions of people exchange music, video, and software illegally. Should publishers sue all their potential customers?

Alternative Legal Techniques

Publishers have other legal tools besides the DMCA. A commonly used technique for protecting digital content is to force the consumer to accept an *end user licensing agreement* (EULA). These are those paragraphs of legalese you have to "accept" before opening shrink-wrapped software or installing downloaded software on your computer.

By using the software, you are technically entering into a contract with the publisher. Although most EULAs are similar, some might contain a few surprises. Many start by saying you have no ownership or control over the product. The publisher then forbids you from using the product in any way other than in the proscribed manner. If the product damages your system, it's not their fault. If it doesn't work, they don't owe you a refund. If you install the software on a Wednesday, they get ownership of your first male child, and so on.

Many people view these agreements as meaningless, but they're not. If you don't agree with them, you can usually return the software for a refund. If you do accept them by opening the packaging or clicking the accept button, the provisions (known as *click-wrap agreements*) are generally binding. In some court cases, the provisions of an EULA were considered inapplicable, but the circumstances have been unique.

The concept behind the click-wrap contract is gaining popularity. It's now being used to further enforce rights on other types of digital content such as music, videos, and documents. Many service-oriented Web sites now require acceptance of this type of contract before providing either free or paid services.

Most users don't read click-wrap contracts; those that do usually lack the specific legal background necessary to interpret them anyway. As a result, users who are part of an organization might unknowingly accept contractual provisions that conflict with company policies. For small businesses, this may be irrelevant, as vendors will probably not find it cost effective to enforce the contract. But medium-sized to large organizations should forbid users from accepting any click-wrap contracts when using company resources. The legal department should evaluate any software or services that require such agreements. If the organization is large enough, it may be able to negotiate with the vendor for a version with more favorable licensing language. Regardless, only knowledgeable and authorized staff should accept these agreements.

Likewise, your organization might decide to use such an agreement when distributing your own digital content. Don't just copy somebody else's contract; get a legal opinion. See if any precedent exists for using this type of agreement to protect content similar to your own. You may save the organization a lot of time and legal fees by using terms and provisions that have legal weight.

The Cumulative Effect

In theory, the passing of new intellectual property legislation should supercede all previous legislation. In reality, gaps or unforeseen situations always occur in which prior law is used to determine a precedent. With all the various layers of copyright and intellectual property law passed since the 1700s, it's still possible to encounter situations where the only applicable prior law is ancient British common law!

Ultimately, this is why some aspects of copyright law are particularly confusing. For example, the DMCA contains provisions concerning the design of boats. Why? Because industrial designs were generally not considered creative works and therefore not protected under copyright law. Similarly, new designs for boats generally cannot be patented because they're rarely "non-obvious" (an invention must be new and non-obvious to qualify for a patent). With no applicable protection, the best designs could be copied by the entire industry with no consideration for the original designer.

Temporary legislation for dealing with this situation has been introduced into the DMCA as an experiment. It has nothing to do with the DMCA per se; it was attached because the proponents knew that the DMCA had a good chance of becoming law and that Congress would consider the attachment harmless. In fact, the additional legislation represents a potentially radical change, which is obscured by the relatively boring and specific application to ship hull design. It creates a prece-

Layers of Laws

Illustration by SageSecure

Laws have evolved over time. Older laws still serve as precedent for new laws, even if they are no longer relevant. This diagram shows how legal issues can "fall through" the historic layers of law. Even today, some cases rely on precedent set hundreds of years ago.

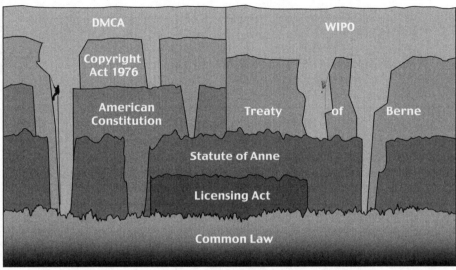

■ **Figure III-1**

dent for an entirely new class of intellectual property protection that has broad ramifications across all industries. Whether it's good or bad is irrelevant. Things like this leave people scratching their heads a decade down the line saying, "How on earth did we get here?" See Figure III-1.

Final Thoughts

Protecting digital content can be difficult and sometimes potentially futile. Nonetheless, few organizations can afford to throw up their hands and surrender their rights entirely. The techniques and technologies we've described can provide a good deal of protection, especially if the organization is vigilant and sensitive to potential exposure. The following chapters will look at some of the specific techniques currently used by content publishers. The technologies and systems that are most compatible with existing legislation consist of digital rights management, copy protection, and digital watermarking.

Future laws may breed new types of rights-protection technologies. The recording and video industry is pushing for laws that will allow them to use aggressive tactics against potential copyright violators. This might include releasing viruses that target and destroy mp3 files or trojan programs that could catalog usage in order to assess penalties. Clearly, a strong amount of public resistance is rallying against this type of legislation, but even a watered-down version would give rise to new types of rights-management tools. Only time will tell what the future has in store for us.

What people think: If content gets illegally released on the Internet, it cannot be controlled.

What we think: Good initial controls will minimize unauthorized duplication and distribution, and good diligence can identify and eliminate sources of unauthorized content on the Internet.

Chapter 6
Reserving Rights: Digital Rights Management

These are the many business strategies and related technologies for controlling intellectual property rights in digital and traditional media.

Technology Overview

The advent of digital devices has created an intellectual property crisis. Is it possible to control the spread of something that can be infinitely copied without loss and at no cost? Even traditional media are not safe. Anything can be scanned or otherwise converted to a digital equivalent. The doom cries of the multibillon-dollar media companies are making Chicken Little seem optimistic in comparison. What's that? Are those war-drums banging in the distance?

Digital rights management (DRM) is a business approach to controlling the distribution of valuable information across all forms of media. The strategy is to manage and enforce intellectual property rights through a combination of technology and law. Creating and maintaining a protected yet accessible environment for valuable content is fundamental to successful digital rights management.

Museums are faced with a similar mandate: protecting valuable artwork without ruining the visual experience. Ropes, panes of glass, and advanced sensors are some of the techniques used by museums. Similarly, digital rights management technologies use encryption and software restrictions to protect content while preserving accessibility. In both cases, the rule is *look, but don't touch or take*.

Gathering revenue is another concern shared by museums and media creators. Museums can charge admission to the entire facility as well as individual exhibits. Patrons can get discounts or privileged access if they have a membership card. Likewise, digital rights management strategies usually incorporate flexible revenue collection systems.

The concept of fair use reproduction gives a vigorous shake to an already complex situation. In the spirit of fair use, many museums allow people to reproduce artwork in limited ways. You can set up an easel and attempt to paint your own copy of a modern masterpiece. You can sometimes use video cameras or take no-flash photographs. You can even purchase a high-quality reproduction from the museum's store (and you can do almost anything you want with your copy, but it would be considered fraud if you tried to pass it off as the original). However, what you can't do is identically reproduce the museum's artwork. Even if it were technically possible to create an exact copy of a painting, the museum physically controls access to the original. In many ways, the physical nature of the medium effectively prevents unauthorized duplication.

If only it were as simple with digital data. One copy is just as valuable as any other. No physical original exists and no physical barriers prevent duplication. One fair use duplication could be used to make thousands of unfair copies.

Most digital rights management strategies solve the fair use problem by ensuring that fair use duplication occurs within a controlled environment. Restrictions are an essential part of digital rights management technology infrastructures. Often the content creator or distributor can fine-tune the restrictions. For example, he or she might allow unrestricted duplication of the first 30 seconds of a song, but only a single copy of the complete song.

Controlling the environment is the most difficult task when building a digital rights management solution. For a DRM solution to be effective, the data must be secure during every portion of its life cycle.

When Fair Isn't Fair

Unfortunately, digital rights management restrictions mean that the content distributor is in charge of deciding what is fair. This could lead to a potential conflict of interests, as the definition of fair may be more restrictive than legally appropriate. For example, in an attempt to prevent the sharing of music over the Internet, some new music CDs can't be played in a PC. But what if the PC is a person's primary sound system? What if you want to make a backup copy of your CD? These are legally valid fair uses of the CD, yet have been prevented by the content distributor.

**Chapter 6
Reserving
Rights:
Digital
Rights
Management**

In order to be effective, protection has to extend beyond the application level. The operating system needs to prevent the interception of unprotected data, and the hardware needs to be secure from end to end. This means that the processor, memory, graphics, audio, and network devices all need hardware-level security. It shouldn't be possible to intercept unprotected information through any component of the computer system.

Once a single PC is reasonably secure, you have the rest of the network to worry about. The application vendors, the operating system vendors, the PC manufacturers, the network hardware manufacturers, and the server software vendors all need to work together in order to make an effective digital rights management solution. That's a pretty tall order. When's the last time you saw that sort of cross-industry cooperation?

How Digital Rights Management Works

Figure 6-1 shows how digital rights management technology interacts with the process of creating a business document and sending it to a colleague.

Security Considerations

Why is all this security necessary? What if the word processor stores the document in an encrypted form so that only someone with the right license and proper reader-software can view the document? Most users would see this as effective security. Unfortunately, it only takes one smart user to break the protection and then post an unprotected version on the Internet.

A skilled hacker knows that a document reader has to do a few things in order to display a protected document. First, the document reader needs to load the protected document, which it must then unscramble. Finally, it displays the document on the user's monitor. This gives the hacker many opportunities to access the unprotected information. For example, when the processor unscrambles the document, it puts the decoded data into the computer's RAM. If a hacker can directly access the document data in memory, the document reader's restrictions can be completely circumvented.

Technology Reality

Reproducing content controlled by digital rights management technologies will always be possible. If you can see it or hear it, you can always take a picture or use another system (such as a video camera) to record what you're viewing or hearing. But it won't be a digital copy, and therefore the quality is generally going to be lower.

Of course, digital copies will also be possible. Hackers will always find a way to crack copy protection schemes. Maybe it is not cost effective to copy the first time, but once cracked, future copies become quite cheap and simple.

A Day in the Life of a Document

Illustration by SageSecure

Normal Document	**How DRM Technology Helps**

 A document is written in a word processor.

 DRM sets restrictions when a document is first created

This program runs on an operating

 DRM integrates closely with the operating system, preventing hacker programs from intercepting the document data

 The operating system controls the hardware in a pesronal computer such as the hard drive, memory,

DRM secures as much of the hardware as possible, further preventing the interception of data by advanced hacker

  When complete, the document may be sent across a network, sometimes repeatedely

  DRM ensures that the document is being sent to an authorized recipient. It may also add a signature to the document, verifying the authenticity of the creator.

 While en route, the document travels through routers, switches, firewalls and other forms of Internet hardware.

The message is encrypted so it can not be read while in transit through the Internet.

 In the end, DRM checks the identification of the associate that receives the message and verifies his authority to read the document. The associate's system is also checked for adequate security measures before the document is unencrypted. Usage of the document is restricted based on parameters set by its creator.

In the end, the associate gets the message and opens it in a word processor. The journey is complete.

■ Figure 6-1

**Chapter 6
Reserving
Rights:
Digital
Rights
Management**

Legal Reality

In many situations, the only way to truly enforce digital rights is through the legal system. Under the DMCA, copying protected content in an unauthorized manner is considered circumvention, regardless of the intended use. The question is how much enforcement do you need? Some people think the DMCA went too far, particularly in the broad nature of its anti-circumvention legislation. Others think it didn't go far enough and have been pushing for even stricter legislation.

One thing is clear: Copyright legislation is in a period of major transition. Many factors are influencing the outcome, including new technologies, global trade issues, and the balance of authority and autonomy for government, industry, and individuals. If history gives any indication of the future, we can expect this transition to last for at least a decade, and possibly much longer. It's unlikely that the results will be ideal for any particular stakeholder, but the situation will at least remain tenable until a new technological revolution changes the way information is distributed.

Consumer Reality

Digital rights management technologies can be perceived as personal invasions to those who are confronted by them as users. These technologies treat all users equally—as potential criminals. This shows a lack of respect, especially when the actual criminals are completely unfazed by the low hurdles in their path. Instead of protecting consumers, the DRM distributors end up abusing them.

One might even hazard that the abuse is intentional. A carefully constructed digital rights management system can be an effective marketing tool by harvesting personal information and preferences. Although many companies might see this as a big win, it can backfire. Would-be customers who don't want to give up personal information might be encouraged to get a "no-hassle" version from an unauthorized source. Nothing encourages the supply of black market goods like a strong demand.

Business Reality

It can be easily argued that the cost of implementing a digital rights management strategy actually far outweighs the potential benefit. On one side are the costs of developing the technology, identifying misuse, litigating and bringing damages, and losing revenue from customers opposed to the oppression. On the other hand, you have the nebulous revenue that could have been earned from "lost" customers, such as the billions in Asia who would have to choose, at retail prices, between hearing the latest boy band and getting to eat for a month.

What about major pirates? Those who copy and resell bootlegs? Like digital rights management is going to stop that! Let's see the *Recording Industry Association of America* (RIAA) go into a major pirating city and arrest every other person they see. It will still have no effect, because even the rabid street dogs have figured out how to burn copies of CDs for profit.

Making the Connection

Cryptography: A critical component to all DRM technologies is the capability to protect the content from unauthorized access via encryption.

Accessing Information: DRM tools need to be integrated at the network level, and access to information containing intellectual property needs to be strictly limited.

Storing Information: A storage method and the storage location are both critical factors that affect DRM.

Best Practices

Many different technologies are used to provide security for digital rights management solutions. Copy protection, encryption, and hardware controls limit access to media. Identification technologies combined with network authentication, for example, compare a user's personal information to a database of authorized license holders. E-commerce systems enable users to access extra content features or additional usage for a payment.

With such a broad base of technology integration, you might think the digital rights management industry would be a vibrant world of exciting opportunities. You'd be wrong. In practice, the world of DRM is defined by the various antipiracy technologies used by of a handful of major publishers such as Sony, Adobe, and Microsoft. DRM software and tools aren't distributed by independent vendors; "The Privatization of DRM" explains why this has happened.

If you are producing content that will be distributed digitally, your DRM options are going to be limited to what is commercially available. "Best practices" aren't really an option here, only the "current practices." The following are some

The Privatization of DRM

A small number of organizations supply the majority of the technology used to create and distribute digital data. Microsoft, Intel, Cisco, IBM, HP/Compaq, Sony, and Adobe primarily control the technology. Disney, Viacom, Vivendi, News Corporation, and AOL/Time Warner control the content and distribution. Also, a significant amount of overlap takes place. For example, Microsoft publishes content and AOL has a significant technology infrastructure.

No single independent vendor could influence all these giants into using a common third-party solution. This, of course, has not prevented a number of companies from trying to create DRM solutions during the dot-com boom. When they all eventually went out of business, the major media giants snatched up their technologies at fire-sale prices. As a result, no significant independent DRM vendors are left.

of the most widely available DRM systems. Chances are you'll be choosing from one of these.

The *Secure Digital Music Initiative* (SDMI) is the most widespread form of digital rights management technology currently available. The initiative was spearheaded by the RIAA. It uses a combination of hardware and software to limit how digital data can be exchanged between compliant devices. The resulting technologies have already become an integral part of many consumer audio, video, and computing devices. Sony has an SDMI-compliant technology called MagicGate/OpenMG. It's built into all Sony devices that record or play multimedia, such as camcorders, minidisc recorders/players, MP players, *personal digital assistants* (PDAs), and PCs. Sony devices can exchange data, when authorized, using their SDMI-compliant MemoryStick. Panasonic has a similar technology called Secure Digital, which is built into devices from competing manufacturers. Of course, even this can be circumvented.

Another major digital rights management initiative is controlled by the *Motion Picture Experts Group* (MPEG), which defines standards for digital audio and video. The MPEG-1 standard is used by CDs and MP3s, and the MPEG-2 is used in digital television and DVDs. *MPEG-21* will be the first multimedia standard that incorporates a complete digital rights management framework. This standard will be finalized in 2009. Many major media organizations, such as the *Motion Picture Association of America* (MPAA), are working with the MPEG group on this standard, but the MPEG-21 standard doesn't address text or Web rights management.

Microsoft has a number of different digital rights management initiatives that may ultimately converge. Their first foray started with tools for controlling audio and video. Their proprietary Windows media format was designed to include optional copy protection. This was trivial for hackers to circumvent and therefore never achieved popularity. Their second initiative was *.NET*, which brings protection to the next level by providing a centralized authorization and license control system (Passport). This also has a number of drawbacks, especially for content distributors who don't want to have Microsoft involved in every transaction.

Intel, IBM, Compaq/HP, and Microsoft have joined together as the *Trusted Computing Platform Alliance* (TCPA). Their purpose is to ensure that the majority of consumer hardware and software supports their planned secure computing infrastructure. This polygamous affair has spawned a love child named *Palladium*. It's a secure operating system that relies on special security hardware. The goal is to give future users a choice of having a "secure" environment that is completely controlled by the TCPA or to have an insecure computer running their choice of DOS, Windows 3.1, or Linux. To call this initiative controversial would be like calling Boston Red Sox fans patient.

If your goal is to secure documents that will be distributed over the Internet, you'll probably want to look into Adobe's digital rights management initiative for PDF documents and e-books.

The currently available solutions are far from perfect or ideal, but they're a start. Companies that need more flexibility and control often implement their own digital rights management systems. This can be time consuming but can also result in a highly effective system that properly integrates with the organization's business processes.

Final Thoughts

The current state of digital rights management reality leaves more questions than answers. Here are a few parting thoughts:

- Will there be compatible standards, or will the major stakeholders continue to move in different directions? Incompatible standards will create security lapses, which will defeat the entire purpose of DRM.

- Who is actually in charge? Who will be the one to monitor the trust infrastructure? Industry? Government? The public? Can any of these three groups actually trust the other two? If, on a social level, no trust exists, how can technology solve the problem?

The law will either make or break digital rights management. A weak law will encourage the creation of circumvention technology. A strong law will punish circumvention. The complexity of the protection technology is irrelevant. If hackers see value in circumventing DRM technologies, they will. Therefore, why is anyone wasting time on a secure infrastructure when the only critical factor is the law?

Chapter 7
Reserving Rights: Copy Protection

Copy protection technology is used to prevent the
unauthorized duplication of digital media.

Technology Overview

The term *copy protection* refers to the wide range of techniques that prevents people from trading, sharing, and using media they have not purchased. The official term for using illegal media is *piracy*. When most people think of software or media piracy, they think about people selling bootlegs of CDs and videos on a street corner in Asia. Few people consider themselves pirates. After all, aren't pirates evil sailors with eye-patches? Yet most people wouldn't think twice about copying a CD for a friend or borrowing the Microsoft Office CD from work to install it at home. Even workplace piracy can unintentionally happen. Many *information technology* (IT) departments install systems using a single set of CDs. Nobody keeps track of the number of installations, and it's not long before the office has more copies than the licenses say were purchased.

Many publishers believe that all this small-scale piracy by typical consumers can add up to real revenue loss, especially because some of these consumers might otherwise pay for the product. Compare this to the less quantifiable revenue loss due to

Asian pirates, who sell to a market of people who can't otherwise afford the product. Although Asian piracy may look threatening on paper, consumer piracy is real. It therefore shouldn't be surprising that the goal of copy protection is to prevent average users from sharing media they have purchased with their friends and coworkers.

> **What people think:** copy protection is annoying.
> **What we think:** copy protection is annoying.

How Copy Protection Works

Copy protection technology can be broken down into three broad categories:

- Media-based technologies rely on the physical properties of the distribution media to protect the content
- Key-based technologies require the validation of a key (such as a serial number) that only an authorized user would have
- Service-based technologies authenticate the individual user before providing access to the content

Media-based technologies are the simplest form of copy protection. Years ago, video games came on cartridges. These were essentially memory chips encased in plastic. Most users lacked the technology to duplicate the cartridges, and this served as an effective copy protection method.

The CD-ROM was the next effective media-based protection. For many years, it was impossible to obtain a consumer-grade CD burner. As a result, it was nearly impossible to duplicate CDs. Thus, CDs seemed like the granted wish of all software development companies: a medium that could hold massive amounts of data yet could not be duplicated at home or the office. Often you couldn't even copy them to your hard drive. The hard drives in those days were too small. For the ultimate in protection, software would require the original CD to be present in the CD-ROM drive in order to run.

Digital media publishers had a big problem when recordable CD drives became mainstream consumer devices. Music and software suddenly were easy to duplicate. As a result, many publishers turned to Secure-ROM, a company that used some novel approaches to protecting a CD. Their process relies on the fact that commercial CDs are physically different from burned CDs. A commercial CD can contain data that can't be burned, such as intentional errors and information on parts of the CD that are otherwise inaccessible. The result is that a burned copy of the original CD is not recognized by the drive as authentic. Ultimately, these media-based approaches don't really work. Hackers are too smart and rapidly find ways to eliminate or neutralize Secure-ROM technology.

When natural defenses are easily surmountable, it's time to put up a locked gate. That's the theory behind key-based copy protection, which generally involves having the user supply information found in the software's packaging or its printed documentation. User-manual protection was a common key-based solution in the early days of copy protection. Activating software required a quote or a word from the manual. This request occurred each and every time the program was launched.

This method has two major faults, however. First, losing the manual meant losing the use of the software. Second, pirates could still share software by simply making photocopies of the manual.

The code wheel was the next evolution in the series of key-based copy protection schemes. Software that used this method came with some form of a cardboard wheel. The wheel would have two or three layers you could line up through small windows in the wheels. Again, the user would be required to enter the secret code upon each use of the software. The wheels were supposed to be difficult to photocopy, but a determined pirate could copy the wheel in little time. Likewise, a legitimate but careless user could easily lose the wheel.

Other interesting attempts were made to better the security system. Dongles, which are pieces of hardware that work in conjunction with the software to secure it, are still in existence today. At first, the dongle seemed like a brilliant idea. After all, it is hardware and therefore expensive to duplicate. Unfortunately, it was relatively easy for hackers to disable the software's reliance on the dongle. Furthermore, the dongle needed to connect to some part of the PC and often interfered with the connections of other peripherals. Finally, if the legitimate owner lost his or her dongle, he or she was really up a creek.

Developers eventually realized that their convoluted key-based systems were hurting the legitimate users and not really diminishing piracy. These irksome protection schemes disappeared in favor of media-based protection until CD burners became widely available. Then they came up with the ultimate key system: the CD key.

If you've installed software from a CD in the past five years, you've probably encountered CD key or serial number protection. Most software comes with a serial number somewhere on the packaging, or via email if the software was downloaded. Most of the time it's printed on the CD case, thus the term CD key. During installation, the user is prompted to enter the serial number. Without the number, the software simply will not install or run.

Software publishers use complex algorithms to generate valid serial numbers. The numbers are usually extremely long and often contain letters, making it nearly impossible to guess a valid number randomly. A second algorithm can verify any given serial number, which means that the CD doesn't need to contain a list of all the valid CD keys (this would be easy to hack), but instead just needs the validation algorithm.

This system sounds pretty well, but it has three fatal flaws. The first is that one CD key can work over and over again. A generic CD key is usually distributed with any pirated software, making it easy to bypass the copy protection measure. Furthermore, hackers are pretty good with math and have long ago figured out how to reverse engineer the original key-generating algorithm from any given verification algorithm. These hackers release *key generators*, which are small programs that spit out valid CD keys on demand. Finally, hackers can alter the software to simply bypass the entire key-checking stage.

Cable TV: Key Versus Service Copy Protection

With traditional cable, every channel is sent to your house. The signal that arrives at your house is scrambled, and the cable box on your TV descrambles the channels you've purchased. The cable company can add or subtract channels from your TV by remotely instructing your cable box. This is key-based copy protection: The cable company has the key that can activate or deactivate your cable box.

The law allows you to buy your own cable box and set it up any way you'd like. The law also requires you to notify the cable company of your settings. It's easy to purchase a cable box that has been configured to automatically descramble all the signals sent to the television. For some reason, most people who buy their own boxes forget to tell the cable companies. It's probably because they're too busy watching every channel for free.

Cable piracy eats into the cable industry's revenue. For years, the cable companies have played cat and mouse with the pirates, trying to detect and damage the descrambler devices. Until recently, they have met with little success.

The cable company changed the game by implementing video on demand, a new technology that only sends the signals you purchase to your home. The signal for *Gone With the Wind* isn't sent to your home until you pay. In theory, this can be done for every channel, not just for movies.

Descramblers can't descramble what isn't there, so the new business model may ultimately thwart cable piracy. Will the pirates find some new way to prevail? As of this writing, no devices are available on the market to enable the theft of digital cable. We will see what the pirates come up with in the years to come.

Service-based copy protection is the newest and most promising option. Users pay for a necessary service associated with the product. This changes the business model. Instead of a one-time purchase, the software is sold for a recurring service fee. The publisher is guaranteed payment up front, usually on a monthly basis. If the payment isn't made, the software stops working.

In practical terms, service-based copy protection means that the publisher needs to control access to critical software functionality. As mentioned in previous chapters, most new game software lets a player connect to the publisher's servers, and he or she cannot access the game without this connection. The publisher can authenticate each connecting user and check that payment has been obtained before allowing the game to run. Thus, copying the game itself isn't enough. A player needs to obtain a valid identity connected to a valid payment system. Most users aren't criminals and will purchase the software instead of stealing credit card information or identities. For most users, this draws a thick line between casual piracy and serious crime.

Security Considerations

When should copy protection be used and what is the correct amount? Is piracy truly harmful or insignificant when it comes to the sale of software products and digital media? Putting copy protection to use comes with many costs. Here are some of the issues to consider before making a decision.

Innocent Casualties

While focusing hard on winning the war against software piracy, many software makers have lost focus of who got them where they are in the first place: the consumer. Copy protection causes one inconvenience after another to the user who legitimately purchases software. The area of serial number protection provides several examples of this.

Imagine this scenario. You arrive home with a brand-new copy of the latest software title. It requires the online registration of your serial number. You break the seal on your software box and enter the new code. An error message from the server states that the serial number is already in use. How can that be? One of two possibilities comes to mind. A hacker's CD key generator happened to randomly generate your serial numbers, or you just bought a used copy of software someone returned to the store. The store then promptly resealed it, presenting it as new, but the original buyer copied the CD and is still using the serial number.

That software piracy is so difficult to stop is bad news for developers. What's been even worse news is that their prevention attempts have caused their legitimate users a great deal of harm. The fact is, only a minority of overall users pirate software. Many users who paid a lot of money for software curse the screen when a legitimate serial number doesn't work. This isn't exactly comparable to the old business adage, "A happy customer is a good customer." Millions of legitimate users are sorely inconvenienced by copy protection schemes every day. Have you tried to reinstall Windows lately and realize you threw out the jewel case that has the CD key written on it? Angry feedback from paying customers about copy protection makes any software maker think twice about this decision.

Lack of Hardware Control

PC hardware gives users a lot of flexibility. The same tools used to protect, encode, encrypt, and secure are available to those who enjoy reverse engineering. The ideal user environment from an antipiracy standpoint is one where the user can purchase, install, and use software, but not manipulate it or its medium. The classic example of such a system is the video game console.

Very few Nintendo, PlayStation, or Xbox video games get pirated. The game publishers rely on the physical limitations of the console system to prevent piracy. Pirating console games often means opening the system and soldering a device to its motherboard. These limitations make it more difficult and expensive to pirate console software than to buy it. Many players would rather not risk breaking their toy.

Most gamers perform a cost-benefit analysis, and software pirates do not want to be inconvenienced.

The real problem is that copy protection does not work. If there is a way to copy something, it will be copied and distributed. And there is always a way because it is easier to break down a building than it is to build one. Even with secure platforms such as game consoles, piracy still exists; it is just less frequent.

False Premise

It has actually been argued that piracy is a good thing. Good? Good for what? Good for sales—or at least some people believe so. Many software companies and industry experts have gone on record claiming that software piracy actually helps sales. There is an interesting logic to this thinking. Piracy always generates an underground buzz about any piece of software. Often this underground buzz translates into a greater volume of total sales.

Another theory is that many people use pirated software to test an application. If they end up using it, then they purchase the legitimate version. The pirate's argument here is, "Why shouldn't I try before I buy?"

Social Rationalization

The original computing community could be safely described as a bunch of digital hippies. People of this lineage would argue today's commercialized software does not fit the vision of the computing community they had 30 years ago. The community they foresaw was about sharing and hobbyists, not about profit and commercialization. Many users still feel this way today, especially with the advent of the Internet. In simple terms, why is software so expensive? Shouldn't it be shared over the Internet so we can all equally benefit from it and advance it further? This explains why piracy is not just a technology problem, but a social problem as well.

Making the Connection

Storage Media: Copy protection is used on many different kinds of media. CDs and DVDs are the most commonly affected ones today, as they are responsible for the widespread distribution of data and video.

Best Practices

Technology companies have been waging war on piracy with only one weapon: more technology. It is only recently that they have considered using a new business model to combat the problem. Changing to a service-based model has proven to be an effective method in preventing piracy and maintaining profitability. Unfortunately, this model is not appropriate for every type of application. Traditional productivity applications (word processors, spreadsheets, and so on) don't really have a need to connect to the Internet and therefore can't take advantage of this type of business model.

Final Thoughts

Whether viewed as a technological, social, or business problem, piracy is still a force in the marketplace. Copy protection has had limited effect on curbing the problem despite its many advances. As new mediums for the delivery of data become prevalent, new forms of piracy will emerge. Will copy protection stay ahead of the game? Only time will tell.

IV
Determining Identity

Summary

It's not enough to have a secure connection between two machines. You also need to be sure that the person or computer you're connected to is who it claims to be. This part discusses the pros and cons of the many available identification systems as well as ideal technology combinations.

Key Points

- Philosophy tells us that a person is more than just the uniqueness of his or her body.
- Technology views a person as a combination of attributes, knowledge, actions, and possessions.
- A digital identity allows the defining characteristics of a person to be rapidly accessed whenever and wherever necessary.
- When properly combined, multiple types of identification technology can improve security.

Connecting the Chapters

Modern identification systems use a combination of technologies, ranging from simple passwords to complex biometric systems (such as fingerprint or retina scanners). This part's chapters explore the most commonly used identification technologies and concepts as follows:

- **Chapter 8, "Passwords,"** examines the words, phrases, or patterns that grant access to a system.
- **Chapter 9, "Digital Certificates and Trusted Authentication,"** covers the electronic documents that verifiably prove the bearer's identity.
- **Chapter 10, "Portable Identifiers,"** discusses the physical items that can associate a digital identity with the bearer.
- **Chapter 11, "Biometrics,"** concerns the technologies that measure a person's vital statistics in order to determine identity.

Introduction to Determining Identity

Throughout history, humanity has been making an ongoing effort to discover itself. We have accumulated and analyzed knowledge and wisdom for thousands of years. For the most part, the goal has been to find answers to two simple questions: "Who am I?" and "Why am I here?"

Sadly, little progress has been made. Although we can provide functional answers to complicated questions such as, "What keeps the earth from crashing into the sun?" we're still at a loss when it comes to those seemingly simple questions. And even then, all we really know is what we've observed, and how do we know our observations are complete? Even worse, these questions beg the shortest and most complicated question of all: "Am I?" You'd think this one would be easy. After all, the only possible answers are yes or no.[1] Unfortunately, philosophers can't agree on either one of them.

The young, brash Western philosophers shout, "Yes, I am!" After all, if there's no self, there can be no identity. If there's no identity, there can be no ownership. If there's no ownership, there can be no capitalism. If there's no capitalism, there can be no Coca-Cola, and we get very grumpy without our daily dose of caffeine. When asked to give a reason for this inarguably practical logic, Descartes succinctly replied, "Cogito ergo sum."

In contrast, the ancient Eastern philosophers believed that the self does not exist. In their view, we are all part of a cosmic unity. The sense of individuality that we call our self is just an illusion of perception. With the right frame of mind, one can see through this illusion and join completely with the singular consciousness. One might achieve such enlightenment by pondering this Zen koan: Who is the difference between one self?

The illusion of reality may be interesting to some, but try explaining it to a hungry person. Eventually, if our individual physical bodies don't get food, we starve and die. In today's world, it's understandable why Western thought is the easier of the two to swallow (sorry). That's a good thing for us, because it's the only mindset compatible with a chapter on identification technology.

Modern society blends philosophical, religious, and technical perspectives on humanity. Most of us Western thinkers assume we have a self, and that it's the only one attached to our body (schizophrenics and those who are possessed might not be so certain). When we see other bodies walking around, we assume they have unique selves too. We call these wandering self-body combinations people. Based on our assumptions, we learn to identify people by the unique features of their bodies. Faces, shapes, voices, smells, movement patterns—these are physical attributes that help us differentiate among people.

A person is more than just the uniqueness of his or her body; people can also be defined by their behavior, their knowledge, the history of their actions, and the items they possess. When this information is shared or well known, it provides an alternate means of identification that can be used in the absence of physical recognition. For example, you may be looking for somebody you've never met. You've been told that she's a tall woman (physical) wearing a leather jacket and a red scarf (possessions). She walks in a funny manner (behavior) and her name is Sarah (knowledge). By combining these factors, you can identify this unknown person with a large degree of certainty.

When you have finished mulling over these deep thoughts about the nature of identity, you may want to drop by your local bank and make a withdrawal. Afterwards, maybe you'll go shopping. During each transaction, you will have a very real need to prove your identity, especially if paying with a credit card. Philosophy won't help you here. Technology has its own way of looking at you that is neither Eastern nor Western. In today's world, your digital identity is just as important as any other self you may or may not have.

Your Digital Identity in General

Philosophy asks, what makes someone what he or she is? Is it his personality or his work? What truly defines a person? What provides every individual on this planet with a sense of distinctiveness?

Identification technology asks, who are you? Tell me who you are. Can I believe that you are who you say you are? How can I trust you? How can I determine your identity with accuracy? How do I know you are not pretending to be someone else? Will anybody vouch for you?

Although philosopher might wax eloquently about the nature of identity, technology does not treat such matters esoterically. Technology is forced to view the defining qualities of a person in a clinical manner through precise and accurate measurements that can be consistently repeated. Computers simply do not have the capacity to factor in the immeasurable. For example, no accurate way exists to measure all the changes in a person's body and behavior due to stress. Computers therefore cannot predict the effect a person's level of stress will have on their voice and body patterns. To a computer, a person under high stress may appear to have a completely different identity.

Using technology for identifying people isn't perfect, but it's a necessary component of a modern society. Technology can provide trustworthy methods for proving identity in a global society where it's often impossible to personally verify if someone is who he or she claims to be. For example, if you purchased a cell phone and a service account, the service provider checked your credit report before they decided to extend you the credit of using their phone service. These typical authentication methods prove that as the global economy grows, the world is getting smaller. Technology provides the necessary means for achieving this.

In another example, when a person becomes financially independent, he or she also begins to build a digital financial identity. Computers record the details of every banking and credit transaction, all of which are collected, organized, and analyzed by several different organizations. In some cases, information is collected from external sources, such as government records. Organizations also exchange information in order to build more accurate and extensive digital profiles. The centralization of this information makes it easy for vendors to obtain digital profiles whenever they need to verify the identity and financial status of a customer. This most frequently happens when a vendor wants to extend credit to a customer.

The Perils of Digital Identity

Most of the time, consumers enjoy the benefits of convenience that come with centralized digital identities. The problems only happen when someone uses the F word, *fraud,* which has become a major problem in the last decade. Criminals have realized it is easier to steal digital money than it is to steal the green variety. The key to virtual robbery is virtual impersonation. Grab somebody's digital identity, and you can take his or her digital assets.

Identity theft is accomplished by obtaining the few pieces of information needed to establish a trusted digital identity. This is not difficult to do. You would be surprised how much of the necessary information is contained in your junk mail. For example, your name, address, and date of birth are often found on common credit card offers, most of which end up in your garbage. Intercepting the right piece of real mail can provide a thief with even more information, such as your bank account numbers and your social security number. Public records can be used to fill in the gaps (see Figure IV-1).

Instant Approval!!!

The credit check process begins when a consumer gives a vendor or store clerk certain bits of critical information about himself. This information is required to verify his digital identity. It usually includes his home address and phone number, date of birth, and social security number.

Once the consumer's identity has been established, a summary score is displayed to the vendor. This score is a dynamic view of the consumer's financial history. The score quickly informs the vendor whether the consumer is a high, medium, or low credit risk. The vendor then makes an instant decision about taking the consumer on as a customer.

This system has its strong and weak points. The speed of information retrieval is a tremendous asset to many creditors. In addition, the credit scoring system provides an accurate evaluation of a consumer's ability to take on additional debt and pay bills on time. But errors do occur. It is unfortunately the consumer's responsibility to ensure that the information is accurate. The trouble is that in many cases, the consumer either has no control over the information or is unaware that the information even exists. It is also difficult, if not impossible, for a consumer to control the way their digital profile is used.

Once these pieces of information are assembled, the holder can begin impersonating the unlucky victim. The information can be used to open credit card accounts, activate cell phones, and even lease automobiles. Identity theft is a federal crime. That said, justice is served far less often than necessary. Identity thieves are difficult to catch, and it's equally difficult to obtain substantial proof for prosecution.

Proving Your Identity

Illustration by SageSecure

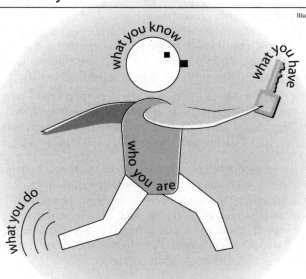

■ **Figure IV-1**

Catching the perpetrators of identity theft crimes is complicated by the fact that most people do not realize they are being victimized until months after the crime has occurred. Usually, the first indication of a problem is a call from an unfamiliar creditor trying to collect money. The victim's response is that the creditor has made a mistake. "I don't have a cell phone account with that company. How can I owe them any money?" The cell phone may only be the tip of the iceberg. Credit cards, lines of credit, and other financial services may have been abused. Crimes may even have been committed in the victim's name, and he or she can look forward to weeks of frustration tracking down and clearing every identity abuse. Years after the fact, he or she will continue to have trouble obtaining legitimate credit due to damage done by the thief.

By the time the dust settles and police reports have been filed, the relatively anonymous criminal is long gone. No perpetrator is apprehended and no record exists for how all this occurred, so who gets stuck with the fraudulent charges? In the best-case scenario for the victim, the credit lender or the vendors take the hit. However, the additional costs of fraud absorbed by vendors each year are ultimately passed onto the consumer.

Digital Identity: The Secure Way

Identity theft is a disaster. What can be done about it? As we've stated before, it's not possible to provide total security for anything, but it is possible to raise the bar way above the heads of most criminals. A good security system is too complex or costly for the average crook to crack. In the case of secure identification, raising the bar involves using a combination of techniques for establishing identity.

Earlier in this chapter, we showed you how a person could be viewed as a collection of physical attributes as well as a set of knowledge, behavior, possessions, and history. These are called *identification factors*, and modern security theory has organized them into the following four categories (see Figure IV-2):

- *What you know* refers to specific *knowledge* that can help someone prove his or her identity. This needs to be knowledge that uniquely relates to a particular individual. The most common example is a password. Other examples include personal history details, such as a mother's maiden name, elementary school teachers, or pet names. Government-issued identification numbers are also frequently used. The lesser known the information, the more secure it is as an identification factor.

  Counterpoint: Many supposedly private bits of data are actually not as private as you'd think. A large amount of personal historical information is publicly available or otherwise easy to discover.

- *What you have* refers to items that are in one's *possession*. These portable identifiers include physical keys, documents, clothing, jewelry, vehicles, and residences. Items that are easy to carry and difficult to duplicate are the most appropriate to use as a basis for identification. Unique documents, such as birth certificates, passports, and government identifications are designed to make duplication incredibly difficult. Credit cards and keys are easier to duplicate, but more convenient for daily transactions.

  Counterpoint: If an item is lost or duplicated, somebody else can use it to impersonate the original owner. Physically protecting the identifying item becomes critical and makes the item less convenient to use.

- *What you are* refers to physical *attributes* unique to one's physical and biological makeup. Examples of these traits include fingerprints, hand topography, hand geometry, and retina/iris patterns. Each one is extremely difficult to duplicate and very specific to a person. Biometric systems are used to record and compare these physical traits.

  Counterpoint: Biometric systems use computers, which can be tricked or bypassed. It's easy to mistake the precision of a biometric system for accuracy. For example, biometric fingerprint scanners may mistakenly grant access to a gel mold of an authorized fingerprint.

- *What you do* refers to unique patterns of *action* that a body generates. This includes handwriting, typing, speech, and movement patterns. These

Identity Theft

Illustration by SageSecure

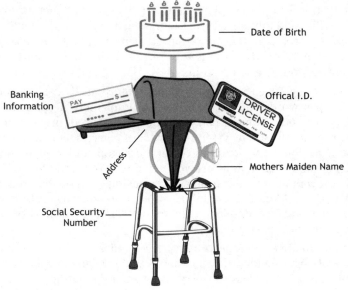

■ **Figure IV-2**

characteristics are far less consistent than direct physical traits, as they are easily influenced by external factors. For example, an awkward writing surface can make a signature unrecognizable, and loud background noise can do the same to a voice. Even under good conditions, the patterns exhibit a wide variance. As a result, systems that capture and compare these traits often have a large tolerance for variations. Such tools are also considered biometric systems.

Counterpoint: Forgers and impersonators take advantage of the tolerance for natural variations, knowing that "close enough is good enough" in many cases. A signature is an example of an easy-to-forge pattern that is commonly used as a critical identification factor in many situations (for checks, credit cards, and so on).

How Many Factors?

A determined intruder can fake any of the four identification factors, but faking more than one factor simultaneously is a significantly harder task. A security system that requires validation of all four factors is difficult to fool. Of course, this assumes that the identification system as a whole isn't the weakest link. To that extent, we can call the security level of the system itself the *fifth factor*.

Ideally, every critical system should require users to satisfy all four identification factors. In reality, this is hard to implement. Each identification factor requires a different type of verification infrastructure. Combining all four into a single system requires complete control over the operational environment. Without total control, compromises are made based on the nature of the environment and lead to weaknesses in the entire identification system.

For example, a high-security system might require an ID card, a voiceprint, a handprint, and a password. This might be easy to implement if you're securing the door to a facility. The environment can be physically secured, preventing intruders from attacking the identification system. But what if you're securing a network with remote users? Are you going to equip every desktop and remote user with a hand/card/voice scanner? If so, how are you going to protect the identification system? What will prevent an intruder from compromising a remote computer and capturing or replaying the identification information? Can you physically protect your remote users from being forced to log in at gunpoint?[2] The identification system can only be as secure as the weakest link.

Designing an effective identification system requires balance and consistency. The importance of ensuring identification needs to be weighed against the practical needs of the business. It's useless to worry about secure identification if an unidentified person can obtain the same information through other channels. For example, if an intruder can gain physical access to the network and critical computers, the intruder can bypass all the identification systems by directly accessing the desired data or observing the data in transit.

Similarly, the use of identification technology needs to be consistent throughout the system. Inconsistencies create opportunities for intruders. One of the most inconsistent systems in common use is the credit card. A credit card with a photograph has all four identification factors: what you have (the card), what you do (the signature), what you know (the card number, expiration date, and billing address), and what you are (the picture). This four-factor system seems secure, until you realize that only one factor is necessary in most situations. If you're at a restaurant or a retail store, only possession is necessary, because most waiters and register operators will not closely check the signature. Online or over the phone, neither possession nor a signature is needed, just the knowledge of the card number and other card details. Criminals only need to obtain the card number, expiration date, and billing address or the card itself to gain full use of the credit account. As a result, credit cards are an extremely easy and effective vehicle for fraud.

In most cases, two properly and consistently implemented factors will provide enough security. A password combined with either a physical item or a biometric

---

[2]Most sane managers probably don't worry about remote users being held at gunpoint. After all, few organizations are involved in anything that would attract gun-toting criminals. Furthermore, intruders rarely need to resort to such extreme measures.

measurement raises the bar adequately against most intruders. A three-factor system, however, that combines passwords, biometrics, and physical identifiers is considered even more secure. "What you do" patterns (other than signatures) are generally used as a fourth or optional factor in ultra-high security systems that have very controlled environments.

New consumer computing devices will make implementing multiple-factor systems easier and more convenient without sacrificing the security of the entire identification system. Devices such as laptops, keyboards, and mice are already integrating fingerprint scanners and card readers. Combination devices are being developed that look like credit cards but can also read fingerprints and generate pass codes. These devices will be responsible for integrating biometrics into many aspects of daily life.

Final Thoughts

The rest of the chapters in this part examine the major technologies used for identification: passwords, digital certificates, physical identifiers, and biometrics. We'll also look at systems for centrally managing and controlling complex identification systems across a networked environment. Understanding these technologies will help you design or improve a comprehensive identification system within your own organization.

Chapter 8
Determining Identity: Passwords

A password is a word, phrase, or pattern
that grants access to a system.

Technology Overview

Passwords are based on the concept of keeping secrets, which can be a great security device. Under most circumstances, you're in complete control of the information you keep in your head.[1]

Everyone knows that keeping a secret is hard. Tell a single person and the secret is out. A secret can become public knowledge within a matter of hours if it's juicy enough.

Two basic types of passwords exist: good and bad. Bad passwords are those that are easily guessed, those that can be compromised through research, or those too hard to remember. Good passwords are impossible for someone else to figure out. Good and bad password systems also exist. Bad systems encourage, or neglect to

---

[1]The Surgeon General would like us to state that drugs, hypnosis, alien abductions, and other unusual circumstances can cause you to lose control of your mind.

protect against, the use of bad passwords. Unfortunately, it's not always easy to come up with good passwords, and only a few good password systems are available.

The majority of computer security systems use passwords that are some combination of letters and/or numbers. These passwords tend to fall into one of four categories:

- **Numeric:** These would be things like your ATM pin number, voicemail pass codes, or alarm system codes (12345 or 1052).

- **Word/phrase:** These consist of "open sesame" challenge-response systems used in spy movies ("The weather is warm for January, is it not?" "Yes, but the flight of the eagle in May is more majestic.").

- **Mnemonic alphanumeric:** These passwords are a combination of letters and numbers meaningful to the user: "fine4you," or "nice2n0u." One technique is to use the first letter of each word from a memorized phrase or song lyric, such as "Mhallwfwwas" (think baaaahhhh).

- **Random alphanumeric:** Smash the keyboard a few times, or get a pet monkey to do it (*2aq!zm3L or s-gjq32#gm1). If the monkey's poundings start to read like *Hamlet*, it's broken and you need a new monkey.

It's possible to create both good and bad passwords that fit into any of those categories. Bad passwords are often the result of poorly designed password systems. As a result, a number of advanced password systems have been designed to make bad password choices impossible. Unfortunately, most of these advanced systems have one or more drawbacks:

- **Visual/pictographic:** A series of images, or locations on an image, comprise the password. In one particular system, the password is based on recognizing pictures of people. Although this is an effective approach, many login systems can't support graphics. As a result, visual password systems are rarely used.

- **One-time passwords:** An algorithm generates a series of passwords that can only be used once. These are great, but they require each user to carry around a list of passwords. This is inconvenient for users that have to log in frequently. In the upcoming chapter on portable identifiers, we describe a hybrid system that makes this system convenient and effective.

- **Personal information systems:** These rely on personal information unknown to others (your mother's maiden name, fourth-grade teacher, or pet's name). The clear drawback is privacy. Most people don't like to give this information out. Nonetheless, many financial institutions use this information as part of their customer identification system.

- **History systems:** Private actions you've taken in the past are used to verify your identity, such as which files you looked at the last time you logged in. It's an interesting concept, but busy users and those who log in infrequently might not remember what they did during the last session.

What people think: It's not worth worrying about the password security on a minor application that is not business-critical.

What we think: People use the same passwords for many different situations, and a hacker can collect passwords on an unimportant insecure system. Often, at least one of these will grant access to a much more important and otherwise secure system.

How Passwords Work

Most passwords are stored in a *password database*. This database can exist on the *local file system* or on a *central authentication* server. Password databases are protected by one or more systems to prevent hackers from obtaining the passwords or editing the file to insert their own passwords.

The box that asks for the password is called a *password dialog box*. The user enters their password, which is compared to the password found in the password database. If the two results are the same, the user is given access.

Well-designed password dialogs will always scramble the password before sending it to an authorization server for comparison. Otherwise, anyone intercepting the communication can see the password in plain text. The password-scrambling process is a form of encryption, discussed in Part 9.

In such a system, passwords in the database are also scrambled. A comparison is made between the scrambled user input and the scrambled password found in the database, not the actual plain text passwords.

Since the password database never needs to be unscrambled, a scrambling method can be used that makes unscrambling nearly impossible. The security benefit is that administrators, users, or hackers can't unscramble the database in order to learn all of the plain-text passwords chosen by the system's users.

Although this scrambling process might seem secure, it actually is not enough. A user's password gets scrambled the same way every time. A smart hacker will capture the scrambled password in transit. The hacker won't know the user's original plain text password, but it doesn't matter. The authentication server is only expecting a scrambled password. By simply sending the captured scrambled password along, the hacker can gain access. This is known as a *replay attack*, because capturing a transmission and replaying it back at a later time can compromise the system.

Protecting against replay attacks requires using a *challenge and response* system. The password server sends a randomized *challenge*, some numeric/textual information that is unpredictable and different for every request. The scrambled password is combined with the challenge, scrambled again, and sent back to the password server (this is the *response*). The password server combines the challenge it sent with the scrambled password found in the database and scrambles the result.

The Challenge/Response Password System

Illustration by SageSecure

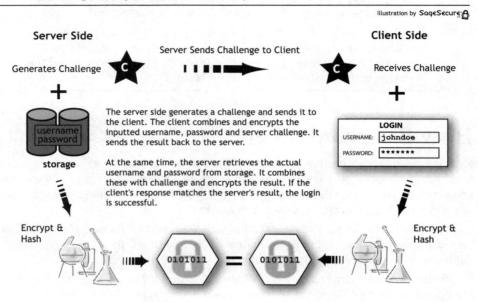

Figure 8-1

This result is compared to the response and should be identical if the password is valid (see Figure 8-1).

The challenge-response system prevents replay attacks because a response is only valid for a specific challenge. Furthermore, scrambling the response makes the password inseparable from the challenge. The next connection will require a different challenge, and any previously captured response will therefore not work.

Security Considerations

We've all seen movies where someone knocks on a door to a villain's lair. A little window in the door slides opens, eyes look through, and a burly voice grunts out, "What's the password?" The hero, watching from around the corner, overhears a henchman's response and uses the same password to gain entry a few moments later. We all sit on the edge of our seats, because we know that an intense bar brawl, gunfight, or chop-socky scene is sure to follow.

This scene illustrates a number of potential problems with passwords. Whenever a password is used, an opportunity is available to intercept it. Overhearing, watching over a shoulder, wiretapping a line, and logging keystrokes are all techniques that can be used to compromise a password. Shared passwords are even worse, because more opportunities for interception are present.

The Dictionary Attack

Passwords that are short or based on dictionary words (in any language) are easy for hackers to crack. Here's how it works.

The computer that controls your ability to log onto the network stores its passwords in an scrambled file. It's usually impossible to unscramble the file by brute force, but hackers have a better technique. They've used the same scrambling system to scramble every word in the dictionary, most common phrases, and all the numbers five to six digits in length. They compare their scrambled passwords to the password file. For example, the word "happy" might get scrambled to "Oxdxx8ffa." If the hackers see 0sdxx8ffa in the password file, they know the user's password is happy.

This technique makes it easy for hackers to reverse engineer bad passwords without actually breaking the scrambling system used on the password file. With even a small amount of users, a hacker is nearly guaranteed to get a few passwords, unless the system administrator implements a system that prevents users from choosing bad passwords.

Passwords tend to be most effective when they're unique to the individual, difficult to guess, and used in conjunction with other authentication techniques. In our spy scene, the hero would have been thwarted if each henchman were recognized by the doorman (biometrics) and had his or her own password. (Of course, this never happens because few movie-goers will believe that a burly, evil doorman can remember hundreds of passwords.) Let's look at some of the issues with passwords a little more closely.

Too Easy to Guess

We all have lots of things to remember. The easiest way to remember a password is to make it something we already have to remember. Birthdays, pet names, and anniversaries—it's easy to use these as passwords because you've already got them memorized. Unfortunately, they're also easy to guess. See "The Dictionary Attack" for more information.

Too Hard to Remember

Ever get an assigned password that is a number or garbled collection of letters with absolutely no meaning? What was the first thing you did? You wrote it down somewhere, because you knew there was no way you'd remember the password. Unfortunately, writing it down just creates another security problem. Now you have to secure a piece of paper. A casual glance by the average office will invariably reveal a number of post-it notes with passwords scrawled about. The same problem is also caused by policies that force passwords to be frequently changed. If passwords are getting written down, something major is wrong.

Input Vulnerabilities

As you know by now, relaying a password is difficult to do securely, regardless of which method you use. Spoken passwords can be overheard. Keypads or keyboards can be watched from a distance, or dusted for fingerprints. Telephone lines can be monitored for dial tones. These are just a handful of ways that traditional passwords can be compromised.

With a computer, even more opportunities for disaster exist. Hackers can use keystroke loggers to capture everything typed into the system or they can invade a computer's cache, a storage area in memory that enables rapid access to information. These cache files can be examined and the password can be stolen in a matter of seconds.

Storage/Transmission Vulnerabilities

Not every application is designed with security in mind, and often passwords are transmitted across a network without encryption. Sometimes they're even stored on the server without encryption. It's not always easy to tell if an application is securely handling passwords.

You might think that it's not worth worrying about the password security on an application that is not business-critical. You'd be wrong. People use the same passwords for many different situations. A hacker will harvest passwords on unimportant insecure system. Often at least one of these will grant access to a much more important and otherwise secure system.

Making the Connection

Cryptography: Passwords are often scrambled when stored and transmitted using encryption to add extra security.

Managing Security: Poor password policies lead to easily compromised passwords. Management needs to understand the limitations of the system users and the restrictions created by business processes. Users need to be taught the importance of password security and how a lazy attitude could hurt them and their company.

Vulnerability Scanning: A common vulnerability analysis technique is to scan systems' password lists for dictionary passwords.

Best Practices

The first step in password security is to recognize and avoid bad passwords. Some examples of bad passwords include the following:

Just numbers: These are too easy for hackers to guess.

Dictionary phrases: These are also easy to crack, as explained in the sidebar.

Overly complex/random passwords: These are likely to be written down or forgotten.

Personal information: Pet names, a spouse's name, and birthdays are all guessable.

Many systems will actually warn users whenever they choose a bad password. Some will even force them to pick a better one, but these systems aren't perfect. They may not be able to tell if a password is overly complex or if it's based on personal information, but they certainly are better than nothing.

A better approach is to encourage a much more sensible and secure system. We call this the "by-the-book" method. Here's how it works:

1. Go to your bookshelf and pick up a book that you like.
2. Open the book, flip through the pages, and stop at an arbitrary point. Don't just pick the page that appears when you first open the book. Many books will open to the same page because of the way the binding has been made.
3. Find a sentence that's easy to remember. If there's nothing you like, choose a new page.
4. Take the first letter of each word to create a password. For example, "Find a sentence on the page that's easy to remember" would be "fasotptetr."

Now you have a password that's derived from memorable words that cannot be guessed. If you need to write something down, write the page number of the book you used (don't write the title or the sentence itself down).

You can use a number of variations to increase the strength of the password. Using capitalization, punctuation, apostrophes, and number substitution can really make things hard to figure out. By using these three variations, our example becomes Fasotpt'e2r. Find a sentence on the page that's easy *2* remember.

If you use variations, be consistent. Don't use caps and punctuation in one password and substitute numbers in another. All your passwords should use a consistent set of variations to prevent confusion.

Simplification by Trust

Wherever possible, we try to make our lives easier by minimizing the details in our lives. Passwords are annoying details, so we want to minimize the number of unique passwords we keep by using a few passwords for everything. This is actually a good strategy, because a few secure passwords are much easier to manage than a lot of insecure ones that need to be frequently changed.

One approach is to use only three personal passwords. The first should be your ultra-secure password. Use the by-the-book method to generate it. You should only use this password on systems you believe are highly secure, such as the login for your office network.

A second password should be used when you have some doubts as to the security of the transmission or storage of the password. For example, accounts on major e-commerce Web sites that store credit card information and other personal details should be protected by this password.

Finally, you simply can't trust certain systems, such as small Web sites that force you to create accounts, free email systems, and so on. For these, use a third, *throwaway* password. When using this password, assume that everyone knows it and anyone can gain instant access to your account.

The same approach can be used when setting master passwords for network systems. The idea is to group your systems into "tiers of trust." For example, your routers, name servers, authentication servers, and file servers all provide critical, interrelated network services. If a hacker gains access to any of these machines, he or she can rapidly gain access to the rest of the network. Once one machine is compromised, the others can often be compromised without using passwords. As a result, it's pointless to give each of these "tier one" machines a separate administration password. Using the same master password makes management easier and doesn't really reduce overall security.

"Tier two" machines include workstations and other systems that have non-administrator access critical network resources. These should all have the same administration password, which should be different from the password used for the highly critical servers. This ensures that a compromised workstation password doesn't give the hacker full access to the more critical systems.

Finally, "tier three" is for those machines that are likely to be compromised, such as those with full access to the Internet. You must assume that hackers will quickly learn the administration password for these systems. Using a separate password here protects your workstations and critical servers.

Advanced Systems

Using advanced systems such as one-time-passwords makes a lot of sense for critical systems that are infrequently accessed. It would be extremely hard for a hacker to compromise a system where the list of passwords is encrypted, burned to a CD, and locked in a safe. An intruder would have to break into the safe and then decrypt the password list. Any intruder willing to go that far will probably find it easier to kidnap the system administrator.

Final Thoughts

In general, passwords are not really secure on their own. It's often easy for a hacker to intercept the usage, transmission, or storage of passwords. Therefore, passwords are best used in addition to some other type of authentication device, such as a biometric scanner or a portable identifier. For example, when entering a *personal ID number* (PIN) on a touch screen, a scan is made of the fingerprints. Access is granted if both the password and fingerprints match. These other authentication systems are described in the following chapters.

A combination system can also be used to spot compromised passwords. A red flag should be raised if a secondary authentication fails, yet the password succeeds. By logging and investigating such failures, a security administrator can detect and deactivate compromised passwords.

Chapter 9
Determining Identity: Digital Certificates

A digital certificate is an electronic document that
verifiably proves the identity of the bearer.

Technology Overview

"On the Internet, nobody knows you're a dog." Peter Steiner's cartoon from a 1993 issue of the *New Yorker* magazine has become an Internet mantra. The anonymous nature of Internet transactions has facilitated boisterous and vibrant social communities among those who would otherwise be shy in public.

Unfortunately, the dark side of the Internet's anonymity is that commercial transactions can quickly lead to fraud. How do you know a "dog" doesn't run the e-commerce site you're about to buy something from? How do you prove your identity to someone over a medium that enables the total manipulation of information? How can you *trust* that what you're seeing is for real?

In the real world, you can prove your identity with a trustworthy photo ID, such as a passport or government-issued identification. These contain a photograph that can be quickly matched against the bearer's face. Antiforgery techniques help make the document itself trustworthy.

Now imagine if these identifications lacked photographs of the bearer. Possession of the ID would be the only important factor in proving identity. If somebody else got hold of your ID, he or she could easily pretend to be you. This is the way many forms of identity used to exist. Years ago, driver's licenses did not contain photographs, but after many incidences of fraud and misconduct, photographs were added to improve security.

The virtual equivalent to a pictureless ID is the *digital certificate*. It contains personal data that can be used to identify the holder. It is issued by a trusted organization, which makes the certificate itself trustworthy. It is assumed that part of the certificate is kept confidential and can only be presented by the legitimate owner. In the "Security Considerations" section, we'll look at this assumption a bit more carefully. For now, pretend that the assumption is valid.

A complete digital certificate system includes people trying to prove their identity, people trying to verify their identity, and one or more trusted third parties capable of performing the verification as well. The entire system is often referred to as a *public key infrastructure* (PKI). The term public does not mean the entire system is designed for public access. Instead it refers to the underlying encryption system that uses a combination of freely shared (public) keys and secret (private) keys.

Two common types of digital certificates exist: *Pretty Good Privacy* (PGP) certificates and X.509 certificates. Both contain personal information about the bearer. Both contain a public key used for verifying and decoding information. The major difference between the two relates to the nature of the trusted third party that verifies the certificate.

Pretty Good Privacy is a popular system for digitally signing and encrypting documents. The *Gnu Privacy Guard* (GPG) is a free, open-source version of PGP (which was originally open source but went through some strange legal issues). PGP establishes trust through an orgy of certificate swapping and signing. As people get to know and trust one another, they sign each other's certificates. The number of signatures and the trustworthiness of the people who sign determine a certificate's overall level of trust. Well-known community members have the most heavily signed and highly trusted certificates. Getting one of these "big cheeses" to sign a certificate automatically imparts a great deal of trust and establishes a stronger case for the certificate bearer's identity.

X.509 certificates rely on a trustworthy central *certificate authority* that can validate the authenticity of a digital certificate. This missionary-style approach to certification has the advantage of convenience: Only one organization needs to verify your certificate before it becomes trustworthy. Obviously, this puts a great deal of responsibility on the shoulders of the certificate-signing authority. They have to prove their trustworthiness to the community or else their certificates are valueless. Verisign is the most prominent Certificate Authority as they validate the certificates used on most e-commerce Web sites.

X.509 and PGP-style certificates have advantages and disadvantages when compared to one another. X.509 certificates can be transparent to an end user and are therefore used heavily in e-commerce Web sites and applications. PGP certificates are often used for securing interpersonal communications because they don't require a central authority's involvement.

> **What people think:** People are unaware they are using digital certificates.
>
> **What we think:** Every time you exchange data over a site encrypted with Secure Sockets Layer (SSL), you are using digital certificates.

How Digital Certificates Work

One of the most common uses of digital certificates is to validate a digital signature. The following steps describe the process of digitally signing a document and validating the signature on the receiving end.

Premise: The sender of a message wants to prove his or her identity to the recipient, and further guarantee that the message hasn't been tampered with en-route.

1. The sender obtains a digital certificate. This contains identifying information, including the sender's public key. The certificate can also include personal identification data such as name and address information. One or more trusted third parties "sign" the certificate and attach their signature to the certificate. The signing process for the certificate is similar to the process for signing a message. Just change the word "message" to "certificate data." Steps 2 and 3 explain how a certificate is signed.

2. A mathematical process known as a *hashing function* is applied to the message. The result is called the hash value. If the message is altered, the hash value of the altered message will not match the value of the original message. For example, a simple hash function might extract every fifth letter of a message. If the message were "The eagle flies at dawn tomorrow," the *hash value* would be "alttr." If the message were changed to "The eagle lands at dawn tomorrow," the hash value would become "aattr." In practice, extracting the fifth letter is not a good hash function because it's too easy to alter the message in ways that would not affect the hash value. Case in point: "Hot bagel flips at nude terror" has the same hash value ("alttr") as the original message. Real hash functions use much more complex systems, and changing just a single letter within the message will affect the hash value.

3. The hash value is encrypted. The encryption is performed using the sender's private key and the public key of the recipient (for additional security). In some cases, a shared secret key is used instead of public or private key encryption.

4. The message is now sent, along with a copy of the encrypted hash value (the signature) and possibly a copy of the sender's digital certificate.

5. The recipient obtains the message. In order to validate the signature, the recipient also needs to have the sender's digital certificate. If it was not included with the message, it can be obtained from a certificate authority or certificate server.

6. When the certificate is presented, the recipient verifies the certificate by using the trusted third party's public verification key to decrypt the certificate signature. The result is the hash value of the certificate data. The recipient runs the same hash function on the certificate data and compares the decrypted hash value with the calculated value. If the two match, the certificate is valid and the identity of the sender has been proven.

7. Finally, the public key included with the certificate and the recipient's private key are used to decrypt the message signature. The result is the hash value of the original message. The recipient runs the same hash function on the received message. If the identifiers match, the recipient knows that the message has not been tampered with during transit (see Figure 9-1).

Security Considerations

Some knowledge is needed to truly understand certificates. Most people don't have enough interest in security to bother learning the intricacies of certificate exchanges, but incorrectly using them can be worse than not using them at all. The false sense of security and trust can be easily exploited.

Theft of a Key

What happens when somebody steals your private key? He or she can sign messages as if they were you. Sure, you can get a new certificate made, but how do you stop people from recognizing the old certificate?

Both PGP and X.509 certificates support the concept of revoking a certificate. Any software that processes a PGP or X.509 certificate is supposed to first check with the signing certificate authority to see if the certificate has been revoked, but unfortunately, this usually doesn't really happen. Checking for revocation can take a while because the major certificate authorities have issued hundreds of millions of digital certificates. As a result, applications such as Web browsers do not automatically check to see if a certificate has been revoked. Although the feature can be turned on, most users are completely unaware of the issue.

Blind Trust

E-commerce Web sites exchange encrypted data with users through a system called *Secure Sockets Layer* (SSL). Rather than verifying the identity of users, this system enables users to verify the authenticity of the site using digital certificates. The goal is to prevent hackers from creating fictitious sites that resemble other legitimate sites yet actually steal credit card information and other personal data.

How Digital Signatures Work

Illustration by SageSecure

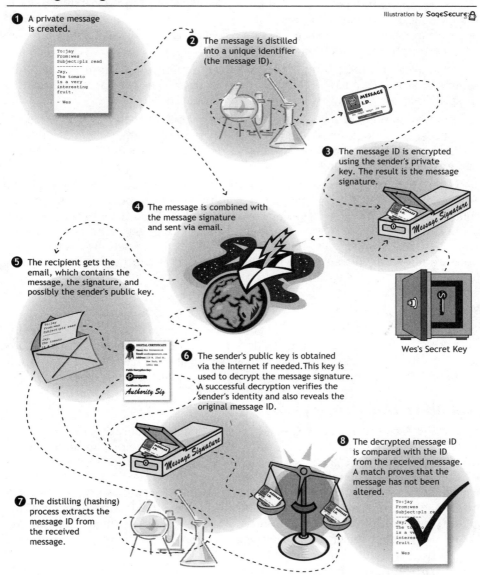

1 A private message is created.

To:jay
From:wes
Subject:plz read

Jay,
The tomato
is a very
interesting
fruit.

- Wes

2 The message is distilled into a unique identifier (the message ID).

3 The message ID is encrypted using the sender's private key. The result is the message signature.

4 The message is combined with the message signature and sent via email.

5 The recipient gets the email, which contains the message, the signature, and possibly the sender's public key.

Wes's Secret Key

6 The sender's public key is obtained via the Internet if needed. This key is used to decrypt the message signature. A successful decryption verifies the sender's identity and also reveals the original message ID.

7 The distilling (hashing) process extracts the message ID from the received message.

8 The decrypted message ID is compared with the ID from the received message. A match proves that the message has not been altered.

To:jay
From:wes
Subject:plz re

Jay,
The to
is a ve
interes
fruit.

- Wes

Figure 9-1

Web browsers have tried to make the digital certificate exchange required by SSL as transparent as possible. To do this, the vendors have preinstalled a set of trusted "root" authority certificates. The certificates signed by these trusted authorities are automatically verified and accepted if valid. The user is completely unaware of the certificate verification process.

A completely transparent yet secure system sounds too good to be true. Sure enough, it is. The fatal flaw is that the transparency comes from assuming that the CA's private verification key will never be compromised.

If a trusted root certificate authority's private verification key were ever discovered, hackers could forge digital certificates for e-commerce sites. Every Web browser would automatically accept these forged certificates and the whole e-commerce system would no longer be secure.

In order to fix this problem, the certificate authority would have to revoke its verification key. But if it revokes the key, all the e-commerce sites with old keys would be invalid. Therefore, the certificate authority would have to reissue every digital certificate it's ever assigned. Then it would have to get every user to update his or her Web browser's trusted root authority certificate settings. Some newer browsers might do this automatically, but the result would be pandemonium for most users.

Monopoly

Unfortunately, because of mergers and business collapses, only a handful of trusted root authorities are available, which leads to the potential for a single point of failure. A hacker can compromise one certificate authority and buckle much of the system. In addition, the current situation could lead to monopolistic pricing conditions.

Exploiting Transparency

Even without compromising a trusted root certificate authority, hackers can still cause mischief. If they have access to a PC, they can add false root authority data to software that automatically trusts the root certificate authorities. Then, they can hack the user's web browser so that when the user goes to Amazon.com, for example, the browser loads the hacker's fake, but identical looking web site. The user's browser accepts the hacker's forged certificate, because the hacker's own certificate authority was used to generate the forgery. The browser trusts the hacker's authority because it automatically trusts anything in its root authority list.

Legal Relevance

One of the big selling points of digital signatures is the concept of nonrepudiation. Only you can present your certificate (assuming you keep it secure). The certificate you present is the recipient's proof that they had an interaction with you. In a court of law, the theory is that you would not be able to deny having that interaction. In reality, it's possible to argue that your certificate was stolen and used by an unknown third party without your authorization.

Making the Connection

Cryptography: Certificates are a part of public or private key encryption.

Virtual Private Networks: Certificates are used to establish trust and secure the link.

Managing Security: Certificates are difficult to use, so only deploy them in appropriate processes. Be careful with transparent certificates used with VPNs and in e-commerce. They can easily overextend trust.

Best Practices

Digital certificates are relatively easy for technologists to understand, but difficult and cumbersome for end users to deal with directly. The following practices will help when using digital certificates in your organization.

Educate users: If your users are going to interact with certificate systems in a non-transparent manner, they will need to be aware of certain issues. Users should know what certificates are, as well as when and why they are used. Users can be easily fooled into compromising the security of a certificate system. If you can't educate your users, you might want to avoid using certificates.

Be careful with transparent deployment: Automatic certificate exchange is useful with systems that have minimal user interaction. Systems such as PGP are difficult to use. Vendors have tried to make software that simplifies the process, but this is generally difficult to do without adding insecurity.

Certificate management: Make sure that critical private certificates are not stored on user systems. Technologies such as smart cards and central authentication can help keep critical certificate data away from the prying hands of hackers.

Legal check: Don't assume that certificates provide additional legal protection, such as proof of a transaction. Check with your legal department and see how relevant legislation treats certificates in practice.

Final Thoughts

Although paper certificates provide a critical form of personal identification, digital certificates are proving their ability to serve as a similarly important form of identification in electronic transactions. Because of their widespread deployment in transparent systems, most people are unaware of the roles digital certificates play in their lives. This may change as people look for ways to directly take digital identity security into their own hands. Understanding and actively managing a set of digital certificates can significantly improve the security of a digital identity, which is especially true when certificates are combined with other identification technologies.

Chapter 10
Determining Identity: Portable Identifiers

Portable identifiers are physical items that can
associate a digital identity with the bearer.

Technology Overview

The earliest portable identifiers were rings, banners, clothes, rarities, and relics. A coat of arms or tartan identified families in some regions. Royalty would have crowns and signets. These items enabled identification even when the face was unfamiliar.

Modern portable identifiers are diverse in form and function. Some are used for generic identification while others enable specific transactions. For example, a credit card is an identifier that is almost exclusively used for purchase. Compare this to a driver's license, which is used to operate a vehicle, get on an airplane, or provide proof of age.

A portable identifier can be read using visual, contact/proximity, active broadcasting, and passive scanning techniques:

- **Visual:** Traditional identifiers such as ID cards and state-issued licenses can be visually inspected. This is frequently observed at fine drinking establishments when some burly guy wearing gold chains asks, "Can I see your ID?"

- **Contact/close proximity:** Magnetic and electronic cards such as credit cards and smart cards are read through physical contact—often by "swiping" the card through a reading device. Some magnetic devices can be detected at a short distance, such as when waved in front of a reader.

- **Active broadcast:** Some electronic devices actively broadcast a signal that can be detected by distant receivers. For example, cell phones and some types of auto alarm systems broadcast a locator signal.

- **Passive scanning:** Other electronic identifiers do not have their own power source. External sensors can nonetheless detect these identifiers from a distance. This is useful when contact is impossible or highly impractical. Many major roadways use passive scanning to identify moving vehicles for the purpose of collecting tolls. A small device mounted on the windshield or kept in the glove compartment is the passive identifier. Implantable passive scan identifiers are now being considered for medical and law-enforcement purposes (see Figure 10-1).

How Portable Identifiers Work

Portable identifiers use a number of different technologies to store and process information. Various types and quantities of data can be stored and retrieved by these handy cards.

Portable Identifiers

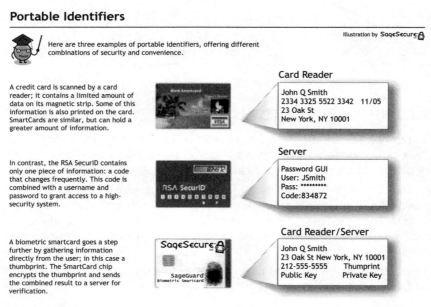

Here are three examples of portable identifiers, offering different combinations of security and convenience.

Illustration by SageSecure

A credit card is scanned by a card reader; it contains a limited amount of data on its magnetic strip. Some of this information is also printed on the card. SmartCards are similar, but can hold a greater amount of information.

Card Reader

John Q Smith
2334 3325 5522 3342   11/05
23 Oak St
New York, NY 10001

In contrast, the RSA SecurID contains only one piece of information: a code that changes frequently. This code is combined with a username and password to grant access to a high-security system.

Server

Password GUI
User: JSmith
Pass: ********
Code:834872

A biometric smartcard goes a step further by gathering information directly from the user; in this case a thumbprint. The SmartCard chip encrypts the thumbprint and sends the combined result to a server for verification.

Card Reader/Server

John Q Smith
23 Oak St New York, NY 10001
212-555-5555   Thumbprint
Public Key   Private Key

Figure 10-1

Bar code, magnetic, or microchip technology can be used to place identification data onto a thin card. These cards a can either be swiped or held in proximity to a reader device. Credit cards, driver's licenses, and telephone SIM cards fall into this category.

Standard magnetic swipe cards (such as credit cards) can contain up to a few hundred characters of textual data. This is usually enough to store a name, the card number, the billing address, and a little extra data. New high-capacity cards can store up to 1.5 kilobytes of data, which is actually enough space to hold a small black and white image. Even more data can be stored on optical cards, which use technology similar to that of CD-ROMs. These were never very popular.

The Smart Card was created for applications that need significantly more storage space. Basic Smart Cards are little more than permanent, reprogrammable memory chips. At the time of this writing, commonly available cards can hold upwards of 64 kilobytes of data. This is much more than what magnetic strips can hold, but is at 5-10 times the cost. New advances in technology are increasing the potential memory capacity even more significantly. Current memory chips used in digital cameras and Mp3 players can hold gigabytes of data on a tiny card. But don't expect to see gigabyte Smart Cards any time soon. The technology requirements for a card that is physically robust are very different than that for delicate digital camera memory.

With a regular memory card, an encryption key can be stored on the card, but the key needs to be transferred to the card-reading device in order to perform encryption/decryption. This creates a security risk if the card-reading device isn't completely trustworthy.

Advanced Smart Cards include a small, low-power processor on the card. If the processor is on the card, encryption keys never have to leave the card's memory. Instead, unencrypted data can be sent to the card and encrypted data will be returned, or vice-versa.

The most prevalent framework for processor-based Smart Cards is Sun Microsystems's Java Card technology. This has a Java Processor on the card that can run limited Java code. The American Express Blue card is a Java Card, although American Express doesn't yet make use of the processor. Finally, the OpenCard framework makes it easy to write software that's compatible with many different types of Smart Cards, including Java Cards.

It's also possible to create a very limited password generation device. Key chains and credit-card sized devices with small digital screens can display part of a password or a response to a challenge for a one-time password. One example is RSA's SecurID system. A small card or keychain-gizmo generates numeric passwords that change and expire every few minutes. These passwords can be used to access a server that supports SecurID. The server and password device rely upon highly accurate clocks to maintain synchronization. The server will not accept expired passwords. This makes it very difficult for a hacker to use an intercepted password—they have only a few seconds to intercept a password, decrypt it and log into a network. Additional layers of security (biometrics, additional passwords, and so on) can be added when necessary.

Security Considerations

A portable identifier that contains valuable information about an individual is a powerful item. With power comes responsibility. Unfortunately, not everyone in the world is always responsible. Imagine the misguided power a portable identifier can have when it falls into the wrong person's hands, be it deliberately or by accident.

The Missing Identifier

The fundamental problem with physical identification devices is that they can be lost, stolen, or copied. People lose things all the time. Ever lose a wallet/purse? Remember the miserable experience of canceling all your credit cards, getting a new license and then checking your credit reports just in case somebody used your information to open new accounts in your name (you did remember to do that, right?).

Hacking

It's possible to duplicate some types of identifiers without ever touching the original. Compromising a card reader is very effective, and often very easy. Small sensor devices can be attached to a card swipe system. These sensors can capture the information from magnetic strips as they are swiped through the reader.

Smart Cards are more difficult to hack—the exchange of data occurs through electronic contact, not magnetically. Furthermore, the exchange is encrypted. Breaking the encryption is a complex process that takes hours. Theoretically, it's possible to build a device that can capture the encrypted transaction for later analysis. Other information, like the amount of power used by the reader and small amounts of electrical radiation escaping from the device, can also be used to break the encryption on a Smart Card. Hacking a Smart Card in this manner requires powerful and sophisticated technology. Of course, if the hacker can get a physical hold on the card for long enough, all of this becomes much easier to accomplish.

Anti-tamper features are built into Smart Cards, but they tend to fail and are often disabled. Even with the features active, it's still possible to access unauthorized information. For example, the chip can be removed from the card and hooked up to common electronics testing equipment that can read data directly from the memory, bypassing all of the security features.

Advanced cards such as Java Cards can be directly attacked by rogue card reader systems. The Java operating system or applications may have implementation glitches that can be exploited. Conversely, a rogue Java Card could be used to attack a card reader, installing a backdoor by exploiting vulnerabilities in the reader system.

Future combination cards will include fingerprint scanning and other user-verification systems to minimize fraud.

Making the Connection

Cryptography: Smart Cards with processors are perfect for cryptographic operations. Data stored on other types of identifiers should generally be encrypted.

Privacy: Portable Identifiers leave trails of information whenever they're used. For example, each time a credit card is swiped, the data on the magnetic strip is captured by the point of sale.

Best Practices

A number of security tricks are used to minimize the risk posed by a lost identifier. The most basic trick is to keep the identifier unlabelled. Password generating devices, magnetic pass cards and other key-type devices should never contain any information about what they open. A simple example is the magnetic hotel key used in modern hotels. The room number is never on the key—if you lose the key, the finder won't know which door it opens. Some hotels don't even print their name on the key, although this is often because the hotel is too cheap to have custom keys printed.

Another important safeguard is the ability to immediately deactivate a lost identifier. This works well for identifiers that communicate with central systems, such as credit cards and passive scan devices. It fails for identifiers that have "off-label" uses, such as driver's licenses. A stolen license can be used to get on an airplane, even if the license itself has been deactivated and re-issued. This is because airlines don't have direct access to the motor vehicle databases (this will most likely change soon).

Smart Cards pose a more difficult problem—critical information can be stored on the memory chip. Changing an encryption key can be a complicated process if the infrastructure isn't designed properly. An institution might deactivate a particular Smart Card, but the information (account number, encryption key, etc.) stored on the card might still be valid.

Combining portable identifiers with other forms of identification technology can increase security. This has been done for years. The original combination token is the photo ID. A driver's license contains authentication information printed on the surface, biometric data (the photograph), and a magnetic or bar code strip with additional information. Some of the newest Smart Card technologies support on-board biometric and password entry systems for added security.

Final Thoughts

Portable identifiers certainly have proven their place in modern society. They have infiltrated the daily lives of many, and provide convenience and added security to those who use them. Nonetheless, the fact remains that when used as a sole system for identification, portable identifiers can do more harm than good. They are best used in combination with other methods of identification security to greatly reduce the risk of fraud or theft. When used properly, portable identifiers are a great addition to personal and industrial security schemes.

Chapter 11
Determining Identity: Biometrics

Biometrics technologies measure a particular set
of a person's vital statistics in order to determine identity.

Technology Overview

The word biometrics comes from the Greek words bio and metric, directly translating into "life measurement." General science has included biometrics as a field of statistical development since the early twentieth century. An example is the statistical analysis of data from agricultural field experiments comparing the yields of different varieties of wheat. In this way, science is taking a life measurement of the agriculture to ultimately determine more efficient methods of growth.

In the most contemporary computer science applications, the term "life measurement" takes on a slightly different role. Biometrics in the high technology sector refers to a particular class of identification technologies. These technologies use an individual's unique biological traits to determine one's identity. The traits that are considered include fingerprints, retina and iris patterns, and facial characteristics. One can see that biometrics is still an appropriate title (see Figure 11-1).

The biological traits used in modern biometric applications are chosen based on our technical ability to catalogue and track them. Some traits are easier to obtain

Biometrics: Taking Life Measurements

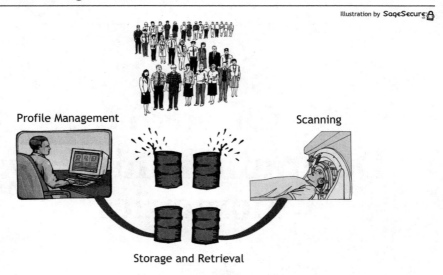

Illustration by SageSecure

Profile Management

Scanning

Storage and Retrieval

■ **Figure 11-1**

than others. Fingerprints, for example, are relatively simple to record and store in a database. They also tend to be less accurate and secure than other more complex biometrics.

Advances in biometric technology are focused on improving the accuracy and security of measurements and reducing the cost to levels appropriate for consumer applications. Simple and low cost systems available today, such as fingerprint readers, will become more reliable. High accuracy systems such as retina scanners will drop in price and will eventually supplement or replace existing systems.

As this book serves proof of, digital security is in ever-growing demand. Complex breaches of security are becoming a worldwide problem. The focus on biometric systems is an industry-wide response to the call for more effective security.

Biometrics, Past and Present

Most people have some degree of familiarity with biometrics, thanks to television and the movies. Hollywood has portrayed biometrics as futuristic technology in science fiction movies, and as elite security technology in spy movies. This has given biometric technologies an expensive and exclusive reputation. Many business owners or executives would most likely say, "We don't need that kind of security; we are not a military facility." Some people don't even think the technology is real, convinced that it's still in the realm of science fiction. As a result, biometric systems have been unintentionally marketed as a very advanced, high-end security technology for many years now.

The difference between today and twenty years ago is seen in both the effectiveness of the technology and the greatly reduced cost. In fact, what one may have only seen in the movies may soon be seen on the front door of your home. Door locks that work using fingerprints or handprints instead of keys are already available at consumer-level cost.

In the coming years a very real and very new market for biometrics will be emerging. Hollywood may have not exaggerated the truth in their movies for a change. Biometrics is truly high tech and, when utilized, gives off an image of an expensive, extremely secure technology. If you have ever had to pass through a retina scanner to get to a meeting, you already know what we mean.

Biometrics is commonly criticized for providing more glitz than security. There can be truth to this claim, depending on how biometric systems are implemented. For example, a retina scanner provides little security if an authorized person holds the door for a stranger standing behind them. Biometrics can only provide effective security when properly combined with other identification factors. Let's take a closer look at how biometrics works to better understand how it integrates within a complete security system.

> **What People Think:** Biometrics is for large and highly secure organizations only. They provide an unprecedented level of security. We can never afford them.
>
> **What We Think:** Anybody can afford to use biometrics. They don't necessarily provide more security.

How Biometrics Work

A biometric device is a combination of a scanning interface and a software system that includes a database and measurement comparison procedures. When a user interacts with a biometric interface the software system will react positively or negatively. A positive response may give the user access to something, or just acknowledge a match in the database. A negative response may deny the person access, or simply determine that the individual has not yet been catalogued. For example, a negative response may tell administrators that the person in front of the hand scanner needs to be recorded for future access.

The first time an individual uses a biometric device, his or her measurements need to be scanned and catalogued. This process is known as enrollment, and serves two main purposes: recognition and authentication.

Recognition systems compare the incoming measurements to every measurement in the database and simply report if a match has been found. These systems are used in numerous applications. Stand-alone fingerprint and handprint scanners check to see if the incoming print is in the "allowed" database. Voice recognition systems match incoming patterns to lists of known words and phrases.

Recognition systems can be set to automatically enroll any unrecognized measurement. They can also be set to reject already recognized measurements. This combination is useful for situations where a person can only participate a single time.

Authentication systems compare the scanned measurement to a particular expected measurement associated with a digital identity. A user first claims his or her identity by either supplying a username or a physical identifier, such as a smart card. The authentication system then retrieves the expected measurements from a database. It then compares the user's measurements to the stored values. If a match is found, the user will gain access according to the privileges specified in his or her digital identity.

There are many different types of biometric devices in use today all over the world. While the main function of all types of these systems is to identify and authenticate, they each perform the task with a unique style.

Fingerprinting has long been used to track criminal and citizens alike. The tip of every finger has a characteristic known as "friction ridges." These friction ridge patterns appear to be similar overall, but no two friction ridges are exactly the same. Police forensic teams have learned to quickly identify identical sets of prints based on patterns within the ridges.

Biometric systems have taken the concept of fingerprinting to new heights. The biometric interface specifically images the ridges of the fingertips using an especially touch-sensitive scanner. The pattern is converted to a digital file and is securely stored in a database or is compared to an already stored image. The same process used by forensic teams is performed in less than a second by a special combination of hardware and software.

Consumer biometric devices have been available to the casual user for quite some time. Fingerprint scanners for PC computers are available as stand-alone devices, and have also been built into mice and keyboards. The fingerprint scanner can be used to prevent unauthorized access to the PC. Some software also uses the fingerprint as an encryption key. This means a fingerprint can be used to protect individual files from prying eyes.

Face recognition software is one of the more recent developments in the field of biometrics. It is far more complex in function, but based on the same principles as fingerprint scanning. Face recognition works by employing a combination of scanning hardware and processing software. The hardware includes discreet high-resolution digital video cameras that can be placed virtually anywhere.

Face recognition software takes the digital image of your face provided by the camera and uses advanced statistics to identify patterns. A common technique is to break the image into a grid, and create a table (matrix) showing the average amount of darkness in each grid region. A mathematical technique called "eigenspace" is used to simplify the table, reducing the image to a set of unique equations. The eigenspace technique essentially treats the image as a topographical map, where facial features are denoted by differences in shading. The resulting equations summarize the relationship between key facial features, such as the distance of nose tip to eyes and cheeks.

Amazingly, these relationships remain constant for any given person regardless of the camera angle, distance, or lighting conditions. Likewise, the relationships are relatively unique—the odds of two people having the same "eigenface" are very small. In many cases, a facial recognition system is more accurate than a human. These systems are far less sensitive to hairstyle, facial hair, glasses, skin tone, and other factors that might confuse a person. They can also match from images that are too small, blurry or distorted for the human eye.

Matching the face is often the easy task. In many cases, face recognition software actually has to perform the more difficult task of identifying the head in a given camera image. In all but the most controlled of environments, the task of "finding the face" is far more difficult due to complex backgrounds and other interfering factors.

One prominent example where this technology has already gone into use is Las Vegas. The phrase "Vegas, baby, Vegas" now has a whole new meaning. Casinos all over town are putting face recognition technology into their already complex and highly advanced security systems. This allows them to quickly identify burned[1] gamblers and escort them out of the casino. With face recognition technology this feat can be accomplished before the individual even gets a seat at the blackjack table.

Before identification technology could capture faces it was able to scan eyeballs. More than a fingerprint, the retina, iris, or cornea of one's eye is able to provide a completely unique pattern by which to identify an individual. The retina, which is located on the back inside of the eyeball, is made up of a series of minute capillaries, which produce a distinct pattern. The cornea and the iris are completely unique in shape and color. An eye scanner can record any combination of this information and store it in a database.

Security Considerations

Biometrics comes with its share of side effects, not all of them being negative. The problems that come with biometric technologies range from nuisances to cost. Planning ahead to incorporate the next generation of security devices is much easier when you know what surprises may lay around the corner.

Cost: As with every new toy, cost is an issue. Not only is the cost of the initial purchase a factor, but so is the cost of upkeep. With biometrics, maintenance can be a real resource problem. Upkeep of almost all biometric systems requires a consistently high level of maintenance and management. Systems that require contact

---

[1]The term "burned" in a gambling town like Las Vegas refers to an individual who has broken casino rules in the past and been caught for it. Once his or her face and habits are connected every casino in a gambling town will quickly ban the individual from gambling. Note the authors only know about this from reading and not direct experience.

suffer from dust accumulation as well as hand cream build up or grease on the sensors. This necessitates routine cleaning as well as diagnostic checks and sensitivity adjustments. Failure to complete the cleaning and tweaking process regularly will most often result in a high degree of system inaccuracy.

Maintenance: Technical problems also plague complex biometric systems. The fact is, they rely on large relational databases to catalogue and store the images they scan. These relational databases tend to degrade over time. They also tend to show a decrease in accuracy and response time as the data pools they store increase in size. The overall result: as the use of the biometric system increases, the overall performance of the biometric system decreases. This in turn lends itself to more maintenance and even greater upkeep costs.

Privacy: Although maintenance, technical problems, and cost may not be a surprise, the following fact is likely to catch most off guard. Biometric systems can often create an invasion of medical privacy for the people required to use them. Because of the constant accurate tracking of certain biological traits, the database is capable of detecting minor changes. These changes can often provide enough information to accurately diagnose certain medical ailments. For example, conditions like diabetes and stroke will change blood vessel patterns in the eye. A retina scanner may start rejecting a user as these conditions mature. Upon examination, the change in the eye pattern will be identified as the culprit. Many consider the capturing of information that can lead to such conclusions to be an invasion of medical privacy.

Health: Another medical related problem with using biometric security devices is the perceived health risks. Many people are not comfortable having their retina scanned, as they feel it may be doing damage to their eye. Additionally, there is the concern of such devices spreading communicable diseases. Germs that cause many ailments may be lying on the interface your body is forced to interact with each day. Whether such health risks are real or not is important in the long-term. In the short-term this may result in user resistance when trying to rollout biometric security for the first time.

Acceptance: With all these issues surrounding biometric security it becomes clear that successful corporate implementation may not be so simple. One component required to achieve success with a biometric security policy is full staff cooperation. Biometrics will only fit into your corporate culture if your employees accept it as part of their routine. People put up with many hassles on a daily basis—traffic jams, lack of parking, and so on. When they finally get to work they may not appreciate a three-tiered biometric security system. It takes the right attitude and adjusted skill level to prevent such a system from becoming a nuisance.

Still, there are components to biometrics that make many people feel it is worth the trouble. For one thing, company staff may feel more comfortable knowing their employer has invested in their safety. This is a big plus in the morale category. Complex-looking security systems give an extraordinary appearance of a secure environment.

Biometrics and Handguns

A company called Bioscrypt[2] has been working closely with Smith and Wesson for several years now. What do a 150-year-old gunsmith and a biometric company have in common? They forged a partnership for better gun control, which has resulted in smart guns. Guns that will actually get to know their owner, and in turn only allow their owner to use them. Supporters of this idea believe it could be the perfect solution to some of the handgun problems seen around the world. No longer will anybody be able to use any gun.

Here's how it works: When the gun is purchased a small biometric device in the handle scans the fingerprint of the owner. Before the gun is permitted to fire the handprint is scanned to determine the user's identification. If gun does not recognize the handler's fingerprint it remains locked and unable to fire. In this way, biometrics is providing a giant leap in handgun safety. If a police officer, for example, has his gun taken, the assailant will be unable to use the gun. A child will no longer be able to find and fire a handgun that was stored around the house.

In the end, you have to ask yourself if your security philosophy truly calls for biometrics. Highly effective security can be achieved without biometrics. That said, a properly implemented biometrics solution could effectively enhance both real and perceived security. As biometrics technologies drop in cost, they will become more viable for small to medium-sized business environments. If these benefits outweigh the costs and potential deterrents, biometrics may be in your organization's future.

Making the Connection

Cryptography: Biometric data is often stored in an encrypted format.

Managing Security: Biometric technologies should only be deployed in manners consistent with an organization's security philosophy. Successfully deploying biometrics requires careful thinking and organizational commitment to security.

Privacy: Biological measurements are considered highly sensitive personal information and therefore the data gathered by biometric devices must be handled delicately.

---

[2] Formerly Mytec technologies, changed name to Bioscrypt after acquisition and merger.

Best Practices

Biometric technology alone will only provide one-dimensional security. When it's combined with other technologies that bolster its weak points, you can have the makings of a secure environment. As we have established, wholly protecting assets always requires following the four categories of authentication methodologies. Biometrics alone only covers two of these categories, "what you are" and "what you do."

At the corporate or government level, biometrics is seen to integrate with technologies that cover the other two categories of authentication. These technologies include smart cards, passwords and *personal identification numbers* (PINs), magnetic stripes, and even physical keys. The highest level of security for these large organizations uses a minimum of three-factor authentication. That is, their security systems integrate three of the four categories of authentication methodologies.

For example, let's follow a mythical employee named Steve as he tries to gain access to a completely secure area of his company. He approaches a door and the door has a complex-looking device requesting input. The device has a chin rest, a keypad, and a card-swiping receptacle that resembles those used in retail stores. Steve places his chin on the rest, at which time his retina is scanned. Upon verification of his retina pattern in the database he is then asked to swipe his magnetic smart card. If the card matches his assigned card code he is the asked by the system to enter his PIN number. With the correct input and collaboration of all three forms of authentication the door releases its lock and allows Steve access.

In this case the security system is requesting something Steve knows (the password), something he possesses (the smart card), and something he is (retinal scan). This is how three factor authentication works, and here are some of the benefits: If Steve drops his card on the ground in the cafeteria and someone else picks it up, it is completely useless. Steve's smart card on its own will not grant anyone access to anything. If someone finds a way to forge Steve's retina pattern they will still be unable to gain access without his smart card or direct knowledge of his PIN number. This offers solid protection against manyw socially engineered attacks as well.

Integrated Biometrics and You

While at one time biometrics was only used in large organizations, practical use for this technology at home has been established. Earlier in the chapter we looked at fingerprint security devices for home computers. Many devices you come into contact with on a daily basis will most likely incorporate biometrics in the years to come.

Cell phones have been a recent example. Many cell phones have voice-activated software built into the handset that allows the user to call whomever he wants by uttering a name into the phone. Voice recognition software has been available for quite some time. This software is "trained" to recognize your voice and language. The result is that it can type out on the screen whatever you say into a microphone. Its goal is to completely replace the need for a keyboard.

Biometrics: Accuracy vs. Acceptance

Illustration by SageSecure

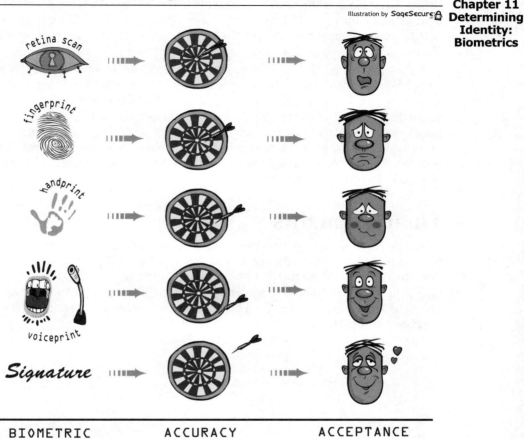

BIOMETRIC   ACCURACY   ACCEPTANCE

Figure 11-2

Advantages to consumer integrated biometric devices are more than just cosmetic. Ultimately, entire systems will improve and in turn make people's lives better. Medical records, financial information, and government records will all be accessible via local computer interfaces. Requests for this information will be granted based on biometric authentication. The result will be fast and deliberate access to collaborated information from anywhere in the world.

In many scenarios this type of technology could save lives. Imagine if a doctor in a California hospital could access a complete compilation of your medical history inside of one minute when you live in New York. If you were to get into a car accident while traveling, the doctor would know about all of your pre-existing conditions and be able to administer the proper treatments in record time. This would solve massive storage problems as well, since analog records of any kind would no longer need to take up space. Biometrics provides the security necessary to make this possible (see Figure 11-2).

Ironically, when the consumers, users, and the biometric industry in general were polled an interesting discovery was made; the comfort level that people have with biometric technologies is inversely related to their effectiveness. This makes sense from a practical standpoint. People are comfortable with what they know. Most people have adjusted to giving their signature as a method of identification. On the other hand, having your retina scanned still seems a little too "Star Trek" for most people today. This information is worth serious consideration before implementing a biometric security device.

Final Thoughts

It may be frightening to realize that databases of faces are being created with each passing moment. Along with the stored faces, habits and identities can be traced over time. Now consider the real life implications of a technology this powerful. It is only a matter of time before a camera can capture your face and your motions and translate them into your identity. This could have massive implications on the future of privacy across the world.

V

Preserving Privacy

Summary

The dark side of authorization and identification is that it relies on information; the more the better. This information can be used to violate the privacy of those you're trying to protect. Furthermore, hackers can more easily compromise your network via social engineering if they have access to personal or private information. Because of this, keeping personal information private is a major need for many organizations.

Key Points

- Privacy is a relative term and it means different things to different people. How much of it someone has depends on perception and environment.
- Privacy is not automatic and takes great effort to achieve and maintain.
- Preserving digital privacy is difficult because information can be obtained quickly and easily. In a digital environment it is often difficult to determine if privacy is being compromised.
- A data trail is inevitably left behind whenever interaction with a computer system occurs, unless the user wipes the trail as they go.
- Organizations have social, moral, and sometimes legal obligations to protect the privacy of personal information that they have collected.

Connecting the Chapters

Various technologies can be used to minimize privacy exposure. Other technologies are explicitly used to collect information and invade privacy without the user's permission. The following chapters explain how to use security tools to increase relative privacy. By managing spam, prohibiting tracking tools, and remaining anonymous whenever possible, a user can reduce the exposure of their private information.

- **Chapter 12, "Anonymity,"** covers techniques for remaining untraceable or without an identity across a large scale network such as the Internet.
- **Chapter 13, "User Tracking,"** examines techniques for tracking the patterns of usage of an individual computer.
- **Chapter 14, "Spam Management,"** investigates ways to keep junk mail from clogging your network and wasting the time of your users.

Introduction to Preserving Privacy

Few topics inspire more debates than the issue of privacy. Scores of organizations exist solely to fight for "privacy rights." Sometimes the corporate world is the enemy, for trying to exploit personal information in order to manipulate customers into buying products. Governments are also targets, for using personal information to stifle freedoms.

What Is Privacy?

What are you thinking? What did you do today? Why are you here? How is your health? Who were you with this afternoon? What did you talk about? How do you feel? What is your favorite color? The answers to those types of questions are personal information. If somebody approached you on the street and started asking those questions, it would most likely be unnerving. Unless deliberately shared with others, information of this nature is considered to be private. Likewise, if someone followed you around and eavesdropped on your conversations, you'd say that they invaded your privacy.

Privacy means different things to different people. The quantity, quality, and relevance of privacy are all matters of relative perception. Some people care significantly about privacy; some care very little. For example, one of the authors is a relatively private person. The other author occasionally runs naked through the streets shouting out his social security number.

In practical terms, privacy is the ability to control access to personal information. You have privacy when you can prevent people from observing your actions and learning what you know. A home is private because the doors and windows can be closed, controlling knowledge about the contents of the house and the actions inside.

Sometimes there's a logical reason for privacy. Privacy can be a defensive mechanism. Certain information can be harmful if it falls into the wrong hands. Information can also be misinterpreted if taken out of context. There are many sound reasons for protecting medical records, financial information, and other types of socially personal information.

Other times, the desire for privacy is purely emotional. Many people are uncomfortable sharing thoughts, experiences, and other personal information. The discomfort has no logical explanation. Even if the information is of no interest or value to a third party, many people will still strive to protect their privacy.

Just as it's impossible to have total security, it's also impossible to have total privacy. The only truly private information is the thought you have but never share. Everything else can be observed, overheard, or inappropriately shared.

The perception of privacy is usually different from the reality of privacy. Ever close a door to have a meeting in "private"? Does this really prevent someone from overhearing? There are a myriad of ways for an eavesdropper to figure out what's happening beyond the door. Much can be learned from observing people before and after they go into a "private" environment, even without bugging the room. Nonetheless, the people in a closed room perceive a greater degree of privacy.

Anonymity can influence the perception of privacy. Some people are comfortable disclosing "private" information anonymously. Perhaps it's because they feel the information can't be used against them personally. Similarly, some people will use pseudonyms when taking actions that could negatively affect their reputation.

An example that illustrates perception and anonymity is the "Tale of Two New Yorkers."

Dave has just moved to New York from a small town in Idaho. Back home, everybody knew him. Although he lived a few miles from his nearest neighbor, the gossip mill in town was very effective. Everybody seemed to know everything about him. If he started dating a girl, the entire town knew within a day.

When Dave got to the big city, he knew nobody and felt like a nameless face in the crowd. Nobody recognized him and nobody knew or cared what he did. He didn't even know the name of the guy living next door, although he could hear him talking on the phone through the wall.For the first time in his life, Dave felt that he had some privacy. Hundreds of people were always within a few feet, but nobody "saw" him or cared what he did. Even if they did watch, he didn't have to talk to them or see them again.

Tom lives next door to Dave. He has been living in the city for 10 years and knew mostly everybody in his building, except for the guy who just moved in to the Robinson's old apartment next door (he knows it's a guy because he can hear him talking on the phone through the wall). Getting to work takes 20 minutes extra because almost everybody en-route wants to gossip about something or other.

Tom feels like he had no privacy. If he starts dating a girl, everyone in the neighborhood immediately knows. Tom fantasizes about moving out west, to a small town where he'd be miles from his nearest neighbor. Nobody would see or care what he did. There he finally would have some privacy!

As the story shows, it's easy to create the perception of greater privacy. It's much harder to increase real privacy. Let's look at where real privacy comes from and how we can get more of it.

How to Achieve Privacy

The way you approach building privacy depends on who you think might want your information. Most people want to keep information from one or more of the following four groups:

Criminals: There are people in the world who will use personal information to harm you, be it physically, financially, or socially. Identity thieves use private information to impersonate you, doing financial harm in the process. Stalkers can use private information to harm you physically.

Society: Some of the things we think and/or do would be frowned on or misunderstood by our societies. We keep these thoughts and actions private whenever possible. Tabloids thrive on exposing "private" aspects of the lives of celebrities.

Organizations: Many commercial and non-profit organizations will use personal information to market their products and services more effectively. They collect vast amounts of profile and preference information in the hopes of identifying a high value market. This information is sold and traded on the open market. Even criminals can use it to target potential victims.

The Government: Some governments like to know what their citizens are doing. Usually, this is justified as the only way to detect seditious or treasonable activities, a necessity for ensuring national security. There have been moments in history where governments abuse their power to violate personal privacy. Past abuses cause some privacy advocates to fear for the future.

Different types of information require different privacy considerations. You probably don't care if your friends and social circle know your mother's maiden name. However, you certainly want to keep that type of information out of the hands of criminals.

For the most part, your personal information starts in your control. You can decide:

- What information you share
- Whom you share information with
- How you share the information

You cannot control:

- Criminal attempts at obtaining private information
- Conclusions that are drawn from observations of you or your actions
- A third party, trusted or not, passing the information on to another
- Key information imposed on you that is ultimately controlled by some other organization, such as a government ID number or a credit card

These ultimately form the core of most privacy violations. While you can't actively stop people from violating privacy, you can protect yourself as best as possible. Four basic ways of enforcing control over private information exist:

Keeping secrets: If you don't tell anybody, nobody will know.

Discretion: Be careful when deciding what to share, whom to share it with and how to share it. Don't let anybody see what you're doing; don't let him or her overhear what you're saying.

Establishing trust: If you need to share a secret, share it with someone you trust. We all learned this lesson when we were young. A secret told to the wrong person became public knowledge within minutes.

Law: Society at large can choose what is considered to be private or not and can mandate protection via laws. Laws can't directly prevent privacy violations, but they can discourage bad practices and offer remedies.

Privacy is not automatic; building and maintaining privacy takes work. Only a few countries guarantee privacy rights to their citizens. Even then, those rights can only offer protection as long as

- Laws exist to preserve those rights.
- Citizenship dues are paid.
- The country maintains its sovereignty.

Privacy isn't cheap. Those countries that offer it can do so thanks to the efforts of people who wish to preserve privacy for all and the lives lost defending the sovereignty of the country in times of war. Even those who never fought for privacy or their country pay the costs of citizenship (taxes, civic responsibilities, etc.).

Where laws end, money begins. Wealth can buy a certain amount of privacy. A house in the hills, tinted windows on the limo, a private jet—these are tools that the wealthy use to increase their effective privacy.

Protecting Digital Privacy

Privacy in the digital world has all the problems of privacy in the real world and then some. The speed and ease at which information can be obtained and shared makes preserving digital privacy a difficult task.

Digital privacy is impossible without digital security. Offline, we can easily keep secrets in our head. The information is safe there until people figure out how to read minds. Online, there's no direct technology equivalent to the human brain. Intruders can easily access storage systems without solid security in place.

Let's revisit the four techniques for control and see what they mean in a digital setting:

Keeping secrets: A digital secret is information that only you can access. Encryption is today's technique for providing that sort of security. Steganography can also be used with encryption to make it hard for others to tell that you have a secret. Both technologies are discussed in Part IX, "Hiding Information."

Using discretion: If you've decided to encrypt your sensitive information, it means you care about your digital privacy. Don't undermine your efforts by sending the information to others in an unprotected format. *Virtual private networks* (Chapter 27, "Cryptanalysis") and encrypted email (Chapter 35, "File Integrity") can be used to ensure discretion during the transmission of information. Equally important is ensuring that the recipient takes the same precautions in storing the information as you have. Digital Rights Management tools (Chapter 9, Digital Certificates") can be used to maintain privacy even after the information has left your direct control.

Another aspect to discretion is ensuring that others are not spying on your actions. Digital footprints are often left behind whenever you interact with a computer system. Privacy control is about erasing those tracks. Some of the technologies we'll discuss later in this part (anonymizers, cookie management, and anti-spyware) can help in eliminating footprints.

Ensuring trust: Do you know whom you're interacting with when you're online? Perhaps your Web browser shows the homepage of your favorite online store, but how do you really know that you've contacted the store's server? What if it's a hacker's server designed to mimic the store? You can't watch the data travel through the network, so you don't know.

Digital certificates (Chapter 13, "User Tracking") use a trustworthy system (public/private key encryption) to prove that someone is who they claim to be. Most people don't realize it, but they're transparently built into the online shopping experience. Web browsers automatically verify the identity of a "secure" site and only complain when the certificate doesn't match. Without using digital certificates, the Internet would be a dangerous place to exchange sensitive information.

Email is a tricky subject for trust. Most people consider their email address to be private information. Spam violates privacy by using your email address without permission. Unlike unsolicited snail mail, spam costs the sender nothing. The people who pay are the ISPs and customers whose bandwidth is being consumed. It also consumes time, as users have to sift through tons of mail in order to find the few important messages. The spam-fighting techniques discussed later in this chapter can help in regaining some email privacy.

Paper Trails and Digital Footprints

You have probably heard the term "paper trail." As much as computers have done for business, they've only been around a relatively short time. For most of the last century, transactions were recorded on paper. Even today, major institutions are not yet comfortable trusting all of their data to digital-only storage. Consequently, many transactions still have a printed record as well as the associated digital information.

Before computers, you could learn a large amount about a person by studying associated paper trails. Purchase histories, passport visas, and train and airplane tickets—all of these paper records could be used to link a person with a place and/or action.

Today, digital systems record your every action online. Credit card companies monitor financial activities. Internet Service Providers and corporations can monitor Internet activity. Market analysis firms also capture Internet data through spyware and devices placed on critical points throughout the Internet. Digital footprints are left everywhere. Even if you don't know where you want to go today, somebody knows where you've been.

The law: Online privacy laws are the subjects of frequent debates and legal battles. In Europe, there are already strong laws protecting the privacy of individuals, both online and off. In America some types of information are protected (medical records and so on), but the legal system has not yet decided how it wants to handle other types of information. The law requires that online organizations post a privacy policy, but this policy can simply say, "There is no privacy here."

Protecting the Digital Privacy of Others

What happens if an organization is collecting information on others? Most companies collect information on their customers simply as a matter of billing. Depending on the industry, the amount of data stored will vary. Doctors and accountants tend to have the most personal information. Consultants and attorneys may have valuable organizational knowledge. Stores can gather extensive preference information based on shopping patterns.

Organizations have social, moral and/or legal obligations to respect private information that has been collected. That information, if carelessly handled, could harm customers.

The relatively poor controls on personal information across the Internet have given powerful ammunition to identity thieves and cyberstalkers. In fact, stalking via the Internet is a growing problem. An amazing amount of information is available via the Internet:

- Telephone, email and mailing address search engines can find contact information for many people.
- Mapquest and other navigation-aid Web sites can provide overhead aerial photographs of an address.
- The contact information for anyone who registers an Internet domain name is publicly available through the "whois database."
- Some companies and organizations make employee information accessible via a public search interface. Some universities make student information available through various online search tools (finger, Web search, and so on).
- Anonymous "detective" agencies advertise thorough background checking for a low, quick and easy fee.

All this and more is available without resorting to social engineering, hacking or more directly invasive techniques.

There are many sad stories of children, women, and men being stalked on the Internet. Anyone who uses public discussion forums, chat rooms or publishes their own Web site is a candidate for cyberstalking. This is a threat to the emotional and physical well being of the person being stalked.

Final Thoughts

Assume nothing.

Trust no one.

Believe.

To the extent that your organization has control over personal information, you should consider the different ways in which the information can be exploited. Whatever is ultimately done with the information, everybody affected should be informed of the protection they do and don't have.

Chapter 12
Preserving Privacy: Anonymity

The ability to remain untraceable or without an identity
across a large-scale network such as the Internet.

Technology Overview

The Internet was not designed with anonymity in mind. Every transaction leaves
a trail of digital footprints. The major Internet routers keep records that can allow
the matching of Internet traffic to specific users. To make matters worse, many applications send personal information all over the Internet. Email, web browsers, and
chat software function as platforms for the exposure of confidential information.
People who care about privacy try to minimize this exposure by using the Internet
anonymously.

How can you obtain total anonymity on the Internet? Put on a disguise. Go to a
cyber café far from where you work or live—preferably in another country. Pay cash.
Browse the Web, but don't log in to any web sites. If you need to log in to a site, create an account that will be used only once. Use a completely random password and
username. Ensure that there is no identifying information in any email messages you
send. Wipe your fingerprints from the keyboard before you leave, or wear latex
gloves. Never come back to the same café again. If anybody looks at you for more
than a few seconds . . . kill them.

Most people consider this approach to be a bit extreme—after all, paying cash is rather inconvenient. They want some anonymity, but without the mess. Ideally, one should be able to browse the Web, make purchases, share files, chat, and send emails anonymously—without having to use secret agent techniques.

Anonymizers are a class of technologies that attempt to provide a certain degree of anonymity in a relatively transparent manner. These software programs use a number of techniques to obscure a user's identity.

Obtaining anonymity on the Internet involves removing all traces of identifying information from communications. This is not easy. For example, an anonymous email needs to have a "from" address that is meaningless and must be sent from a machine that has no relationship to the sender. The message contents also must not reveal the identity of the sender, or must be encrypted.

Most commercial anonymizer

Anonymous Services

One of the benefits of the dot.com boom/bust is the relatively wide availability of free Web-based email. With minimal effort, anyone can set up an untraceable email account. The only trick is finding these service providers. These email accounts are useful when you need to provide an address for web site registration, or if you need to post an address somewhere on the Internet. People can contact you, but they can't figure out your identity.

Another popular service is PayPal®. This service allows two parties to make a reliable, yet anonymous financial transaction. Each party creates a PayPal account (which may or may not be anonymous). PayPal brokers the transaction. Neither party needs to know the other's identity; they just need to know if the transaction is valid. PayPal provides anonymity by acting as a trusted go-between.

---

services operate on the principle of a *trusted intermediary*. In this model, the user makes a connection to the service provider that is not anonymous. The service provider then acts as an anonymizing intermediary between the user and other parties. The user has to trust that the provider won't compromise his or her anonymity.

A web proxy anonymizer is an example of a trusted intermediary. Users first browse to a proxy anonymizer Web site and enter the destination web address into a form on the site (this step isn't necessary if the browser has been configured to automatically use the proxy). The proxy site retrieves the page and displays it to the user. The user never directly visits any web sites other than the proxy site. The web proxy site "promises" to remove any identifying information about the user from their records. Of course, this is impossible to verify; it's purely a matter of trust.

Clearly, the trusted intermediary model is not ideal. A better solution is one where trust is unnecessary. It should be impossible for any intermediary to know or discover the identity of either the sender or recipient. The technology exists to make this happen, and it's known as a *mix-net*. When used properly, a mix-net can provide someone with a chance to remain completely anonymous on the Internet.

David Chaum, a leading cryptologist, developed the basic concept behind mix-nets in the early 1980's. His focus was on anonymous email. Chaum referred to the email server in his model as a *mix*. Derivative email systems built based on Chaum's theory are known as *anonymous remailers*.

How a Mix Works

In Chaum's original mix server design, users send encrypted messages to a mail server. The server decrypts the message and waits until a bunch of other messages accumulate in its queue. It then sends out the messages in a different order from how they were received. An attacker cannot correlate the incoming encrypted messages with the outgoing messages; the contents look different and the order is all *mixed* up.

A major flaw in Chaum's mix model was that there was only one mix. If a hacker compromised the remailer, the anonymity provided by the system would likewise be compromised. The solution is to chain a bunch of mixes together into a mix-network (mix-net). A hacker would have to compromise all of the mixes used in order to defeat the shroud of anonymity provided by the mix-net.

A mix-net takes the remailer concept to the next level. The mix-net sends messages through a collection of mixing servers, which obscure the original source and final destination of a message to an observer. Across mix-nets, servers are considered to be *hops* on a path that lead to an ultimate destination.

Cryptography is used to protect confidentiality of the message and routing instructions. Every mix-net server has a public and private key pair; all information exchanged is encrypted. Eavesdroppers can't read the message or the routing instructions. They can't tell which, or how many, relays the message has passed through. The only thing a relay server knows is the address of the next relay server.

Since Chaum's original work, several approaches to mix-networks have been attempted. All approaches have the same goal in mind: to maximize anonymity across large networks. Routing mix-nets and permutation mix-nets are two approaches that have been frequently used and well documented. Though commonly used in different environments, these approaches illustrate the progress and limitations of mix-nets today (see Figure 12-1).

Routing Mix-Nets

The routing approach to mix-nets has taken over the more traditional methods. In this approach, each mix-server takes information from clients or from other mix servers and then forwards that information in a more processed form to some other server. Routing mix-nets are currently in widespread use because it is relatively easy to modify a mail server or router to strip headers. This means that most networks already have the hardware in place to create a mix-net.

The routing approach relies on multiple source and destination mix-net servers. Each server will receive messages in the form of encrypted packets. The server, using public key cryptography, decrypts the message. Inside are another encrypted

How Routing Mix–Nets Work

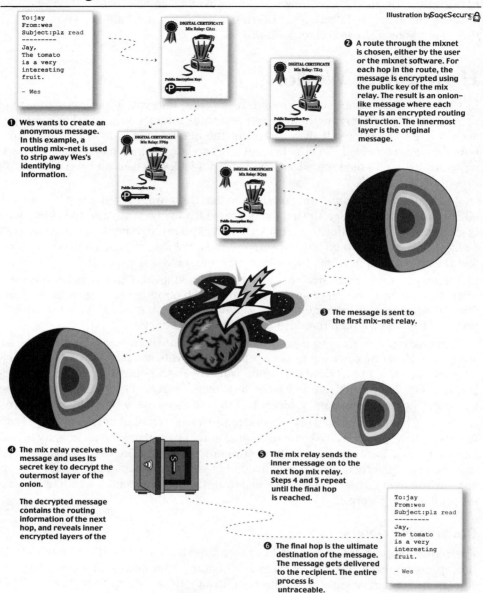

To:jay
From:wes
Subject:plz read

Jay,
The tomato
is a very
interesting
fruit.

- Wes

Illustration by SageSecure

❶ Wes wants to create an anonymous message. In this example, a routing mix–net is used to strip away Wes's identifying information.

❷ A route through the mixnet is chosen, either by the user or the mixnet software. For each hop in the route, the message is encrypted using the public key of the mix relay. The result is an onion-like message where each layer is an encrypted routing instruction. The innermost layer is the original message.

DIGITAL CERTIFICATE
Mix Relay: CA11
Public Encryption Key:

DIGITAL CERTIFICATE
Mix Relay: TX13
Public Encryption Key:

DIGITAL CERTIFICATE
Mix Relay: FF69
Public Encryption Key:

DIGITAL CERTIFICATE
Mix Relay: BQ93
Public Encryption Key:

❸ The message is sent to the first mix–net relay.

❹ The mix relay receives the message and uses its secret key to decrypt the outermost layer of the onion.

The decrypted message contains the routing information of the next hop, and reveals inner encrypted layers of the

❺ The mix relay sends the inner message on to the next hop mix relay. Steps 4 and 5 repeat until the final hop is reached.

To:jay
From:wes
Subject:plz read

Jay,
The tomato
is a very
interesting
fruit.

- Wes

❻ The final hop is the ultimate destination of the message. The message gets delivered to the recipient. The entire process is untraceable.

■ **Figure 12-1**

message and the address for the next mix-net hop. The inner message is then sent along to the next hop, and then the process begins again. This is like peeling away the layers of an onion. The outer layers are stripped away, meaning that a server can't trace the data any further than one hop back. The more hops/layers used, the less likely it is that an observer will be able to track a message from source to destination.

- **Benefits:** This is the most common approach used when establishing mix-nets and the most straightforward to set up and maintain. Mix-nets can be "piggy backed" onto mail servers, for example, because many mail servers are capable of stripping headers out of packet data. Mail servers are well understood by network administrators and therefore more easily manipulated into the form of a mix-net.

  Once servers are configured to strip headers of incoming messages, public key cryptography is fairly easy to add. A mail server set up to strip message information and encrypt all message instructions can also be easily replicated and distributed to multiple locations. In short, the means and methods involved in the routing approach are fairly attainable and can be quickly implemented on multiple host networks.

  Another benefit to the routing approach is that mailed instructions are encrypted using public key cryptography. The routing instructions for a specific mix-net are encrypted with the public key of that particular mix-net. The result is that each mix-net can only read the encrypted instructions destined for it, and not for any other mix-net in the chain of hops.

- **Detractions:** The major drawback to routing mix-nets is that they rely on routing. In other words, each mix-net server is required to analyze traffic that it receives, and actively strip headers from certain packets of that traffic. This can put an immense strain on server hardware, especially if it is also performing other critical services such as *Simple Mail Transport Protocol* (SMTP) for email. Finding an effective and practical balance in the implementation of a routing mix-net can be time-consuming and costly.

There is another more traditional approach to mix-nets known as permutation mix-nets, or list-based mix-nets. Routing mix-nets cannot collaborate with one another as traditional mix-nets can. So the following is a key point of trade off in the world of privacy and anonymity: increased reliability at the expense of inter-server communication. In other words, you lose all the benefits the mix-net can provide as a direct result of inter-server communication.

Permutation/List Mix-Nets

Also known as the list approach, this mix-net method can be powerful and effective. All the messages that enter a mix-net are what make up the list. Messages that make up the list can be linked back to senders because when senders directly pass a message on to a mix-net the communication is not anonymous. The mix-net takes a whole list of messages it has received from various senders and permutes them. The new permuted list is sent out of the mix-net at the other end. The permuted list now provides anonymity for all the original senders because it is impossible to match an outgoing message to a message on the original incoming list. As a result, no one can match a message to a sender, with the exception of the original sender.

Benefits: The permutation approach offers the possibility of many mix-net servers being able to work together knowingly. Because permutation mix-nets do not rely on secret encrypted instructions, each server can collectively work on the process of discretion. This approach is more pure and theoretical than practical. In other words, it is the ideal way mix-nets should be designed and implemented, but is not realistic given most network structures and set up costs.

Detractions: The permutation approach to mix-nets is not very practical, and is mostly put to use in academic circles.

Security Considerations

Why does anyone care to remain completely anonymous on the Internet? One common opinion is that the need for total anonymity online is a result of two situations:

- Someone is up to something illegal.
- Someone is paranoid about privacy.

Few businesses have a legitimate need for anonymity. If anonymizers are being used, it's probably because somebody is doing something wrong within the organization.

Proxy anonymizers, such as Web and email proxies, can be used to get around an organization's security devices. Some organizations block access to sites such as Hotmail or Yahoo mail. If a user can reach an anonymous Web proxy, they can use the proxy to reach the blocked sites. Monitoring logs for evidence of proxy use is a good idea for any organization that wishes to restrict Internet access.

If an organization does have a legitimate need for privacy, it should consider using a mix-net. Unfortunately, mix-nets are not simple to create, they require a good deal of equipment and expertise. Most users will need to find a public mix-net, or one created by another organization, but finding such a mix-net can be difficult. Mix-net users obviously want to remain anonymous, so they're unlikely to discuss the existence of their system in a public forum.

Once a mix-net has been found, a certain amount of trust is needed. Is the mix-net real? Is the owner of the mix-net compromising anonymity by running rogue servers? Have hackers taken over the mix-net machines? Mix-net researchers are looking for ways to verify the integrity of a mix-net, minimizing the amount of trust necessary.

Making The Connection

Cryptography: Mix-nets rely on encryption to intercommunicate securely. Without encryption, eavesdroppers could read the message and the routing instructions. Encryption also ensures that a mix-net server can only read its own routing instructions. The server can't read the routing instructions for any of the other mix-net servers.

Connecting Networks: Mix-nets are the most effective when multiple mix-net servers are used to strip a message of identifying information. This requires a chain of mix-net servers to be used over multiple "hops." Setting up mix-net servers properly to receive, interpret, and send information to other servers on other networks is vital to the efficiency and success of anonymity.

Best Practices

Anonymizers are not commonly used in business today. Most businesses are struggling with basic security and other general technology issues. A business rationale for using anonymizers within an organization rarely exists.

Anonymizers are generally more applicable to individuals or smaller groups and organizations that hold privacy in an extremely high regard. Clandestine agencies may also use anonymizers to make their communications untraceable.

Final Thoughts

Anonymity in the digital age is extremely difficult to achieve. This is really not different from the physical world and the lack of anonymity most people contend with in everyday life. There are many different "digital footprints" that individuals leave behind when they use computer technology. Mix-nets are one tool people can turn to if remaining anonymous is of extreme concern.

Chapter 13
Preserving Privacy: User Tracking

The capability to track the patterns
of usage of an individual computer.

Technology Overview

Few technologies cause more debate and consternation than those that track the activities of computer users. On the one hand, there are many legitimate reasons for user tracking such as compliance, performance analysis, troubleshooting, and policy enforcement. On the other hand, users see tracking as an invasion of privacy—Big Brother looking over their shoulder. To make matters worse, a large degree of confusion and misunderstanding exists about the nature and capabilities of the many different user-tracking technologies.

Many organizations have an internal need for user tracking. In many countries, companies have the right to monitor workplace activities. Compliance laws may even require the auditing of employee activities. Scores of software systems are available for tracking and controlling computer tasks such as web browsing and email. The software can be used to identify employees who are misusing company resources or spending too much time on personal activities.

A large number of organizations also track the activities of their customers. This can help to better predict and control inventory, improve service, and allow targeted marketing. In the physical world, credit cards and frequent shopper cards can link purchase histories with customer names and addresses. In the digital world, far more information can be gathered. It's even possible to get detailed information about potential customers, even if they never make a purchase!

Digital user tracking took an age-old marketing concept and made it available to businesses at light speed. No longer are mail-in and phone surveys needed, (not that phone solicitations have subsided). Instead, customer's choices can be "watched" as they are made. Even better, some of the technology needed is essentially built into the process of web browsing.

The Tracking Debate

For the most part, internal workplace auditing is seen as a necessary and understandable evil. But many people see customer tracking in a completely different light. Privacy advocates feel that customers should always be aware of when and how they're being tracked. A person should never be tracked without his or her permission. Some advocates feel that this is enough: if a person knows that they will be tracked, they can choose to not do business with a company. Others feel that tracking is wrong unless a person is asked about it and agrees to it ahead of time. Furthermore, if a person refuses to be tracked, he or she should not be denied service.

This chapter focuses on the two most common digital user tracking technologies: cookies and spyware. Both are tools that can be employed by organizations to collect information about Internet and computer usage. Organizations that collect the data can analyze it for their greater purpose, which is usually determining what customers want.

Cookies were originally created to simplify the process of creating Web-based applications such as online shopping systems. They enable a web site to store a small amount of information on the visiting user's computer. This information can help a web site keep track of the user's progress through the site, which is necessary in order to have features like shopping carts.

Through a creative use of cookies, it's possible for certain companies to monitor a user's browsing habits across a wide range of web sites (but only on web sites that have actively chosen to use this service). These companies can recognize users as they travel from one site to another. The collected data can then be used to create profiles of users, which can provide valuable market research information. In most cases, a web site will hire a tracking company in order to obtain the market research data.

The danger is that the data gathered by these "third-party" cookies might be combined with personally identifying information. The result could be a personal profile created without the consent or knowledge of that respective person.

Although few companies, if any, are doing this, privacy advocates are deeply concerned about the possibility.

Spyware is really more of a business model supported by a technology concept, rather than just a technology. Many vendors of popular "free" software distribute spyware with their products. In exchange for getting the software for free, users agree to allow their computer habits to be monitored. This information can help software vendors make a better product. In some situations, the information obtained from spyware may be valuable to other companies and can be resold.

Spyware is a much more sophisticated concept than cookies: spyware programs can actively gather information about computer users. A spyware program is often installed on an operating system like any other piece of software. In their most sophisticated forms, spyware tools can gather information about how a computer is used even if the computer is offline! Later, the information is sent to a database where the offending organization can use it as they wish.

How Cookies and Spyware Work

A cookie is simply a piece of text that a Web server can store on a user's hard drive. Cookies enable web sites to store information on a user's machine and gain access to that information when needed. The data that are stored in a cookie file are known as *name-value pairs*.

In many cases, cookies benefit the web site operator and web site surfer. A web site may store information about a username and even a password that an individual used to gain access to a site. A cookie may also be used to store information about how a web page was set up or accessed. The use of cookies breaks down into three main categories:

Online ordering systems: Cookies are used to remember what a person wants to buy. For example, if a user shops at a web site, filling up a shopping cart or a wish list, but then logs off, later, when the user returns, all the items on that wish list will be remembered and remain in the shopping cart.

Site personalization: Just as it sounds. Cookies are used to remember preferences that a user sets at a web site. How a forum is used, how items are displayed, lined up, or any visual preference the site allows.

Web site usage analysis: Cookies can be used to identify repeat visitors to a web site. Some types of analysis software can create "profiles" based on accumulated browsing patterns. This can be a helpful tool for Web site designers or web masters as they can use the information to help improve the site.

Cookie data are simply name-value pairs stored on a hard disk by a web site. The web site stores the data on the local system only to retrieve it at a later date. A web site can only receive the data it has stored on a system that has been to the site. It cannot look at any other cookie or any other file on a system.

Many people use Microsoft's Internet Explorer to browse the Web. Since Internet Explorer is tightly integrated into Windows, the cookies are stored in a Windows directory called */cookies*. Take a look in this directory on your own computer to get an idea of how many individual cookies have been set as a result of browsing the Web with Internet Explorer.

You can see in the directory that each of these files is a simple and quite normal text file. They can be viewed with any simple text viewer, and read in plain text. The file name of each cookie will reveal the particular web site that placed the cookie there in the first place. Both before and after a cookie is placed on a hard drive, the data contained within the file moves around a great deal. Here is a closer look at how cookies operate:

1. The URL of a web site is typed into a browser and the browser sends a request to the web site for the page.
2. When the browser sends the request to the site it simultaneously looks on the local system for a cookie file. If it finds a cookie already exists for the requested site, the browser will send all of the name-value pairs in the file to the site's server along with the URL. If it finds no cookie file, it will send no cookie data at all.
3. The requested web server receives the request for a specific page and the cookie data (if it existed). If name-value pairs are received, the web site will use them.
4. If no name-value pairs are received, the site still has information. It now knows that the user has never been there before. The server creates a new ID for the user in the database and then sends name-value pairs to the requesting machine in the header along with the web page it sends. The browsing machine stores the name-value pairs on its hard disk.
5. The web server can change name-value pairs or add new pairs whenever the site is visited and a page is requested. The cycle repeats itself upon each visit, whether there is a unique or a repeating viewer. Figure 13-1 illustrates this process.

Spyware

Spyware, unlike cookies, is actually capable of collecting information. Cookies are severely constrained in their capabilities, but spyware can gather information on any aspect of computer use. A web site can gather limited demographic and statistical data automatically provided by the Web browser and Internet protocols, and read cookies set by its own domain. Spyware can observe and disclose any data that is stored, enters or exits a computer. This information can then be used for just about any purpose, or sold to the highest bidder.

As much as current spyware modules are capable of infringing upon privacy, they have the potential to do even more. Spyware exists as an independent, executable program on your system and has the same power potential as any program. The following is a list of some of the capabilities of spyware software:

How Cookies Work

Illustration by SageSecure

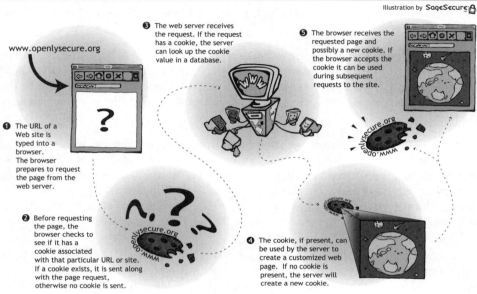

❸ The web server receives the request. If the request has a cookie, the server can look up the cookie value in a database.

❺ The browser receives the requested page and possibly a new cookie. If the browser accepts the cookie it can be used during subsequent requests to the site.

www.openlysecure.org

❶ The URL of a Web site is typed into a browser. The browser prepares to request the page from the web server.

❷ Before requesting the page, the browser checks to see if it has a cookie associated with that particular URL or site. If a cookie exists, it is sent along with the page request, otherwise no cookie is sent.

❹ The cookie, if present, can be used by the server to create a customized web page. If no cookie is present, the server will create a new cookie.

■ **Figure 13-1**

- Keystroke monitoring, arbitrary file scanning, random selected application monitoring, reading of cookies, interfacing with system Web browsers to determine browsing habits, and monitoring various aspect of general computer system usage.

- When an attempt is made to uninstall spyware, it can notify the creator it is being removed or tampered with by the user.

Spyware programs will contact "home" on occasion to send collected information back to a central database where it can be analyzed. All the information obtained by the spyware can be used by the spyware author for marketing purposes, or sold to other companies for a profit.

Security Considerations

The following three types of cookies exist:

Session cookies: These are not considered a privacy threat because they are deleted when the browser window is closed. They're often necessary for the operation of e-commerce Web sites.

Persistent cookies: These raise privacy issues because they are stored for a potentially lengthy period of time. A persistent cookie can enable a Web site to build profiles on repeat visitors. This can also be accomplished without cookies by requiring users to log in.

Third-party cookies: These are persistent cookies that are set by a Web site other than the one that is being viewed. A third-party tracking company can use these cookies to build up user profiles across multiple sites. These are considered to be the most significant threats to privacy. Privacy advocates refer them to as "spyware cookies."

As a point of comparison, persistent and third party cookies are quite similar to many forms of privacy invasion that are considered parts of normal living. Retail merchants use credit cards and reward club cards to track customer spending patterns. Customers can avoid this by paying in cash. The downside is that you have to carry lots of cash everywhere. Likewise, web users can turn cookies off, but this will make it hard to access some web sites. In both cases, the added privacy comes at the cost of convenience.

Spyware poses a greater privacy concern than cookies. The biggest problem with spyware is that it is often placed on a host machine without the knowledge of the user. The concept of spyware did start as a business model; the exchange of functional software for user statistics. However, in time, spyware programs were being used without the exchange of anything. In other words, spyware is discretely installed on a system in exchange for nothing, and without the permission of the user.

Because a spyware program is an independent executable program it has extensive privileges. A multiuser operating system may somewhat impede a spyware application's abilities. However, spyware can still have all of the same privileges of the user who installed it. Imagine the possibilities when someone installs spyware with administrator privileges on a network.

Some spyware modules contain many additional insecure features. This may include the ability to automatically update the spyware program, or install sister programs without the user being aware of the change. This opens the door for further abuse of the system by malicious attackers who can exploit this feature of spyware by uploading and running any program they wish on a remote system!

Spyware is a much more dynamic tool with which to track information. Cookies may keep track of browsing habits specific to one or more web sites, but spyware can watch much more. Spyware can also track information when a user is offline, collecting it and sending it as soon as a user connects to the Internet. In this way, spyware is spying in a full time capacity. It does not need to be connected to a web server or database to collect and store information.

Regardless of the degree of severity, spyware and cookies both have an impact on privacy and, in turn, on the overall security of a network or computer system. Cookies and spyware are out there, and depending on the philosophy and policies of the organization, can be ignored or eliminated.

Making the Connection

Hiding Information: Cryptography and related tools can be used to lock and hide important data on a local computer system. Data that is encrypted cannot be read or interpreted by spyware programs that may retrieve it on a system.

Detecting Intrusions: Intrusion detection software can notify users of cookies and spyware that exist on their system. Running intrusion software regularly is a good way to prevent unauthorized access to private data.

Best Practices

It is up to the user or the organization to take a stance on privacy. Roaming the Internet and downloading applications with no scrutiny will always yield a high count of cookies and spyware on a host machine. If a firm stance to protect privacy is chosen, spyware, persistent cookies, and third-party cookies will be avoided whenever possible. Here are some things that can be done to find the right balance:

Cookie management: Modern Web browsers provide sophisticated controls for handling cookies. Most browsers differentiate between temporary cookies, persistent cookies, and third-party cookies. Users can choose to accept or deny all cookies. Users can also choose to accept or deny cookies on a site-by-site basis.

Cookie control software also exists for systems that have older browsers or for users who want even more control over cookie information. These programs provide all of the functionality of the most sophisticated browsers, and then some.

When a Web site attempts to set a cookie, it is not always optional. Many popular sites require cookies. In cases where the site insists on setting a cookie, there is a simple decision to be made: accept the cookie or never browse the site. This means that denying a cookie, or to be even more extreme denying all cookies, can take your Web browsing experience nowhere, fast.

One solution is to set the browser (or use an external program) to delete all persistent cookies on a regular basis. This allows a user to accept cookies in order to browse a site, knowing that the cookie will be promptly deleted once the browsing session is finished. Another good recommendation is to never accept third-party cookies. These cookies are almost exclusively used for tracking purposes. For many people, these settings provide a good balance between privacy and usability.

Detecting spyware: There are software products available, many for download, that are specifically designed to find, identify, and remove spyware. These programs need to be updated frequently. Frequent updates allow the removal of the latest and greatest spyware applications that may have been installed on the host machine. It is also important to be aware that new spyware is developed often, sometimes weekly. In addition, spyware makes its way onto a system with Internet access very often. Running spyware detection tools daily would not be too extreme.

Deleting spyware: Once spyware is found what is to be done with it? That depends on the user and how he or she feels about privacy invasion. All spyware detection tools provide the ability to automatically delete detected spyware. Some utilities go a step further and provide resources and information for what a particular spyware application is capable of doing. Some people may have a greater threshold for spyware than others. If you want to know what the authors do, the answer is simple: We delete all of it.

Final Thoughts

Cookies and spyware have been grouped together in this chapter because they both have an impact on privacy. Despite their public association, cookies and spyware really have little to do with each other in technology terms. Cookies are text files that store basic information for Web sites. Spyware are programs that can monitor every mouse click and keystroke when installed and executed on a computer.

Ultimately cookies and spyware have far less of an impact on privacy than government policy and legislation. If a government doesn't want to permit the use of tools that invade the privacy of its citizens, it can make technologies like spyware illegal. Unfortunately, functioning in today's society means giving up a certain amount of privacy. Many everyday activities involve interaction with tools and technologies that can invade the privacy of individuals far more effectively than cookies or spyware.

Chapter 14
Preserving Privacy: Spam Management

Everybody likes getting mail, until they see that it's junk mail.
This chapter discusses how to keep junk mail from clogging
your network and wasting the time of your users.

Technology Overview

Jerry Seinfeld once asked,"What is mail, really?" His punch line was that mail boils down to two things: bills and junk mail. Why get excited about mail? It is not as if anybody actually sends you anything interesting like letters anymore. Sadly, traditional mail, or snail mail, is mostly a lost art. A typical trip to one's mailbox reveals nothing besides bills and solicitations. There is little reason to send letters through the postal system when they can be sent for free over the Internet and reach their destination almost instantly.

Email has taken over snail mail as a routine form of written correspondence. As the medium has grown in popularity it caught the attention of the snail mail solicitors. Now digital junk mail, known as spam, has followed the paper letter writers to their new medium. The result is that the average person's daily email contains 10 credit card offers, 8 young girls who show it all, 15 guaranteed ways to make up to $1,723 per week from home, 3 herbal solutions for impotency, 2 pharmaceutical solutions to impotency, and a mechanical solution to impotency that implies certain physiological similarities between men and bicycle tires.

Why Spam Happens

Spam is the digital version of junk mail, which is overzealously referred to as "direct marketing." Statistics show that for every 100 flyers, brochures or other marketing drivel mailed out, there will be an average of one to two responses. If a hundred thousand are mailed out it could yield a few thousand customers. Bulk mail is cheap, but not that cheap. It can be an effective strategy for certain products, but for others the cost of the mailing might exceed the realized profits.

Email changes the cost-benefit equation. Because email is "free," marketers can send out millions of electronic advertisements at little to no cost. Even if the response rate is much less, the massively wider reach can mean even more total responses. And there's little direct economic downside to spamming.

Many people think spam can do more harm than good for the compa-

Everybody Pays for the Internet

Dad always said nothing is free in this world. For quite some time he seemed to be wrong, as the Internet appeared to provide an abundance of free services. However, the Internet is not free. Somebody somewhere pays for the infrastructure of connectivity. The major telecom companies provide much of the bandwidth used to get data from one system to another. Spam takes up a large amount of this bandwidth. Somebody has to pay for the bandwidth used, so the telecoms pass these costs on to *Internet service providers* (ISPs), who then pass the costs on to corporate customers and consumers in the form of higher rates. As usual, it appears Father knows best.

nies that send it out. This is because users get angry when they receive spam and may choose to not do business with the spamming company out of spite alone. The proponents of spam marketing respond by explaining that even bad press can create good overall exposure. The sentiment is that ultimately people will remember the product, but forget that they first heard of it from a spam message.

How Spam Happens

In order to spam mailboxes, two things are needed: a list of email addresses, and a mail server that will permit the sending of spam. Getting the addresses isn't all that difficult. Many companies are in the business of selling direct mailing lists. Some go as far as targeting the email addresses based on demographics. How do these companies build their lists? The following are some of the more common techniques:

Harvesting: Search engines crawl the Web, looking for email addresses in Web pages. The addresses are automatically extracted and added to a database. Does a Web site have your email address on it? If so, it's probably in a spam database. Email addresses posted to newsgroups or Web-based forums are at risk too. These discussions are often archived, and the archives are available as Web pages. Email harvesters love list archives because they have tons of addresses and are often in a consistent format making them easier to farm.

Trial and error: Some spam systems will test a number of common email addresses at every domain on their list. This includes common names, administrative accounts (webmaster and root), marketing accounts (sales, info, and support) and such. Many email servers will send a "not found" response if the address is invalid. The spam system can determine valid addresses by testing tons of addresses and marking those that don't generate a "not found" message.

List exchange: Ever provide an email address to register for a Website or an e-commerce site? That address is stored in their database. It is common for these sites have privacy policies that say "we will never share your information . . . except with our partners." Since a "partner" can be anyone, this gives the site freedom to exchange "private" information with "partner" marketing companies. Once the marketing company has the information, it could end up anywhere.

How Spam Management Works

The best solution to the problem of spam is installing a spam filter. This is software that is designed to block spam. Some spam filter packages can also help manage email in general.

Four basic types of spam filters exist: blacklists, whitelists, pattern matchers, and adaptive/heuristic systems. The most effective spam blocking systems use a combination of these techniques.

Blacklists: Various antispam organizations maintain lists of mail servers that relay large volumes of spam. Blacklist spam filters check the source of an incoming message. A message is blocked if it originated or passed through a known spam relay.

The problem with blacklists is that they block some ISPs entirely because many ISPs have poorly configured mail servers. As a result, mail messages from users of that ISP are discarded. This can be a real problem if a client or important contact is using a blocked ISP, as real messages will never be received.

Whitelists: The opposite of a blacklist, whitelists specify nonspam addresses. If the address is not on the whitelist, it is marked as spam. Many whitelist packages will automatically synchronize a user's address book with the whitelist.

The obvious problem with whitelists is that a lot of desired email will get dropped. Anyone making a first contact by email will not be in the whitelist. For business users, a pure whitelist solution is completely impractical. It's most useful for children. Parents can control the whitelist, ensuring that only friends and family can reach the child via email. This can help protect against online stalkers, and is guaranteed to eliminate spam that would be damaging to the child.

Pattern matching: Spam tends to be very simple and repetitive. For example, any email message that has the words: hot, free, teen, or sex and a complicated web link is likely to be spam. Either that or you've got some strange friends.

Pattern matching is highly effective against certain types of spam, but the database of patterns needs to be constantly updated to keep up with changing spam patterns. Ultimately if the list of patterns grows too big it will create performance issues on the mail server.

Another problem with pattern matching is that it does occasionally wipe out desired messages. Advanced systems assign a confidence factor to each message, only trashing those messages that pass a certain threshold. This sort of system does require ongoing user tuning and monitoring.

- **Signature databases:** An advanced type of pattern matching, signature databases take a "fingerprint" of a message and compare it to an online database of known spam messages. If the fingerprint matches then the message is marked as spam. These spam systems often integrate with email clients, allowing users to update spam message fingerprints to the database. This is how new spam signatures are identified and added.

- **Header analysis:** Spam messages usually forge the mail headers in an attempt to prevent anyone from tracing their source. Some spam tools use pattern matching to identify forged headers. The forged headers look a little different from real headers. For example, in some cases the apparent route taken by the message is invalid.

Adaptive systems: These are the most complex antispam systems. Adaptive systems attempt to learn from the mail recipient. When a spam message shows up, the user marks it as spam. The system then tries to analyze the message, looking for distinct patterns and other identifying criteria specific to that message. In the future, messages matching these patterns are automatically filtered out. Over time, adaptive systems get better at sorting spam from good messages using this artificial antispam intelligence. A system called "Bayesian Filtering" is one example of a popular adaptive technology used in a number of spam filtering packages.

Security Considerations

In general, getting rid of spam is purely beneficial. Even positive side effects exist, such as the fact that spam filters often catch email viruses (they're just a different type of spam). That said, there are a few things to watch for when installing spam filters:

False positives: Spam filters that are too aggressive or too broad can prevent important messages from getting through. It's better to let a few spam messages through than to "over filter" and inadvertently delete "ham" (desired email).

Processing overhead: Any organization that has several users, or maintains a mail server that handles large volumes of mail, may find that spam filters tax their mail server. In general, email is not a performance intensive application, so it is common for mail servers to operate on older, slower hardware. With spam in the picture, mail servers can quickly run low on processing power or drive space.

One solution is to separate the mail server functionality onto different physical servers. The existing server can be used to only handle incoming and outgoing mail. Mail delivery systems (such as POP and IMAP) and spam/virus filtering functionality can be handled on a separate, more powerful machine.

Making the Connection

Internet Services (email): Understanding how email works is necessary to better understand why spam happens and how to protect your network against being a spam relay.

Firewalls: These enable less secure internal servers that spammers can't reach. Your external mail server can be configured to only talk to the proxy, eliminating its effectiveness as a relay.

Viruses and Trojans: Virus emails are a special type of malicious spam. A good virus Trojan horse detection system will help eliminate this ugly monstrosity.

Best Practices

A number of effective techniques minimize spam besides the use of spam filters. Most of these techniques can be used alongside spam filters, or can provide some protection in situations where filters can't be applied.

Make sure your own primary mail servers have not been blacklisted. The major blacklist services have a tool for determining if a mail server has been blacklisted. It is a good practice to do this periodically, even if there is no reason for your server to be blacklisted. A malicious person could create a situation that would result in an "innocent" mail server or email address being added to one or more of these blacklists. Many ISPs use these blacklists to block server access to their networks. Once a mail server is on a blacklist most of the mail it sends, whether actual spam or not, will be blocked. All of these services do provide methods to get removed from a blacklist if a server was added in error.

It is a good idea to download and apply mail server patches. Serious spammers will look for mail servers with known security flaws. These are mail servers that they can really "own." With a large enough selection of insecure servers, spammers can effectively avoid getting blacklisted and send out as much spam as they desire.

The following is a list of other quick tips to help in the war against spam:

1. As mentioned earlier, email addresses are harvested from web pages. It is a good practice to never post "real" email addresses on a web page.

2. If you do need to put your email address online, you can obfuscate the address by adding stuff that might confuse a harvester, such as M*dot*Johnson*removeifnotspam*at*example*dot*com.

3. Be careful. Harvesters are smart; they can match for typical obfuscation patterns such as me.at.example.dot.com, or menospam@example.com. Get creative.

4. Always keep a "garbage" email address for situations that require providing an address to a non-business or family contact. Businesses might want to issue a number of addresses to each employee: one for internal, one for family/friends, one for vendors, and one for customers. Garbage addresses are also useful for filling out forms on the Web. This way, spam collects at an email address that is not important.

Final Thoughts

Future spam fighting will happen on a number of fronts. New email technologies might make it far more difficult for spammers to operate. For example, mail servers might need to be "registered," just like domain names. Registered servers would only exchange mail with other registered servers. Spam complaints against a server could result in the loss of its registration.

Legislative action is another very effective weapon against spam. In Europe, it is now illegal to send unsolicited commercial email. Spam may slowly disappear if enough nations follow suit and adopt antispam legislation and follow it with law enforcement. There may always be some spam, but if the risks outweigh the rewards most commercial spammers will be forced to stop.

"There's an interesting article in the Washington Post (March 12, 2003) which includes an inside look at AOL's spam control center in Northern Virginia. The story reports that roughly 40 percent of all e-mail traffic in the US is now spam, up from 8 percent in late 2001 and nearly doubling in the past six months; that AOL's spam filters now block 1 billion messages a day; and that spam will cost U.S. organizations more than $10 billion this year from lost productivity and the equipment, software and manpower needed to combat the problem." —From slashdot.org

VI
Connecting Networks

Summary

The tools used to build and connect networks are not always secure. Some are notoriously insecure (wireless). This chapter discusses issues and solutions for securing the basic components used to build and connect networks.

Key Points

- Networking allows computer systems to share information across vast distances.
- Network hardware is used in two different ways: connecting computers together and connecting networks together. Critical differences exist in the security implications.
- Specialized hardware enables networks to become large and complex by routing and controlling massive amounts of data.
- Securing data over networks is far more complicated than keeping isolated data secure. Securing data over wireless networks is more complicated than doing so over traditional wired networks.
- Numerous common security problems are the direct result of poor network design.

Connecting the Chapters

A number of technologies can be used to connect networks. Each technology has its security benefits and detriments. Although the tools themselves are relatively simple, many ways to mix and match the components are available. The interactions between devices create additional, unique security issues. In order to effectively evaluate the security of a network design, one must understand the components and how they interact.

- **Chapter 15, "Networking Hardware,"** describes how hubs, routers, and switches are the three basic "joints" used to connect computers and networks together.

- **Chapter 16, "Wireless Connections,"** explains the use of radio waves and signal processing to connect systems and networks without the encumbrance of wires.

- **Chapter 17, "Network Lingo,"** covers the basic languages that network systems and applications use when interacting and transferring data.

Introduction to Connecting Networks

In its very brief history, computer science has gone through some very rapid evolution. One of the most noticeable advances is the invention of networking, which greatly expanded the ways in which computers could be used. Computer users were no longer limited to the resources available on their systems, but instead could tap in to the resources of any system connected to the network.

Before networks, transferring information between computers was a physical process. The data would be written to some form of removable storage and carried from one machine to the other. Today this sort of file transfer is affectionately called "sneakernet."

The limitations of storage media directly affected the process of transferring information between machines. Punch cards were among the earliest "removable storage" systems. They weren't particularly convenient or space efficient. This chapter would probably take up a few boxes of punch cards. Transferring significant amounts of data involved using a forklift. Floppies were far more convenient, but still could only hold a relatively small amount of data. Tapes held a lot more, but took forever to access. Still, for many years tapes were the storage system of choice for data that needed to be moved between machines.

Many applications that exist today were inconceivable before the network. Collaborative environments, audio/video conferencing, email, to name a few. These applications fundamentally rely on the ability to move information rapidly between two computers. Email would be pointless if you had to stick a bunch of punch cards in an envelope.

Networks allow computers to take on specialized roles. Systems can focus their processing and storage resources on specific tasks such as handling files, running applications, or transmitting email. These focused systems are called network servers. They centralize organizational resources across a manageable number of powerful computer systems. End-user systems, called workstations, can become less complex as more functionality is shifted to the central servers. This makes managing the workstations easier, which can be a real benefit in large organizations.

One Computer, Two Computer, Red Computer . . .

In the beginning, computer networks were simple and quite limited in scale. This meant that computers connected to each other were usually in the same room, or maybe within the same building structure. This type of network is known as a *local area network* (LAN). Its purpose is to provide nearby users the ability to share information with each other quickly and easily.

Local area networks are powerful. They are capable of storing large amounts of information and providing fast access to the information. All of the information and services needed by an organization can reside on a LAN. Access to a LAN is generally provided through a computer system called a client or workstation. Many successful organizations have just a single LAN. One unfortunate problem with LANs is that they're "local," and therefore rarely extend beyond a single facility or even a single area or room.

Local area networks are limited in scope to the resources available on them. What if there was needed information available on another LAN? How could one get to that information? Could a network be networked to another network?

Sometimes, people in that local office need access to information stored in another local corporate office. Maybe they need financial information specific to that office to do a comparison of office profitability, for example. Whatever they need, in order to get it they need access to a remote LAN. The only way to get that type of access is by having a network connection from one LAN to the other LAN.

Wide area networks (WANs) are the result of two or more networks being connected to each other. On a WAN, an end-user in one location has all the resources of the local network and the additional networks with which it is connected. This allows for a level of communication between isolated locations that greatly expands organizational resources.

Many different examples of local and wide area networks exist. One example is a corporation that has multiple offices with a separate network in each office. Each office network has its own LAN. Data important to the local office resides on servers connected to the office LAN. This information would all be accessible to anyone who works in that local office. If the individual LANs were connected into a WAN, users in one office could obtain critical files from the other branch offices.

In some cases, a single office might have multiple LANs. Think about a company with a number of independent business units, such as a major bank. For security and functionality reasons, each unit might have its own LAN. A WAN could connect each unit to centralized shared resources while preventing the separate LANs from inter-communicating. This might be necessary to prevent the currency traders from communicating with the investment advisors (which would violate various trading regulations).

Specialized Networks Need Specialized Hardware

Whether a situation calls for a LAN, a WAN, or both, certain technologies are used to make sharing information across computers possible. These technologies are the devices that connect computers to computers and networks to networks. The larger the network needs to be, the more specialized the hardware becomes. Simple and relatively inexpensive devices can be used to build LANs. More complex and intelligent devices are needed to handle large LANs and WANs. Critical junction points in extremely large networks require very advanced equipment that can cost millions of dollars.

The Internet itself is little more than the largest WAN ever created. It's truly just a network of networks; an endless, ever growing mesh of many smaller WANs and LANs across the globe. It is made possible through the use of specialized pieces of hardware known as *routers*, which receive, interpret, and distribute information between networks.

Major Internet networks that connect a large number of smaller networks together are called *backbones*. They have been given this name because they carry the brunt of the Internet's traffic. Major and minor networks join together at gigantic data access points along the backbone. Enormous volumes of network traffic often pass through these points to reach their ultimate destination.

As networks grow larger, the hardware used to connect them together becomes more specialized. For example, Internet backbones have very large routers, capable of processing enormous amounts of data. These routers receive and distribute millions of bits of data every minute. The routers communicate with each other in order to determine the best way to pass information along from one network to another.

In facilities called *exchanges*, multiple backbones are joined together by the most powerful and specialized of routers. These routers can switch data from one backbone to another. This allows data to reach destinations that are not directly connected to a particular backbone. It also allows data to travel to its destination via more efficient routes.

A true Internet Backbone requires the capacity (bandwidth) to carry an enormous amount of data to many different geographic regions. The data is usually carried along cables, although it also travels via satellite. It's not cheap to run a cable from New York to Los Angeles. Because of the immense cost of deploying connectivity, it isn't merely any company that can afford to become an Internet Backbone. Only large communications companies can afford to run copper and fiber cable all over the globe. These large communication companies have control over many of the Internet's backbones.

Some networks don't qualify as backbones, even though they connect many large and small networks together. *Internet service providers* (ISPs) are a typical example. These companies buy up a large amount of bandwidth from a backbone provider and often are directly connected to the backbone. Their business model is to provide high bandwidth access to the Internet for their customers for a monthly fee. They maintain rooms full of routers that provide an intermediary connection between their customers and the backbone.

How does all this tie together? Air travel provides a good analogy. Getting from point A to B via the air travel network involves a process similar to getting data around the Internet.

Airline companies have a network of destination cities and routes. Major air carriers are the equivalent to Internet backbones. Their routes cover most of the world. Smaller regional carriers are more akin to ISPs. If you live in a small city, most of your traveling will require taking a regional carrier to a larger airport where you can switch to a major carrier. These larger airports are like exchanges. You can then take a major carrier from one large airport to another. You may even have to switch back to a regional carrier to reach your final destination.

Although some major destinations can be reached with a direct flight, others require changing planes. Direct tickets are often more expensive than discount tickets that can involve switching carriers and making additional stopovers. This is less efficient in terms of time, but carrier saves money by filling less valuable seating space. In the same way, routers can move traffic from one network to another based on cost and time efficiency considerations.

Networks: Power and Peril

Wide area networks such as the Internet open up new realms of possibilities for information exchange. This brings even more power to individual users as well as entire organizations. Of course, with great power comes great responsibility and cheesy movie references. In the case of networks, responsibility means keeping important data safe.

Keeping the data on computer networks safe is no easy task. It is certainly a more arduous and difficult task than keeping data on isolated computer systems safe.

The more isolated a computer system is, the fewer options there are to gain access to its data. Networking opens up a multitude of channels and potential backdoors that intruders can use to gain access to a computer's data.

Personal digital assistants (PDAs) are an example of computers that are not traditionally connectable to networks (newer PDAs can use wireless technologies to directly connect to the Internet). Most PDAs are capable of storing a good deal of information, and some of it may be very personal and private. If all of that information is entered directly into a PDA without wireless access and the PDA is always kept with the owner, there is little to no opportunity for someone to gain access to the information. The only access point is the PDA itself, which unless lost, can only be used by the owner.

Now take that very same PDA and synchronize it with a desktop computer on a corporate network. All of the information stored in the PDA gets duplicated on the office workstation. To make matters more complicated, the office workstation is probably connected to the office LAN and also to the Internet. The information stored on the PDA went from having one access point to many access points. The data would be far more secure if it were only synchronized to a computer with no network connection. It would be even more secure if it were never synchronized at all. Many people would feel, however, that this would defeat a major purpose of using a handheld device.

Networking in general involves a complex set of decisions about security. The general problem with securing data on a network is the innumerable amount of available access points to the data. The more "connected" a computerized environment gets, the less secure it becomes. In contrast, the more connected the environment, the more convenient working with data becomes.

Much like the PDA scenario, what conveniences are organizations willing to sacrifice for the sake of security? Certain types of networking and networking architecture offer convenience that is too good to pass up. Wireless networking technology serves as a perfect example.

Wireless networks give users all the power of networks in an ultra-flexible environment. Computers can be set up anywhere without having to worry about cabling constraints. This is particularly useful for laptops, which don't even need a power source. Many organizations have been drooling over this technology for quite some time, as it frees up many of their traditional spatial limitations. Unfortunately wireless networks are even more insecure than traditional wired networks, because they offer yet another unprotected access point; the airspace around the network.

Wireless networks are little more than radio waves traveling through the air. In a wireless network, data is flying through the air all around the office. A love letter to your significant other sent through a wireless network connection spends time floating in front of your face. This means that the information is out there, just waiting for someone to grab it and look. As a result, the use of wireless networks poses an additional risk to an already fundamentally insecure situation.

Connecting Correctly

The biggest security problem with most networks (the Internet included) is that they were not designed with security in mind. In fact, most networks are not designed with anything in mind except for solving the immediate needs of an organization.

CEO Steve: "Bob, I need to get this file from this office to that office by 5 p.m. tomorrow."

Networking Bob: "Okay, Steve, I'll get the two locations networked together fast."

Little planning goes on in traditional work environments when it comes to networking. For this reason, network planning is often an afterthought. An examination of the average network tends to reveal that networks are created and then grown as needed, without a long-term plan for their superstructure.

CEO Steve: "Bob, we should have central copies of all the data from our branch offices."

Networking Bob: "Okay, Steve, I will implement database and file replication immediately."

If Networking Bob reacts quickly to the needs of his organization he may receive praise. Even if he takes his time to properly set up remote database replication, chances are he will not be willing to alter the underlying network very much. Sometimes when the carpet is lifted there are just too many dust bunnies to clean. Other times, the dust bunnies decide to attack the rest of the network, choking basic functionality as they get sucked into the server air vents. The fact is, when restructuring a network that has "grown out" from the ground up, things are going to get worse before they get better. Many organizations do not want things to get worse at all, even for a minute.

Proper network design is a topic worthy of a book of its own. We can't hope to do the topic justice in this limited space, but we will point out a few key things that you need to keep in mind (see Figure VI-1).

A good network design starts from the following questions:

- Which network services are necessary to support organizational business processes? For example, a shipping company needs to be able to track items from pickup to delivery. This means that the processing office, the delivery office and the trucks/boats/planes all need to be able to send tracking data into a central tracking system. That requires a network. The design of the network is going to be heavily influenced by the specifics and requirements of the business process.

- What sort of security is necessary to satisfy business needs? Our shipping company probably doesn't want thieves knowing about the contents and whereabouts of every package. Therefore, tracking information needs to be securely communicated and handled.

- Is there a conflict between necessary services and security needs? In other words, does the business need to use the network in a way that is fundamentally insecure? Some elements of processing and handling the shipping company's tracking data might be unavoidably insecure.

- Is there enough money in the budget to provide the needed services at the necessary security level? The example shipping company can't afford a multi-million dollar security system. Some of the less critical security techniques might need to be dropped.

- Is it possible to use traditional risk management techniques to cover up areas where the security needs can't be satisfied? Insurance and other risk mitigation techniques can minimize the exposure in the event of a problem. The network doesn't *have* to solve every problem.

If a network is already in place, then other important questions also need to be considered:

- What is the minimum amount of change necessary to achieve business and security goals?

- Is the network providing extra, unused services? It is usually the case that a network offers many more services than are actually needed by an organization. If excess services are available, trimming down the fat is the next step. Services create access points to networks, the fewer services, the fewer holes that exist for an outsider to enter and exploit.

- How are the actual network devices connected? The best way to get an overview of the relationship between network devices is by diagramming a network on paper with flow chart software. At a glance this will summarize what hardware connects to what, and how information flows along pathways in and out of a network. A network map of this sort is often called a topology, or sometimes incorrectly called topography.

A Typical Insecure Network Design

A common network design mistake is overextending trust. Here, the firewall protects the internal network from the Internet. That's good, but what's protecting A's servers from A's workstations or workgroup B? Nothing.

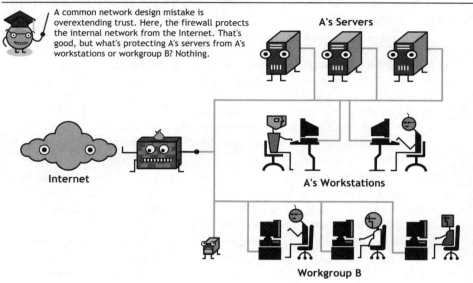

A More Secure Network Design

A better solution is to use the firewall to segment the network into three groups: A's servers, A's workstations and B's workgroup. Some firewalls have extra network ports, allowing them to create these new segments. Multiple firewalls can achieve the same effect.

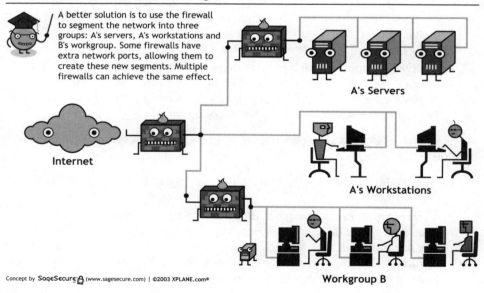

Concept by SageSecure (www.sagesecure.com) | ©2003 XPLANE.com®

■ **Figure VI-1**

- Is the network segmented into functional units? Keep in mind that network areas don't always need to openly communicate with each other to provide services. Very often restricting what hardware can communicate with what hardware will offer tremendous improvements in security. This process of segmenting network communications is often referred to as subnetting. When a network is subnetted it relies on connecting hardware to control communications. This is why connecting hardware is such an important tool in network security.

Final Thoughts

Designing a new network, or restructuring a network that has been in existence for a long time may not be an easy task. It certainly is not an enviable task. It will, however, force any company to take a hard look at how convenience and connectivity relate to their overall data security and integrity. Connecting hardware plays a special, centralized role in any network and if used properly can help guarantee improved security and performance.

Once a network has reached a certain point of complexity, the process of design becomes closely intertwined with the process of hardening. Choosing the type of networking equipment and the ways in which the network will be connected is part of network design, but using hardware to control access to individual servers and services is really part of network hardening, which the next chapter covers. To make an analogy: if design is the foundation and structure of the house, hardening is putting the locks on the doors and windows. A house can't stand without a good structure, but even the most well built house needs a door, and that door needs a lock.

Chapter 15
Connecting Networks: Networking Hardware

Hubs, routers, and switches are the three basic "joints"
used to connect computers and networks together.

Technology Overview

When many people think of big networks they also think "complicated." This is
for good reason, as big networks are undoubtedly complicated. They contain millions
of nodes connected by hundreds of thousands of specialized network devices, not to
mention a whole lot of copper wire. If someone told you that giant networks, such
as the Internet, were simple and manageable beasts that anyone could understand
you might second-guess their sanity. It takes an engineering degree and years of
experience to even begin to understand the complex issues that major network
carriers face.

Networks can become complicated, but the fundamental concepts are actually
pretty simple. In fact, powerful and effective network strategies can be created after
understanding just a few basic networking tools and concepts.

Hubs

The simplest type of networking device is the hub. It's the networking equivalent of a basic power strip. Intelligence is not present in a hub; it just ties all the network wires together. Data sent down one wire goes into the hub and travels to every other wire. It's a simple and effective way to connect computers together.

The downside to hubs is that, like a power strip, the devices plugged in share the available resources. If one device on the hub uses too much bandwidth, the other devices can't communicate.

In the most basic type of network found today, nodes are simply connected together using hubs. As a network grows, there are some potential problems with this configuration:

Scalability: In a hub network, the limited shared bandwidth makes it difficult to accommodate significant growth without sacrificing performance. Applications today need more bandwidth than ever before. Quite often, the entire network must be redesigned periodically to accommodate growth.

Chaining limits: When a signal goes into a hub, it gets "copied" and sent to every other port on the hub. The copying process is done with simple electronics (resistors, capacitors, and so on) and simply replicates and boosts the signal. But this introduces a certain amount of distortion. If a hub is connected directly to another hub (chaining), the second hub increases the distortion even further. In general, a third hub raises the distortion to the point where the data becomes garbled. As a result, there can be no more than a maximum of two hubs between any two computers on a network. This makes it difficult to use hubs for networks with hundreds of nodes, as most hubs support no more than 32 connections.

Distance limits: Ethernet signals degrade over distances larger than 100 meters. This, combined with the chaining limits, makes it difficult to use just hubs for networks that occupy a physically large space. Other devices, such as "repeaters" must be used to extend cables for longer distances.

Latency: This is the amount of time that it takes a packet to get to its destination. Since each node in a hub-based network has to wait for an opportunity to transmit in order to avoid collisions, the latency can increase significantly as you add more nodes. Or, if someone is transmitting a large file across the network, then all of the other nodes have to wait for an opportunity to send their own packets. You have probably seen this before at work, when attempting to access a server or the Internet and suddenly everything slows down to a crawl.

Network failure: In a typical network, one device on a hub can cause problems for other devices attached to the hub due to incorrect speed settings (100 Mbps on a 10-Mbps hub) or excessive broadcasts. Switches can be configured to limit broadcast levels.

Collisions: Ethernet uses a process called *Carrier Sense Multiple Access with Collision Detection* (CSMA/CD) to communicate across the network.

Under CSMA/CD, a node will not send out a packet unless the network is clear of traffic. If two nodes send out packets at the same time, a collision occurs and the packets are lost. Then both nodes wait a random amount of time and retransmit the packets. Any part of the network where a possibility that packets from two or more nodes will interfere with each other is considered to be part of the same collision domain. A network with a large number of nodes on the same segment will often have a lot of collisions and therefore a large collision domain.

Collisions can have serious repercussions. If too many occur it is possible that the network breaks down. Each time two frames collide they need to be retransmitted. Each frame ends up appearing on the network twice. If the volume of traffic is high at a time when frame collisions occur, the entire network can seriously deteriorate. An increasing amount of the total network bandwidth is taken up by frame retransmissions. As a result, an increasing amount of legitimate traffic becomes involved in the collisions. Depending on the severity of the collisions, as much as 70% of total network capacity may be rendered unusable.

Segmenting: Although hubs provide an easy way to quickly add computers to a network, they do not break up the actual network into discrete segments. Every computer connected to the hub can talk to every other computer. Splitting up computers into functional groups is often desired for both management and security reasons.

Switches

Think of a hub as a four-way intersection where everyone has to stop. If more than one car reaches the intersection at the same time, they have to wait for their turn to proceed. Now imagine what this would be like with a dozen or even a hundred roads intersecting at a single point. The amount of waiting and the potential for a collision increases significantly. But what if you could take an exit ramp from any one of those roads to the road of your choice? That is exactly what a *switch* does for network traffic. A switch is like a four way intersection where each car can take an exit ramp to get to its destination without having to stop and wait for other traffic to go by.

Switches are a fundamental part of most networks. They make it possible for several users to send information over a network at the same time without slowing each other down. Just like routers allow different networks to communicate with each other, switches allow different nodes of a network to communicate directly with one another in a smooth and efficient manner.

Several different types of switches and networks exist. Switches that provide a separate connection for each node in a company's internal network are called *Local Area Network* (LAN) switches. Essentially, a LAN switch creates a series of instant networks that contain only the devices communicating with each other at that particular moment.

A vital difference between a hub and a switch is that all the nodes connected to a hub share the bandwidth among themselves, while a device connected to a switch port has the full bandwidth all to itself. For example, if 10 nodes are communicating using a hub on a 10-Mbps network, then each node may only get a portion of the 10 Mbps if other nodes on the hub want to communicate as well. But with a switch, each node could possibly communicate at the full 10 Mbps. Think about our road analogy. If all of the traffic is coming to a common intersection, then each car has to share that intersection with every other car, but a cloverleaf allows all of the traffic to continue at full speed from one road to the next.

In a fully switched network, switches replace all the hubs of an Ethernet network with a dedicated segment for every node. These segments connect to a switch, which supports multiple dedicated segments sometimes in the hundreds. Since the only devices on each segment are the switch and the node, the switch picks up every transmission before it reaches another node. The switch then forwards the frame over the appropriate segment. Since any segment contains only a single node, the frame only reaches the intended recipient. This allows many conversations to occur simultaneously on a switched network.

Switching allows a network to maintain full-duplex Ethernet. Before switching, Ethernet was half-duplex, which means that data could be transmitted in only one direction at a time. In a fully switched network, each node communicates only with the switch, not directly with other nodes. Information can travel from node to switch and from switch to node simultaneously. This greatly enhances network efficiency over time.

Fully switched networks employ either twisted-pair or fiber optic cabling, both of which use separate conductors for sending and receiving data. In this type of environment, Ethernet nodes can forgo the collision detection process and transmit at will, since they are the only potential devices that can access the medium. In other words, traffic flowing in each direction has a lane to itself. This allows nodes to transmit to the switch as the switch transmits to them. Transmitting in both directions can effectively double the apparent speed of the network when

Bridging Networks

A network bridge is the same concept as any bridge in the world. It is designed to connect two points together. Many different types of hardware can accomplish bridging. A firewall can act as a bridge, a hub or a switch can act as a bridge, and even a router can act as a bridge. The act of bridging networks requires connecting two network segments together. Sometimes bridging is performed transparently. A transparent bridge enables the bridging device to remain anonymous on a TCP/IP network. This technique can be useful with firewalls. A firewall can be set up as a transparent bridge and dropped into a network with total invisibility. In this mode, the firewall's network interface cards do not maintain IP addresses. As a result, the firewall cannot be directly interfaced with via the network. It is more difficult to compromise a firewall that cannot be detected.

two nodes are exchanging information. If the speed of the network is 10 Mbps, then each node can transmit simultaneously at 10 Mbps.

Most networks are not fully switched because of the costs incurred in replacing all of the hubs with switches. Instead, a combination of switches and hubs is used to create an efficient and cost-effective network. For example, a company may have hubs connecting the computers in each department and then a switch connecting all of the department-level hubs. In recent years prices of switches have become more reasonable and higher incidences of switches in small networks are being utilized.

Switches solve almost every "hub" problem we mentioned above. They're inherently more scalable. They do not have the same type of "chaining limitations" and they can be used like repeaters to overcome distance limitations. They are less susceptible to latency, network failures, and collision problems. They can solve most network efficiency problems. But they aren't the solution to every networking problem. They are still LAN devices, only designed to connect a single network. When it is time to connect whole networks together, a *router* is called in to do the work.

Routers

Routers are specialized devices that send or "route" packets of information to their destination along one of many potential pathways. Complex algorithms are often used to figure out the best path possible by balancing distance, time, availability, and cost issues. While the function performed by routers is complex, inside the box they are composed of a combination of hardware and software much like a personal computer. The difference is that routers are streamlined to perform specific functions constantly, over very long periods of time. They use special operating systems and high-performance hardware in order to minimize the chance of failing.

Routers, like switches and hubs, act like the "copper piping" of a computer network. What all of these devices have in common is their ability to receive signals from computers or networks and then pass those signals along to other computers. The router, however, stands out from these devices, as it is the only one that examines each bundle of data as it passes and decides where to send it along. To make these decisions, routers must receive two kinds of raw data: addresses and network structure.

When you send email to someone how does the message know to end up on your recipient's computer rather than on one of the millions of other computers in the world? A good deal of the responsibility falls on the shoulders of routers. This is because routers are devices that allow messages to flow between networks, rather than within networks.

A router has two separate-but-equal job functions. First, routers ensure that information goes only where it's needed and nowhere else. This is critical for keeping large volumes of data from clogging the connections of "innocent bystanders." Second, the router makes sure that information gets to the intended destination in the most direct and efficient way possible.

In its most basic capacity, a router can be used to connect two separate computer networks, passing information from one network to the other. It also can protect the networks from one another, preventing the traffic on one from unnecessarily spilling over. As the number of connected networks grows, the routing decisions become more complex, but the basic operation and function of the router remains the same.

If you have enabled Internet Connection Sharing between two Windows-based computers, one of the computers (the computer with the Internet connection) is functioning as a simple router. In this instance, the router does so little—simply looking at data to see whether it's intended for one computer or the other—that it can operate in the background of the system without significantly affecting the other programs you might be running.

Slightly larger routers, the sort used to connect a small office network to the Internet, will do a bit more. These routers can make some basic security-related decisions when handling incoming and outgoing traffic. The volume of traffic processed is usually higher than a computer running Internet Connection Sharing can efficiently handle. As a result, most offices use stand-alone router devices rather than software running on a workstation or server.

The largest routers, those used to handle data at the major traffic points on the Internet, handle millions of data packets every second. These routers are large stand-alone systems that have far more in common with supercomputers than with office servers (see Figure 15-1).

Routers, Switches and Hubs

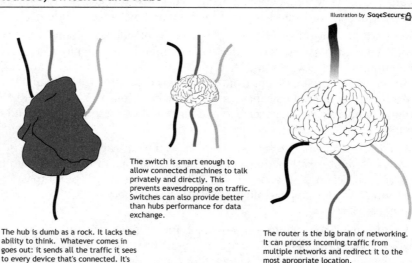

Illustration by SageSecure

The switch is smart enough to allow connected machines to talk privately and directly. This prevents eavesdropping on traffic. Switches can also provide better than hubs performance for data exchange.

The hub is dumb as a rock. It lacks the ability to think. Whatever comes in goes out: it sends all the traffic it sees to every device that's connected. It's trivial to eavesdrop on any computers connected to a hub.

The router is the big brain of networking. It can process incoming traffic from multiple networks and redirect it to the most apropriate location.

Figure 15-1

How Routers Work

Routers need some way of differentiating local traffic from traffic that needs to be routed to another network. This problem is solved with a subnet mask, which is a portion of the data's destination address that is similar to a postal code. The destination route is determined by the sender's subnet mask, which contains a source and destination address. If the traffic is intended for a remote network, a router will point it towards its destination.

One of the tools a router uses to decide where a packet should go is a configuration table. A configuration table is a collection of information, including the following:

- Information on which connections lead to particular groups of addresses
- Priorities for connections to be used, based on cost measurements called "metrics"
- Rules for handling both routine and special cases of traffic

A configuration table can be as simple as a half-dozen lines in the smallest routers, but can grow to massive size and complexity in the very large routers that handle the bulk of Internet messages. Processing a large configuration table requires a lot of processing power. As a result, routers come in a range of sizes, varying based on the amount of networks they need to route between and the amount of traffic they need to handle.

Let's look at an example network scenario that uses a medium size router to handle Internet routing. In this example, the router is connecting a 50 node local network with the Internet. The office network connects to the router through an Ethernet connection, specifically a 100 base-T connection.[1] There are two connections between the router and the *Internet service provider* (ISP). One is a T-1 connection that supports 1.5 megabits of data per second. The other is a DSL line that supports 128 kilobits of data per second.

The configuration table in the router tells it that all out-bound packets are to use the T-1 line, unless it's unavailable for some reason. If it can't be used, then outbound traffic goes on the DSL line. The DSL line is used as a backup against a problem with the faster T-1 connection. The router ensures that human intervention is not required to make a switch to the DSL line in case of trouble. The router's configuration table automatically handles the routing of packets until the T-1 line becomes available again.

In addition to routing packets from one point to another, the sample router has rules limiting how computers from outside the network can connect to computers inside the network and how computers inside the network appear to the outside world. While most companies also have a special piece of hardware or software

---

[1]100 base-T indicates a connection throughput of 100 megabits per second. This architecture uses a twisted-pair cable. This cable is similar to your phone cable connecting your phone to your wall jack, only it contains eight twisted wires instead of four.

called a firewall to enforce security, the rules in a router's configuration table are equally important to keeping a network secure.

Security Considerations

Hubs are bad. Switches are good. Need we say more? There is no beneficial reason to have hubs in your network other than cost. As soon as you can afford it, replace your hubs with switches. The performance and security benefits alone make switching to switches an efficient choice.

Few network administrators know how to perform maintenance tasks on router hardware. Most routers run very custom operating systems with complex command line driven interfaces. This often means you need an engineer who has specific training in order to perform maintenance and fix routing problems. Many companies end up calling in router experts in the same way they'd bring in electricians.

Those who have the special knowledge of how to program your router are not being paid to think about security strategy. In many cases a router is the main public gateway to a corporate LAN. As such, it needs to be tightened down as much as possible. As your routing needs change, so does the exposure of your network to the networks it is connected with.

Consulting with a routing engineer regarding your security policies is imperative to maintaining a safe networking environment. With the proper specific request made, a good engineer can tighten down your router appropriately and ensure the routing tables are not pointing in any wrong directions. Items that need to be secured in particular include the routing tables and the system address tables.

Growth goes hand-in-hand with networking. Once a network has been implemented the eventual obsolescence of its initial hardware is almost inevitable. As networks grow, the capacities of the connecting hardware can create bottlenecks. Where a fast Ethernet hub used to suffice, a switch may now be needed. Collisions may be slowing down your network traffic for months before it becomes known. Planning for growth and monitoring network choke points is critical in the prevention of network aches and pains.

As a final note, most mid to high-level switches and routers provide the ability for remote management. This means that any computer system that can reach the device via the network can change its configuration settings. The management interfaces usually come with a default password set by the manufacturer. If the network administrator doesn't change the password, hackers can easily gain access to the equipment.

Making the Connection

Managing Security: The design of a network directly affects the ability to effectively manage and monitor network systems.

Outsourcing Options: Many companies lack the resources in-house to properly design networks and configure network devices.

Detecting Intrusions: Many of the tools used to analyze networks for intrusions are very sensitive to network design.

Best Practices

Imagine a small company that designs logos and marketing material for magazines. This small, fictitious design firm has 15 employees, each with a computer. Seven of the employees are graphic designers, while the rest are in sales, accounting, and management. The designers must transfer very large files back and forth to each other as they work on projects. To do this, they use a local area network connected together using a 16-port hub. This type of simple local network is known as a peer-to-peer network because it does not use a central server.

When one designer sends a very large file to another, the process utilizes most of the network's bandwidth (capacity), slowing the network for the rest of the company. This slowdown occurs because each information packet sent from a workstation is viewed by all the other workstations on their local network.

During a particularly busy week, the network bogs down to a crawl. Desperate for a solution, the designers bring in a consultant. The consultant looks at the network, rips out the hub, puts a switch in its place, and leaves a bill for $2,000 on the table as he walks out the door. Now when two designers exchange files, the rest of the network stays clear of traffic. The result? No more network performance problems.

A year later, the design firm has grown a bit. They have a dozen more employees, spread out over a few departments. One day, a sales employee gets an email virus from a friend who works at a telemarketing company. A few minutes later, the virus spreads across the network, destroying some critical files on a designer's computer. The CEO, in a panic, calls the consultant again.

The consultant walks in with a big smile on his face. He knows that this time, there's going to be some real work to do. He recommends splitting the network into two smaller networks. One network will be for the designers, the other for the rest of the company. Computers on the designer network will not be able to access other company computers except in very limited ways, and vice versa. Furthermore, both networks need to share some of the same resources, such as a full time Internet connection. To do this, a router is necessary. A router can control the communications between two networks while ensuring that the networks are connected to the outside world.

The router is the only device that sees every message sent by any computer on either of the company's networks. When a designer sends a huge file to another designer, the router looks at the recipient's address and keeps the traffic on the designer network. When one of the designers queries a file from the company's sales database, the router forwards the message between the two networks. When either network wants to access the Internet, the router forwards the traffic out through the Internet connection.

As the design firm continues to grow, it will probably add more servers. Eventually, there may be enough servers to warrant having a server room. For better management and security reasons, the servers may ultimately be placed in their own network. The router can be used to carefully control access to the server network on a department-by-department basis. The router can even control access to individual servers and services.

Final Thoughts

As networks grow and evolve, it's important to periodically re-evaluate the design of the network. Does the current design properly serve the needs of the organization, or has it evolved itself into a corner? Often networks are littered with the remnants of temporary solutions used to put out fires. Cleaning up the litter by redesigning small or large portions of the network can provide major benefits in terms of security, performance, and manageability.

Chapter 16
Connecting Networks: Wireless Connections

Wireless technologies use radio waves and signal
processing to connect systems and networks
without the encumbrance of wires.

Technology Overview

Wireless communication has to be one of history's most reinvented wheels. The first wireless technology was spoken language. This was good, but the range was limited and privacy was difficult within the effective range. Drum beats and smoke signals were invented to help increase the effective communication range, but they virtually eliminated privacy.

In contrast, wire communications rely upon a particular conveyance media. Couriers were the earliest "wires," physically conveying information from one place to another. This was far more private than drum and smoke signals, and could be used over greater ranges. The downside was that couriers were relatively slow and could be easily hampered by terrain, weather, and other adversarial forces. Carrier pigeons improved the speed and range of couriers, but at the expense of security and reliability.

The first modern form of wireless communication was the radio signal. Scientists had already figured out that sound traveled as waves, and that these waves could be transmitted across electrical wires. As scientists began to learn about the concepts of electricity and magnetism, they realized that the same information could be sent through the air as electromagnetic *radi*ation, thus the name *radi*o. Thankfully the names radiomatic and radiotron didn't stick. Society would have never progressed past the 1950s.

The first consumer devices capable of broadcasting radio signals were ham radios. Then came infrared devices and cordless phones, which initially had very short ranges. Infrared devices were particularly limited: they required a clean line of sight to the receiver. It wasn't long before cellular phones arrived. The first cellular phones had much greater range than any of their predecessors. Today's cellular and cordless phones are even more advanced, with greater range, clarity, and privacy features.

Wireless Networking

The most popular type of wireless network technology today is known as Wi-Fi, which stands for *wireless fidelity*. It refers to any device that communicates using the 802.11 group of standards (802.11a, 802.11b, 802.11g, and so on). There are other methods of wireless networking besides Wi-Fi, but they are designed for very small, localized, or home environments and will not be discussed here.

Under 802.11b, the most popular consumer-level wireless fidelity standard, devices communicate at a speed of 11 Mbps whenever possible. If signal strength or interference is disrupting data, the devices will drop back to 5.5 Mbps, then 2 Mbps and finally down to 1 Mbps. Though it may occasionally slow down, this keeps the network stable and very reliable.

The following are some advantages of 802.11b wireless networks:

- They are usually reliable.
- They have a long range (300 meters) in open areas.
- They can be integrated into existing wired-Ethernet networks.
- They are backwards compatible with older 802.11 devices.
- They are as fast, or slightly faster than, 10baseT Ethernet networks.

On the other hand:

- The 802.11b wireless standard is notoriously insecure on many levels.
- Range can fluctuate depending on interference (in dense office environments range can drop down to 25 meters or less).
- Prices for low-end wireless devices have dropped, but high quality wireless networking components are still expensive.
- Superimposing multiple wireless networks can result in sudden and severe degradation.

How Radio Works

The key principle behind the radio signal is the concept of a wave. When you go to the beach, you see endless examples of waves. Radio waves are not very different from ocean waves, except for the salty water thing. If you had to measure the waves in the ocean there would be two things you could record: the height of each wave (amplitude) and the time between each wave (frequency). The number of surfers on the wave is irrelevant unless you happen to be a hungry shark.

Information can be transmitted by *modulating* (manipulating) the shape of waves. The signal starts off as a simple wave with a constant frequency and amplitude. The wave can be manipulated in three basic ways:

- Creating pulses by turning the wave generation device on and off. This is called *Pulse Code Modulation* (PCM). Morse code relies on this principle. It uses groupings of long and short pulses to represent letters.

- Varying the amplitude while holding the frequency constant. This is called *Amplitude Modulation* (AM).

- Varying the frequency while holding the amplitude constant. As you might guess, this is called *Frequency Modulation* (FM).

Can you spot the familiar acronyms? The radio can receive amplitude and frequency modulated signals, which it converts to music. As everyone who has used a car radio knows, AM signals sound lousy but travel a long distance. FM signals sound good, but go out of range much faster. Most modern music stations broadcast using FM signals because of the extra quality. AM is used for talk radio, which doesn't need the high fidelity. As a result, FM and AM might as well stand for Funky Music and Awful Monologues. See Figure 16-1.

Why do signals go "out of range"? Picture an empty balloon. Notice that the rubber is pretty thick. Now blow up the balloon. The rubber gets thinner and thinner until it finally breaks and the balloon pops. The strength (power) of the signal is like the

Wave Concepts

Illustration by SageSecure

The Basics:

A square wave is the simplest wave. It's created by alternating instantly between two tones (hi and low). The diagram shows a single "oscillation" of a wave.

The amplitude is the height of the wave. The frequency is the number of oscillations that can pass by in a second.

A sine wave smoothly alternates between two tones. Sine waves are used as the basis for most wireless transmissions.

Modulations:

Pulse Code Modulation (PCM) involves turning the wave signal on and off, creating "pulses". Information is transmitted by varying the length and space between pulses (like morse code).

Amplitude Modulation (AM) conveys information by increasing and decreasing the amplitude of a sine wave. The frequency remains the same.

Frequency Modulation (FM) conveys information by increasing and decreasing the frequency of the wave while keeping the amplitude the same.

▨ **Figure 16-1**

Power, Frequency, and You

All radiation can be harmful if the amount of energy is high enough. A microwave oven operates at 2.4 GHz and can cook food in seconds. Wait a minute . . . 2.4 GHz? That is the same frequency as new cordless phones, Wireless LANs and Bluetooth devices! How come we're not all crispy fried by now?

Two basic differences exist between a microwave oven and a cordless phone. The first is frequency. Microwave ovens use very specific frequencies that cause water molecules to resonate. Cordless phones do not focus these frequencies. Secondly, a microwave uses a very high power signal and shielding to contain the radiation within a small space. At those high power levels, enough energy is transmitted to heat the water molecules within the food. Phones use a much smaller amount of power. Even if phones used the same frequencies as microwave ovens, the energy of a phone's transmission is not high enough to do any water heating or cellular damage.

Regulatory bodies have determined safe levels of transmitting power for various frequencies. The higher the frequency, the more dangerous the radiation and therefore the lower the allowable power level.

balloon's rubber: There's only a limited amount of it. Close to the transmitter it's pretty strong, just like when the balloon is empty and small. As the signal travels outwards, the power is spread out in every direction. The further away it gets, the more spread out it gets. Eventually, it's so thin that receivers can no longer detect the signal. That static you hear when you drive out of range from your favorite radio station is the sound of the signal's balloon bursting. The more power, the further the range (more rubber lets you blow a bigger balloon).

Many of today's wireless technologies are variations on the concept of FM radio. The "static-free" nature of frequency modulation makes it a better choice for conveying information than amplitude modulation. The main difference between most FM wireless technologies is the "band" of frequencies in which the devices operate. The *Federal Communications Commission* (FCC) in the United States allocates frequencies for operation to minimize interference between different kinds of radio devices.

Current wireless communications systems run on very high frequencies. Safety regulations require an extremely low transmitting power for these frequencies. Radio interference and physical obstructions further reduce the signal strength. The resulting signal is very difficult to detect beyond a limited range.

In order to increase the range of wireless communication devices, advanced signal detection and processing techniques are necessary. The most popular solution used in today's wireless devices is called *spread spectrum* technology.

The concept behind Spread Spectrum technology is to spread radio energy across a wide frequency spectrum, reducing the power at any one frequency. This also reduces interference and makes eavesdropping more difficult. In the United States, spread spectrum technology is required by law for any frequency in the 2.4 GHz band.

**Chapter 16
Connecting
Networks:
Wireless
Connections**

How Spread Spectrum Works

Imagine you and a friend are standing on opposite ends of a noisy room. You want to say something private to your friend, but he can't hear you over the ambient noise and neither of you can move across the room. The "high power" solution is to write the message in large letters and hold the sign over your head. Unfortunately, this is not very private. A "reduced power" solution would be to write the message on a piece of paper and ask somebody to deliver it to your friend. The problem is, that person will see the message and/or he or she may forget to give it to your friend. A better option is to write each word on five pieces of paper. Hand the first word to five random people wandering through the room and ask them to deliver it to your friend. Now do the same for the second word, and so on. Chances are, at least one copy of each word will get through to your friend. Furthermore, nobody else in the room will have the complete message. You used the same amount of "reduced power" to send the message, but with more security and reliability. This is the basic concept behind spread spectrum technology.

Two types of spread spectrum systems are in use: *Frequency Hopping Spread Spectrum* (FHSS) and *Direct Sequence Spread Spectrum* (DSSS).

- **FHSS:** A wide frequency band is divided into a number of smaller bands. The transmitter and receiver initially start communicating over one of these smaller bands. At specific intervals, the transmitter and receiver both change to a new band, selected by using a pre-arranged system.

- **DSSS:** A wide frequency band is divided into a few narrower bands. Each device uses one of the narrower bands. The device then transmits data over multiple frequencies within this band simultaneously. Wi-Fi uses DSSS because of the higher data rate it can achieve.

How Wireless Integrates into Your Network

There are three types of wireless network components: access points, end points, and bridges. An access point connects a wired network to a wireless network. It's a device that broadcasts the network signal to the surrounding environment. End points are the wireless adapters that are found in PCs and laptops. These are just like traditional Ethernet adapters, except they have antennas instead of jacks for an Ethernet cable. Bridges are used to extend the range of a wireless network. They receive signals from a nearby access point or bridge and rebroadcast them (see Figure 16-2).

Integrating Wi-Fi

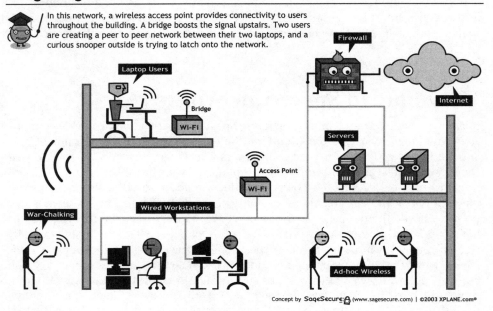

In this network, a wireless access point provides connectivity to users throughout the building. A bridge boosts the signal upstairs. Two users are creating a peer to peer network between their two laptops, and a curious snooper outside is trying to latch onto the network.

Concept by SageSecure (www.sagesecure.com) | ©2003 XPLANE.com®

■ **Figure 16-2**

When a wireless end point is within range of an access point, it will attempt to connect to the access point. Most wireless connections support some type of encryption and/or authentication system. In many cases, possession of a valid encryption key is the only form of authentication necessary. Some access points support additional types of authentication for added security.

It's also possible for two or more wireless end points to create a peer-to-peer network without an access point. This is called an ad hoc wireless network.

Security Considerations

Wi-Fi networking technology is pretty simple and easy to use. It takes little effort or understanding to add wireless connectivity to a network . . . in an insecure manner. It takes a significant amount of knowledge and effort to add Wi-Fi support in a way that doesn't significantly reduce overall network security. Unfortunately, the ease of Wi-Fi means that people who don't understand the security issues can still successfully create such networks by installing a Wi-Fi device onto the network. Let's look at some of the reasons why Wi-Fi can be more trouble than it's worth:

Range: The biggest problem with wireless communications is that the data ultimately travels everywhere. Even though the box says that your wireless LAN card has a 100-meter range, a properly connected tin can will boost the range to a kilometer! The reason for this is that the LAN signal is still traveling well past the 100-meter range; it's just much weaker. The can acts as an antenna, increasing your LAN card's sensitivity and enabling it to pick up the weaker signals.

With the right tools, even extremely short-range signals can be seen from a distance. That's the theory behind TEMPEST, a technology that can show someone what's on your computer monitor from across the street. It works by using an incredibly sensitive antenna to detect the electrical radiation coming from your video card. Sci-fi? Hardly.

Signals get weak, but they keep going until interference distorts the signal beyond comprehension. The signals from the first radio broadcast ever are still traveling through the universe and have already reached star systems roughly 100 light-years away. In fact, the SETI program is relying on this very fact to detect other life in the universe. They hope to eventually detect faint radio signals generated by other civilizations. If E.T. can eavesdrop on your data, what chance do you have to protect it against the average hacker who's much closer and much more likely to be listening?

Interference: The other evil side effect of the range problem is interference. Two signals broadcasting at the same frequency will distort each other. As a result, pandemonium can occur when multiple radio devices are broadcasting near each other at similar frequencies. The problem is that many wireless technologies use similar frequencies. For example, Bluetooth, Wi-Fi and newer cordless phones all operate at the 2.4 GHz range of frequencies. When these devices are close enough to one another they can create interference issues. Unfortunately, the likelihood of unavoidable proximity problems is high, because these devices are becoming fairly ubiquitous within the home and office environments. Even unrelated systems can cause interference. Microwave ovens, some types of lights, and various kinds of machinery can "leak" radiation in the 2.4 GHz range, causing interference.

There are a number of technological solutions for mitigating interference, including: use of different frequencies, use of spread spectrum technology, and intelligent avoidance algorithms. Newer Wi-Fi technology is capable of operating in the 5 GHz range, which will help to avoid interference with other 2.4 GHz devices (until more devices move to 5 GHz). *Bluetooth* uses spread spectrum to minimize interference among other Bluetooth devices, but as the signal hops around it will hop in and out of bands used by Wi-Fi networks. As a result, a Bluetooth device may cause a nearby Wi-Fi network connection to perform badly or completely fail.

Newer Bluetooth and Wi-Fi systems are being designed with intelligent broadcasting systems for avoiding interference. These system monitor interference levels and adjust the way in which data is transmitted to minimize conflicts. The problem is that "intelligent" algorithms make already complex systems even more complex.

Protocol weaknesses: Both Bluetooth and Wi-Fi technologies have flawed underlying security models. Bluetooth security relies upon its short range, spread spectrum signal hopping, and encryption. As we've discussed, range can be boosted with the right equipment. Similarly, it's not too difficult to reassemble a spread spectrum signal, especially if "intelligent" algorithms limit the valid spectrum channels due to interference on other channels. Finally, the encryption system is initialized with a *personal identification number* (PIN) code that is often set to "0000", making it trivial for hackers to forge a connection.

Wi-Fi also has serious security problems. Its encryption system, known as *Wireless Encryption Protocol* (WEP), is relatively easy to crack within a short period of time due to inherent flaws in the protocol design. To make matters worse, most consumer-grade Wi-Fi products initially have encryption turned off. As a result, the vast majority of Wi-Fi access points offer full, public access to anyone within range (1000 meters with an antenna) of the network. This gets to be particularly problematic when an employee sets up a Wi-Fi access point on the office network. Suddenly, your nicely secured internal LAN has a giant hole offering hackers carte blanche access to your entire network and its bandwidth.

Exposure: Just because you can't see them, doesn't mean they can't see you. When you activate a wireless connection with another machine, hackers can see both machines. Even though you're not technically on the hacker's LAN, there's nothing to stop him or her from broadcasting signals at your computer. He or she may even be able to establish an ad hoc wireless network connection with your system without your knowledge. Even if you think you're protected from unauthorized connections, remember that a software bug on your system could let the hacker in through a back door while you're busy having what you think is a secure conversation through your *virtual private network* (VPN).

Unsafe wireless: The availability of free Wi-Fi in metropolitan areas makes it really tempting to hop on. But how safe are these LANs? If a hacker gets into your system when you're on a public Wi-Fi LAN he can control your system when it's back at your office network. Using a free Wi-Fi LAN is like having casual sex without a condom (assuming "like" is used in the broadest sense possible).

What people think: Wireless connectivity is a very advanced concept. It is difficult to imagine that hackers can grab data out of thin air.

What we think: Wireless connectivity is here to stay, but has major security issues that need to be addressed immediately. Current use of wireless networks is extremely risky.

Making the Connection

Hardening Networks: The tools used for hardening networks can help isolate Wi-Fi network vulnerabilities from affecting the rest of your network.

Identifying Vulnerabilities: Only highly secured systems should use Wi-Fi networks—testing the systems for vulnerabilities before going wireless can help prevent hackers from using Wi-Fi to bypass normal security measures.

Part VI   Connecting Networks   **197**

**Chapter 16
Connecting
Networks:
Wireless
Connections**

Cryptography: WEP is an encryption system for Wi-Fi data; understanding encryption is necessary in order to successfully employ WEP on a Wi-Fi network.

Best Practices

It is imperative to tighten the security on any wireless access point, end point, or bridge. The following are some guidelines for successfully tightening a Wi-Fi network:

- At the minimum, WEP should be enabled. Don't make a hacker's job any easier by giving him an open invitation to your network.
- Change your WEP keys frequently, or buy newer equipment that supports automatic changing of WEP keys. Even if hackers crack the key, they won't be able to get back in without cracking the key again.
- Check your end points. Is it possible to initiate a connection with the end point?
- Connecting to the access point doesn't necessarily have to grant access to the network itself. There are a number of security technologies that can be used to add security beyond the measures provided by the access point including additional authentication systems, firewalls and VPNs (described in the next chapter). Assume that a dedicated hacker will gain access to your Wi-Fi network. It shouldn't matter.

What security is available for Wi-Fi end points that will be used on public networks? The best you can do under Windows is to run a software-based firewall. Unfortunately, hackers can get around these fairly easily. A much more secure option is to have a multi-boot system; using one operating system for public Wi-Fi and the other for private work. The authors use a highly secured operating system (OpenBSD) when connecting to public Wi-Fi networks. We boot into Linux/Windows when connecting to the Internet over Ethernet at the office. In the unlikely event that a hacker breaks into the system, he'd have to go through a number of unobvious, extra steps before being able to access work-related files. It would be nearly impossible to do this without being noticed.

If a multiboot system is out of the question, then do not trust any system that has been on a public Wi-Fi network (or any non-office network, for that matter). Constantly scan for viruses and trojans. Consider using a rollout system that rebuilds the hard drive from a known clean image on a regular basis. Just don't assume that the system is "probably safe," unless this type of risk is compatible with your security philosophy.

When deciding on whether to go wireless, it is important to weigh the benefits against the security risks. There are certainly a large amount of security costs, so if the benefit list isn't exceptionally long, waiting may be a good idea. Wireless technology does seem like a fun high-tech toy, but that is not a reason for implementation. Think of it this way . . . the world has run with wires for this long . . . so it can wait a little longer for security's sake.

Final Thoughts

Many people think that no wires are better than wires, or that wireless technology is more advanced than traditional wired technology. This is not always accurate. Historically, communications technologies have wavered back and forth between wires and wireless. Originally television broadcasting was wireless, transmitted over radio waves to antennas on the roof of homes. Then, cable came along, offering better quality and more channels. Soon after, the satellite dish emerged and threatened to eclipse the cable TV business. It offered more choices, more channels and a more advanced interface. Now digital cable offers as much or more than satellite television without the requirement of additional hardware and still maintains its reception during bad storms.

Likewise, wireless networking has some downsides that make traditional wired networks preferable in certain situations. Security, range limitations, interference, and health concerns are among some of the negative aspects of wireless networking.

That said, wireless networking technologies offer very convenient options for staying connected on the move. As time goes by it may be inevitable that wires become things of the past, especially if the security concerns are adequately addressed. Of course, the next "wired" technology might change everybody's minds yet again. Only the future will tell.

Chapter 17
Connecting Networks: Network Lingo

Networking protocols are the languages
that network systems and applications use
when interacting and transferring data.

Technology Overview

What we're about to do in this section is downright evil. We're going to try to explain how the Internet works, from the most basic level. And we're going to do this in just a few pages, which is a most dastardly and nefarious thing to do. Each paragraph here is covered extensively in books that are hundreds of pages long. We've boiled these concepts down to the bone. Actually, past the bone. You're looking at marrow soup here.

Why did we do this? Because it helps to have a basic idea of what's going on in those wires. Beyond the basics, we couldn't even begin to get into any of these concepts without creating a book that would give you a hernia to carry. And frankly, our medical liability coverage isn't that great. If you already fully understand TCP/IP networking, then feel free to gloss down to the "Security Considerations" and "Best Practices" sections.

> **What people think:** Core networking protocols are like Greek to me. I just want to download a file from the Internet.
>
> **What we think:** Understanding the protocols that make networking possible empowers the average user to make better decisions about data.

Ethernet

Many different technologies are available for creating networks. One of the most popular and flexible is called Ethernet. It is a basic system that moves information between two networked computers. Each computer has a network adapter (it looks like a telephone jack on the back of a PC or laptop, only it is larger). This adapter has a unique Ethernet hardware address, which is a fairly large number. Theoretically, every Ethernet adapter in the universe has a different hardware address. The hardware address comes from a chip located on the Ethernet card.

Information travels across an Ethernet network in packets, or frames, which are small parcels of data. Each parcel has a "from" and "to" hardware address, followed by some data to transmit. Packets are usually small; the data in a Web page might get spread across many packets.

Ethernet is nothing more than a basic delivery system. Think of it as a mail tube. Put the letter in and hope it gets to where it's going. The tube itself will not guarantee that the information will arrive at its destination. Ethernet itself does nothing to direct information from one network to another. What Ethernet actually does is pretty basic. It is known as a "link layer" technology, because it deals directly with the hardware that links machines together.

The Internet Protocol (IP)

This is the fundamental technology behind the Internet. One of the main purposes of *Internet Protocol* (IP) is to create a hardware-independent

Encapsulation

A packet generally has two fundamental parts: a header and data. The header contains the from and to addresses, and some additional information that helps when moving the packet around the network. The data can contain anything, including another packet. Think of a packet as a Russian doll. The header is the artwork on the outside. The data is the doll shell. When you open up one shell, another smaller doll is inside. That smaller doll can contain an even smaller one . . . and so on. Encapsulation is the process of putting a packet of one type inside another.

An IP packet sent over Ethernet is actually encapsulated inside the Ethernet packet. So, when you "open up" the data part of an Ethernet packet, you'll find an IP packet, which has a header and its own data. As you'll soon see, the data part of the IP packet can be opened up as well.

system for moving data from one machine to another. Like Ethernet, IP data is also sent in packets with from and to addresses. IP has a separate addressing system from Ethernet, one that is hierarchical in nature. Although IP data is commonly sent across Ethernet networks, it can operate with many other types of link layer systems.

Address Resolution Protocol (ARP)

The question is, how does one computer on a *local area network* (LAN) figure out which Ethernet address goes with which IP address? After all, IP packets need to be stuck inside of Ethernet packets, and Ethernet packets need a hardware address to get from one machine to another.

The *Address Resolution Protocol* (ARP) enables a computer to shout out, "Does anyone on the LAN know the hardware address for this IP address?" The computer with the matching IP address then stands up (digitally speaking) and says, "Yeah, that's me, over here. Inspect my Ethernet address if you don't believe me." Now the first computer can put together a proper Ethernet packet and send the IP packet along inside.

Transmission Control Protocol (TCP)

Neither Ethernet, nor IP ensures that data gets from one machine to another. Packets are just sent off into the void, with good wishes and tearful goodbyes. These are considered unreliable protocols. They never call; they never write.

The *Transmission Control Protocol* (TCP) is a system that works with IP to provide some communication reliability. It works by sending acknowledgements back and forth. If a packet is sent and the corresponding acknowledgement is not received, the TCP system guesses that something went wrong and tries again. TCP data is also sent in packets. These packets are encapsulated in IP packets (which are encapsulated in Ethernet frames).

TCP is considered a "stateful" protocol. In order for two computers to have a TCP conversation, they need to establish a "connection." This is done through a process called a three-way handshake. The first computer says, "Hi, I want a connection." The receiving computer says, "Okay, you can have a connection." Finally, the first computer says, "Okay then, let's talk." All three parts have to happen in order for the connection to go through. Once a connection has been made, information in the TCP header allows both sides to recognize future packets associated with that connection. This means that the connection process only has to happen once, as opposed to once for each packet.

Imagine that you have five or six programs all communicating over TCP. How does one program know that it should answer an incoming connection request? If this were a phone system, each program would have an extension (the main number would be the IP address). In the TCP world, that extension is called a *port*. It's a special identifier that applications can use to figure out which traffic is important.

User Datagram Protocol (UDP)

The problem with TCP is that the back and forth required to create a stateful connection involves a lot of extra data. That extra data can bog down certain types of communications, especially those that don't care if a few packets get lost. Streaming audio, video, and video games are all examples of systems that don't usually care if packets go walkabout. These types of communications rely on such a large number of packets for fluidity that a few lost packets go unnoticed and have little effect on the overall communication.

UDP is a "connectionless" or stateless protocol. There's no handshake and there's no acknowledgement. The major advantage it provides over directly sending out IP packets is the capability to use ports. It also has a few other features, such as a header field to support basic data integrity checking.

Internet Group Management Protocol (IGMP)

Sometimes you want to send a message out to many computers at once. Instead of having each computer establish a separate connection with your machine, you want to send out one message that essentially "carbon copies" all the other computers that want to listen. Multicasting is a technique that enables limited broadcasting over a network. A collection of machines sharing data is called a multicast group. Creating and joining a multicasting group is accomplished through *Internet Group Management Protocol* (IGMP) messages.

Routing

If you've been observant, you've noticed that thus far we've left out a pretty major detail: explaining how the Internet actually works. We've said that Ethernet addresses are only useful on a LAN, and that IP doesn't really care about packets getting anywhere either. So the real question is, how does a packet get from your PC to some computer out on the Internet? Damned if we know, but drop us a line if you ever figure it out.

Routers are special computers that connect multiple networks together. When a packet comes in from one network, the router needs to figure out which other network should get the packet next. It does this by consulting a routing table. This is a chart that can be used to make decisions based on the packet's destination IP address. Routing tables for most corporate routers are pretty simple. The complex routers are those at *Internet service providers* (ISPs) and major network exchange points, where multiple large network providers all interconnect.

Internet Control Message Protocol (ICMP)

Sometimes things go wrong. Machines go down, networks get disconnected, and routes can change. The ICMP exists to help machines tell each other about problems or major changes. It's a simple set of commands that can be sent inside of an IP packet. For example, a router can send a "host not found" error back to a computer

that's trying to make a connection to a nonexistent machine. This can save the computer from wasting time trying to establish an impossible connection.

ICMP can be used as a basic routing protocol if a network only has a handful of routers. One router can use ICMP to tell another that a route is no longer valid or that a better route is available. For reasons we'll talk about later, most routers do not use ICMP when making routing decisions.

Real Router Protocols

The big routers that run the Internet need to talk to one another. They keep each other posted on the best way to get data from one side of the Net to the other. Some of the early routers used a fairly simple system called the *Router Information Protocol* (RIP) for intercommunication.

When a router using RIP first connects to a network, it looks for other routers and asks them for their routing tables. It then puts all the tables it finds together and uses the combined table as its own. At regular intervals, the router will broadcast its routing table to its neighbors.

The problem with RIP is that security is not integrated into its design. Specifically, RIP will accept routing updates from any source. In addition, it does not check the integrity of the routing tables it receives. It is easy to send false routing tables to a router running RIP as its information protocol. In response to this, RIP-2 was created, which is a slightly more advanced version of the RIP protocol. RIP-2 includes some basic security measures; specifically, it is able to support encrypted authentication.

When routing tables start to get large, all the broadcasting starts to create substantial network overhead. A better protocol, called *Open Shortest Path First* (OSPF), was created to fix many of the problems with RIP in more robust environments. Most major routers now use OSPF, as it is much more efficient and secure than RIP. OSPF can handle encrypted authentication and can measure the effectiveness of a route based on the quality (cost) of the connection from end to end. As a result, OSPF can pick an optimal route in situations where connectivity problems are many routers away.

One other protocol is associated with routers, but this one is associated with the biggest and best routers in town. The protocol is the *Border Gateway Protocol* (BGP) and is designed for the largest routers with the heaviest of loads. Generally, routers that are responsible for the backbones of the Internet use BGP.

Dynamic Host Configuration Protocol (DHCP)

The *Dynamic Host Configuration Protocol* (DHCP) is a frequently used networking protocol. Unlike some other protocols that have been discussed, it can be used by a variety of systems in a variety of environments. Whether used by a router at an ISP junction or a small server in a small company, DHCP is used to manage IP addresses. A network administrator defines a pool of IP addresses that are automatically assigned by the DHCP server to any client device that needs one. With such a system, new devices added to a network do not need to be manually configured for connection. In addition, large networks that have more workstations than IP addresses can share IP addresses over many workstations. Only when a workstation is being used will it request an IP address; otherwise, the IP address will go back into the DHCP pool to be reassigned to another computer.

Domain Name Service (DNS)

When you open up a Web page and type "www.openlysecure.org," your computer needs to find the IP address of the Web site's server before it can send a Web request out to the server. The *domain name service* (DNS) is used to look up the IP address associated with a particular domain name, such as www.openlysecure.org. It's essentially a giant phone book, matching names to numbers. Hundreds of thousands, if not millions, of domain name servers exist throughout the Internet. However, all the data can be traced back to only a handful of master (root) servers at the end of the line.

When you look up an address, your PC connects to its local DNS server (most likely at your ISP or inside your company LAN). This server contacts the root server and asks for the IP address of the DNS server responsible for the domain you want. Your local DNS server then contacts the domain's DNS server directly and obtains the correct IP address. Along the way, your local DNS server and your PC will save (cache) the resulting IP addresses for faster future reference. When a *uniform resource locator* (URL) is typed into a browser for the first time, generally some sort of delay will take place before the Web site is located. That delay represents the time it takes the DNS server to find the IP address associated with the domain name. Once a domain has been located, little to no delay should occur when visiting the site for a second time due to the caching process.

Network Time Protocol (NTP)

Many network applications require a certain degree of synchronization in order to operate properly. For example, certain advanced authentication systems need to be precisely synchronized with an external clock. The *Network Time Protocol* (NTP) makes it easy for computers to keep their clocks precisely adjusted to international "precise" clocks.

NTP is designed to prevent everyone from trying to contact a handful of precise clock servers. Instead, a few NTP servers are considered top-level and can talk directly to the precise clocks. Each of these top-level machines has a number of second-level machines. The second-level machines have third-level machines, and so on. Most machines can connect at the third or fourth level. This ensures that the machine's time is constantly within milliseconds of a precise clock's time.

Simple Network Management Protocol (SNMP)

Switches, bridges, routers, print servers, and other basic network devices often need to be configured. The problem is, they do not provide a keyboard or monitor output to allow direct control. Usually, hooking up a special serial cable between the device and

a PC enables a direct link. In some scenarios, physically connecting a serial cable to network devices may be impossible. For example, on a large network with several pieces of equipment, making widespread changes would be difficult if it involved hooking up PCs to each piece of network equipment. To solve this problem, *Simple Network Management Protocol* (SNMP) was created as a basic system for remotely managing network devices. With SNMP, a single computer or terminal can remotely manage basic networking settings on a network hardware device. Most mid- to high-level networking devices provide some degree of configuration management via SNMP.

Security Considerations

We've looked at the basic protocols involved in moving data around a network. Now let's look at the security issues that are inherent in each protocol:

Ethernet: It's possible to forge Ethernet packets. This means that hackers can obscure their trails or impersonate other systems at the lowest levels of communication.Hackers will often attempt to capture raw Ethernet traffic. This gives them the most information possible, as all of the data traveling across the network is contained within the Ethernet packet. Once all the data is collected, it can be sorted at a later date to look for information that will be critical to breaking into additional systems.

Internet Protocol: IP can be exploited in a number of ways. Because IP has no authentication or verification, it's trivially easy to forge packets that appear to come from a different IP address. The IP header also contains options for controlling the route that packets take through a network. Hackers can use this to divert information to machines that they control. A combination of IP abuses can allow a hacker to intercept traffic on a switched network, something that is theoretically not supposed to be possible.

ARP: An attack known as *ARP cache poisoning* tricks one computer into thinking that a hacker-controlled computer is the router or some other important server. It occurs when the victim computer receives a false ARP reply that maps the hardware address of the hacker computer to the IP address of the machine it wants to impersonate (in this example, a router). This updates the ARP cache of the victim computer with the newer incorrect mapping. Now, all the packets destined for the router get sent to the hacker-controlled computer. Often, this machine has been set up to analyze the packets and forward them to the actual router. In essence, the victim now thinks the attacker's machine is the router. In many cases, this sort of attack is difficult to detect.

TCP: *Session hijacking* exploits a fundamental vulnerability in the design of TCP. It occurs when an individual attacking a system interrupts communication between a client and a server at a precise moment. By doing this, the attacker fools the client into thinking it is connected to the server, even though the server has no acknowledged connection. The server resets and looks for another connection from the client, whom the attacker is now able to replicate, thus taking control of the client/server session. This method of attack is a form of session hijacking.

UDP: The Fraggle attack is a denial of service attack where an invalid, spoofed UDP packet is sent to a network broadcast address. The machines on the network all send an error message back to the spoofed address. The packet is usually broadcast over a number of highly saturated networks. These are networks where most of the IP addresses in the subnet are assigned to active computer systems. The result is that a flood of error packets simultaneously arrives at the spoofed address, potentially denying service to whatever machine is there.

Routing: Routers are dream targets for hackers. Controlling a significant router is like controlling the transit system for a major metropolitan area. From a router, a hacker can usually launch any number of complex attacks on the network. Routers that are poorly configured can be real boons to hackers. Even if they can't get in, they may be able to abuse the router to help in attacking somebody else's network.

Routing is a complex art. If your router isn't connected to some major infrastructure, its routing tables should be statically created by hand. Otherwise, you're giving hackers a chance to impersonate a router and mess with the routing tables.

DHCP: DHCP runs on a leasing principle. This means that the IP addresses it controls are "lent out" to the clients that request them. After they have been used for a certain period of time, or if the client no longer needs them, they are returned to the address pool. If malicious users gain control of a DHCP server, they could then lease out IP addresses to their own systems and latch onto that network. Not only would this give them full client access to a network, but it would also disguise their location. Anything that was done from that point forward would look as if it originated from the attacked network.

ICMP: ICMP is extremely easy to exploit, as it was not designed to be secure. Many denial of service attacks have leveraged ICMP in the past to create massive amounts of Internet traffic. Smart hackers can also use ICMP to alter and control routing tables across connecting devices such as routers.

DNS: DNS is a hacker's dream. Grab a DNS server and you can get your hacked machine to impersonate anyone. A hacker can also redirect users to alternate versions of sites that are under his control. For example, a fake AOL, Yahoo, or MSN homepage could show up when the user types the Web address into the browser window. This fake page would capture the user's name and password and then act as a middleman, passing and capturing subsequent communications. The user would see no problem; he'd be successfully checking his email, but the hacker would be intercepting every message and would have permanent access to the user's mailbox.

The recent versions of Microsoft's server operating systems (Windows 2000/XP) desperately want to run DNS services, even if you don't want them to be your DNS servers. This creates a potential security problem: A hacker can take over a Windows machine and reconfigure the DNS any way he want. He can then convince other machines on the LAN that the rogue DNS machine is the real one.

NTP: Many network services rely on being able to get an accurate sense of time. Some services are so sensitive that they require computers to be synchronized to within less than a second of one another. Systems running such services often use NTP to stay in sync. Although difficult to accomplish, in some situations a hacker might be able to impersonate an NTP server and slowly force system clocks out of sync with one another. This could be used to disable or exploit applications sensitive to the time.

Random numbers are frequently used to provide security when something "unguessable" is needed. The reality is, these random numbers are not truly random. Many applications use the time of day (down to the millisecond, and sometimes with even more precision) as a "seed" when generating random numbers. Using the same seed gives the same sequence of random numbers. If a hacker can accurately predict or control the system time by exploiting NTP, he can figure out the seed and therefore the resulting random number, effectively guessing the unguessable.

SNMP: If SNMP is active and not properly configured, hackers on your network can directly access and control many of your network devices. Very often SNMP devices are never configured and have manufacturer default passwords that hackers know. It is important to either properly configure or disengage this protocol on all devices to avoid this problem.

Making the Connection

Hardening Networks: Hardening technologies help mitigate the security risks inherent in these core networking protocols.

Vulnerability Scanning: Most vulnerability scanning will detect insecure protocols running on a network.

Best Practices

In general, disable any services you're not using and block any protocols you don't need. Most firewalls will block any externally originating data that isn't from a trusted machine or wasn't specifically requested by a machine from within the network. That's great for external protection, but many hacks come from within. It's much more difficult to protect against vulnerabilities in the core protocols if the attacker is already inside your network. Keeping that in mind, here are a few specific notes:

Ethernet: Most operating systems handle the creation and reading of Ethernet packets extremely efficiently. The code for handling Ethernet has been extensively tested and scrutinized over years of use. This doesn't mean problems will never happen, but they're highly unlikely at this point. As a result, few applications use Ethernet directly. Programs generally interact with the network at a higher level. There's little advantage to directly creating and reading Ethernet packets, except for the purposes of intrusion detection and malicious hacking.

Intrusion detection devices will generally look for Ethernet packets that might have been directly created by malicious programs. These may have forged addresses and other attributes designed to fool or compromise a network system.

Routing protocols: If you have a serious enough infrastructure to warrant high-end routers, make sure you have some qualified individuals taking care of the configuration and maintenance. Good credentials for this type of work usually include an engineering background with years spent programming routers in large network environments. Audit the router security at least once a year, more frequently if your network is rapidly expanding.

DHCP: This is a great tool for almost any networked environment. At the very least, it will guarantee less grunt work for network engineers. You will never need to manually enter IP address information on a client again. In addition, DHCP can assign other critical information to clients. Commonly assigned information includes DNS server addresses, gateway addresses, and other pertinent network data. It can be especially useful when combined with network address translation to assign private IP addresses to a private network. Be sure to set shorter lease times when using DHCP on a network that has more computers than free public IP addresses. Lastly, DHCP makes a wonderful management tool to watch over a network. At a glance a DHCP server will show an administrator how many clients are connected, what IP addresses they are using, and how long they have been on.

SNMP: Most machines don't need to send or receive SNMP data. Only authorized monitoring and administration machines should be able to send and receive SNMP traffic. Routers and firewalls can be used to enforce this type of policy. Some SNMP devices can also be configured to only allow connections from specific IP addresses. In general, if you're not using SNMP, disable it wherever possible.

Final Thoughts

Today's networks are pretty complex. This section surveyed the core protocols that make networking possible. We only covered the basics, and merely scratched the surface at that. Although complete knowledge of these technologies is only necessary for experts, it's important for anyone dealing with security to have a basic understanding of the key concepts. At the least, this chapter should have given you an idea as to why it's so hard to secure networks: A lot of activity is going on and the whole thing rests on a relatively insecure foundation. If you need to know more about any of these protocols, the companion Web site (www.openlysecure.org) has plenty of links to more resources.

VII
Hardening Networks

Summary

It's not always possible to secure everything. Often many segments of a network connection will be out of your control, especially if you connect to the Internet in any manner (and who doesn't nowadays?). Nonetheless, a number of technologies are available to secure the portions of a network that you *do* control. This part of the book discusses technologies available to harden your network against attacks.

Key Points

- Network hardening and network design are very closely intertwined processes.
- Network hardening compensates for practical network design compromises that real networks need to make.
- No amount of network hardening can compensate for a poor network design.
- Some hardening can be done by removing insecure systems and services, while other hardening relies on adding security-related hardware and software.
- Network hardening technologies can do more harm than good if not properly utilized.

Connecting the Chapters

The tools presented in the following chapters provide security for different aspects of network communication. Firewalls and network address translation protect the entry points to a network. Virtual private networking secures data traveling between networks. Traffic shaping ensures consistent availability of high priority network resources. These tools, when used in combination, can provide great protection of network data.

- **Chapter 18, "Firewalls,"** covers devices that can restrict information traveling in and out of a network.
- **Chapter 19, "Network Address Translation,"** explains a technology that can convert *Transmission Control Protocol/Internet Protocol* (TCP/IP) addresses from one subnet to another.
- **Chapter 20, "Virtual Private Networks,"** looks at using encryption to create a secure network connection between two systems over an insecure network.
- **Chapter 21, "Traffic Shaping,"** examines a system for controlling access to bandwidth in order to improve data security and bandwidth efficiency.

Introduction to Hardening Networks

A fine line lies between network design and network hardening. The line often gets blurred by vendors and technology writers (like us). There's no definitive agreement on where design ends and hardening begins. Some experts don't even see a difference between the two processes. For the sake of everyone's semantic sanity, we're going to start by clarifying what *we* mean by design and hardening.

Network design is about deciding what you want your network to do. The design process involves choosing which network services to provide, and creating a network infrastructure to support those services. Good network design provides the foundation of a secure network. The quality of the design also affects efficiency, productivity, speed, longevity, and maintenance needs.

Network hardening is the other side of the coin—it's about making sure a network *only* does what it was designed to do—and nothing else. Hardening involves using a combination of tools and techniques that can control access to services and protect machines that can't effectively protect themselves.

The fact that networks are "hardened" doesn't mean that they start like Jell-O. A well-designed network is actually pretty solid and difficult to attack. Network hardening is much more like hardening steel, a process that makes something already strong even stronger.

In practice, network design and hardening are very closely intertwined. Design decisions affect the choices available for hardening, and hardening technologies have direct implications on network design. For example, adding a hardening device to a network might require a change to the network topology. The new topology might affect the functionality of other hardening devices, which could prompt a network redesign.

The process of redesigning a network is often confused with hardening a network. A network redesign happens when structural changes affect the flow of information. An example is separating a network into smaller, independent networks. The sidebar shows some ways in which this type of network redesign can help improve security.

Managing Internal Threats

Most network threats come from the inside. A user in one department often exploits knowledge of the network and the business to launch attacks on other parts of the network. Splitting (segmenting) the network is one technique used to minimize these threats. It prevents users from accessing machines in unrelated business units. At the most, they'll only be able to access the systems within their own unit.

The best way to deal with internal threats is awareness. An employee who is attacking the network probably has other work-related problems. His manager and co-workers are in a good position to spot such issues. By limiting the threat within a business unit, accountability is directly in the hands of those affected.

Ideal Versus Reality:
The Need for Hardening

An ideally designed network would be incredibly difficult to attack from either the inside or the outside. But ideal designs are not practical. Their restrictions create far too much inconvenience for the average business.

The practical needs of business often require making compromises in network design, balancing functionality against security. Consequently, most real networks contain insecurities, even those designed with security in mind from the start. Common vulnerabilities include services that don't need to run and systems that can be exploited because of hardware or software flaws.

An ideal network would never interact with untrusted systems. In the real business world, that's not practical because people often need to:

- Access external untrusted resources from within the network.
- Access internal resources from machines on external untrusted networks.
- Exchange information quickly and readily within the organization.

This functionality comes at the cost of weakened security. Uncontrollable and untrusted elements will have pathways into the network. Internal threats may have greater potential for damage.

Network hardening technologies compensate for these and other common design compromises (try saying that three times quickly). They can do the following:

- Provide a central choke point where both internal access to the outside and external access to the inside can be controlled (firewalls).
- Restrict access based on per-user authentication (proxies).
- Prioritize the use of shared network resources (traffic shaping).
- Keep data secure when it travels across insecure networks (*virtual private networks* [VPNs]).

There's No Point in Closing the Barn
Door After the Horse Has Left

No amount of hardening can make up for poor network design. A lock on the door is pointless if another door or window in back is open. A good design limits the possible entrances to those that can be effectively managed with hardening technologies. A poor design either results in too many entrances to protect, or doorways that are so wide that they become fundamentally unprotectable.

So what is a poor network design? Ultimately, it's any design that permits a user to have far more access to resources than necessary. It's like parenting: letting a child eat all the candy he or she wants would be considered poor parenting. Not allowing any candy is impractical (unless both parents are deaf), but good parents control

Table VII-1.   Network Practices

Design Issue	Bad	Good
Need to exchange files	Employees make files and directories on their system available on the network.	A central file server and sharing environment controls who gets to access the files and can provide additional protection against viruses and data destruction.
Need Internet access	Unrestricted access to the Internet from the primary desktop.	Unrestricted access from isolated machines or highly controlled desktop access.
Need remote access	Allow remote connections to desktops or file servers (programs like PC Anywhere).	Implement alternate remote access solutions that are separate from the primary internal network, such as a secure Web-based email system.
Business is growing	Machines are all added to the same network.	The network is segmented based on business organizational structure and functional needs.

access to sweets and junk food. Similarly, network resources are like candy. Internet access, email, and file sharing resources can all be used in excess. Network gluttony can lead to viruses, trojans, and other intrusions.

Table VII-1 compares a few good and bad network practices. There are many others, but this should illustrate the underlying distinction.

Out with the Bad, in with the Good

Hardening involves removing the insecure elements from a network and then implementing additional security. It is more important to remove insecurities than it is to add hardening devices. Fundamental vulnerabilities in the network will undermine the ability to further lock down the network through hardening.

Entry to a network is often gained through services that are either fundamentally insecure or have not been properly configured. Shutting down or reconfiguring services of this nature is an example of removing the insecure. Equally important is the removal of any extraneous software and hardware, regardless of the security implications. If it is not absolutely needed, remove it; this is a good rule of thumb in any design or redesign stage. Simply think of a network as a house, the bigger the house, the more doors and windows through which an intruder can enter. Don't build a mansion if the situation calls for a three-bedroom colonial.

Once unnecessary hardware and software are removed, there will still be security vulnerabilities. Now it's time to bring in the heavy artillery. This is when the VPNs, firewalls, and traffic shaping devices come into play. These, and other hardening technologies, can effectively fence in your network from outside threats. They can protect your data when it's in transit. They can even be used inside the network to protect critical systems from internal threats.

More Harm than Good?

Network hardening technologies are powerful tools. However, these tools can be self-defeating if not properly set up. A poorly implemented hardening solution can even reduce the security of the network.

> **What people think:** We have bought a (insert latest must-have security product here), so we are ok.
>
> **What we think:** It's not what you have, but how you use it.

One of the most commonly misused hardening tools is the firewall. The firewall has gained tremendous popularity in recent years. Who would have guessed that such a complex technology would be mass marketed to the general public? There are many people who can't set the clock on their VCRs yet will happily discuss the merits of their home or office firewall. Well, at least they can't record reruns of Miami Vice.

Firewalls have been marketed as simple "black box" solutions to network security problems. People buy them and plug them into the network, hoping that hackers will be held at bay. Often, these firewall devices are rendered useless due to poor placement and/or poor configuration.

Figuring out the appropriate place on a network topology for a hardening device is not always an easy task. Sometimes there's only one possible choice, but usually there are many options. Firewalls, for example, can be placed right before a router, right after a router, right after a switch, right before a critical server, and so on. Sometimes more than one of the same type of device is necessary to properly address the security needs. It's not unusual for relatively small networks to have two or more firewalls.

Even if the placement is right, improper configuration can destroy the usefulness of a hardening device. A firewall that's merely thrown on the network won't do much good. Understanding how to configure and maintain a firewall requires an advanced understanding of network technology. Likewise, it takes a good amount of business understanding to integrate a hardening device within a network effectively. This creates a real problem, as all of that knowledge is rarely sitting in front of the keyboard when it comes time to make configuration choices.

Adding multiple hardening devices to a network design complicates the entire picture. Not only do these devices need to properly integrate into the network, they need to properly integrate together. Some technologies go well together. Firewalls work well with *Network Address Translation* (NAT), and they work well with VPNs. You would think that this means NAT and VPNs work well together. In fact, VPNs and NAT can interact horribly (described in the section on NAT).

Final Thoughts

The remainder of this part covers the most commonly used network hardening technologies. They provide the ability to regulate, manipulate, and track network traffic. Although these tools are often used to protect the perimeter of a network, they can also be effectively deployed within the network to provide internal control.

Chapter 18
Hardening Networks: Firewalls

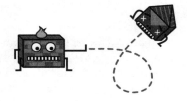

Firewalls are devices that can restrict
information traveling in and out of a network.

Technology Overview

Don't you hate it when you are at a party, meeting someone, getting to know them better, and all they want to do is talk about firewalls? How their network gets scanned by hackers 200 times a day, how they've survived worms and viruses, and how good they are at security? Aren't they special?

Thankfully, party talk is not always that bad, but the term *firewall* has nonetheless broken the geek-speak barrier and obtained social buzzword status. Firewalls have also entered the world of the business mainstream. Businesses and home computer users have become comfortable with the concept of a firewall and its perceived role in the network. Perhaps too comfortable.

> **What people think:** Firewalls solve all of our security problems. We are ahead of the game (sophisticated) because we have one or more firewalls.
>
> **What we think:** Firewalls have become a defense that most perpetrators can circumvent with great ease.

One technology that has greatly contributed to common knowledge of firewalls is broadband (high-speed) Internet access. Broadband is now available to people in their homes throughout many major cities around the globe. Before broadband, it could be argued that home computers were rarely on the Internet because modem connections were typically short. Home modem users didn't feel like they were exposed to hackers (although they actually *were* primary targets). However, a broadband connection is always on, and therefore creates a large stationary target for hackers. Home users now see themselves as potential targets (which they still are). As a result, firewalls are now marketed to both home and corporate consumers as necessary security solutions.

What Is a Firewall?

Not all firewalls are created equal. Software firewalls and low-end "firewall appliances" (such as those built into cable/*Digital Subscriber Line* (DSL) modems) are radically different from the type of high-end firewalls sold by major commercial vendors. The difference is so significant that they should actually be called by different names. Nevertheless, since the public is comfortable with the "f" word, it has been applied to nearly every type of security networking product.

Most firewalls have one thing in common besides their name. They all perform *packet filtering*. A packet filter is simply specialized software that filters the information traveling between an outside network (such as the Internet) and a private internal network or computer system. To oversimplify the process, if incoming data does not meet certain criteria then it is not permitted to pass. More specifically, packet filters have the capability to accept or reject individual *Internet Protocol* (IP) packets based on information contained in the header of the packet. If that statement is confusing, skim over the few paragraphs on IP in Chapter 17, "Network Lingo."

Early packet filters treated each packet as a separate entity. They had no ability to deal with groups of related packets. For example, when a Web browser makes a request, a Web server sends many packets in response. Packet filters saw no relationship between the response and the request. This was a problem, because administrators wanted the ability to say, "If the outgoing request passes our filter, then all the related responses should be automatically allowed through the filter."

Stateful inspection is a concept that was added to packet filters to solve this problem. It compares certain parts of the packet to a database of trusted information. Information traveling from inside the firewall to the outside is monitored for specific defining characteristics, and then incoming information is compared to these characteristics. If the comparison yields a reasonable match, the information is allowed through. Otherwise, the packet is discarded. This procedure allows for a dynamic, on the fly, approach to firewalling.

Packet filtering software usually runs on routers, because firewalls often sit between two or more networks. Another related technology often integrated with routers and packet filters is *network address translation* (NAT), which is described in the following section of this chapter. In fact, most packet filters are capable of performing routing and NAT. There are some exceptions, though, such as software-based packet filters and *bridging firewalls*. Zone Alarm and Black Ice Defender are two examples of PC-based packet filters.

For many years, packet filtering, routing and network address translation were the only things that firewalls could do. Stateful Inspection provided a breath of fresh air, but it still left much to be desired. One glaring issue was that a packet filter has a blind eye toward the content of a packet. A malicious web packet looks the same as a harmless one. This is because firewalls were not supposed to look at the actual application data within each packet. That's a job for a different type of solution, called a proxy server, described in the sidebar.

Firewalls

Illustration by SageSecure

Firewalls attempt to close all access points to a network. Decisions are made at the front door as to who can come in and who will be turned away.

Firewall rules can specifically allow or deny certain types of traffic. Incoming traffic is examined to see if it meets the criteria of the rules.

Of course, there are always other ways in besides the front door.

Rules of Entry
No Spoofing
No Telnet
No FTP
Keep State

Firewall

Vulnerability

■ **Figure 18-1**

Proxy Servers

A function that is often combined with a firewall is a proxy server. The proxy server acts as a middleman between a client and a server. It understands application data and can therefore make intelligent decisions about information.

When a computer protected by a proxy server makes a request to a remote server, the proxy server intercepts the request and inspects or manipulates the data. It then forwards the request on to the remote server (assuming it doesn't deny the request). When the remote server sends a response, the proxy server intercepts the information. Once again it has a chance to inspect, manipulate and potentially deny the data. The net effect is that the remote computer never comes into direct contact with anything on the protected network other than the proxy server.

Proxy servers can be used to control outgoing access as well as incoming access. For example, a web proxy server can require a password to access a general Internet web site. It can also prevent access to web sites that contain objectionable keywords, such as "sex," "x-rated," or "Rush Limbaugh."

Proxy servers can also make Internet access more efficient. Frequently requested data can be stored on the proxy server for rapid retrieval. Web proxies store frequently accessed web pages and images, speeding up retrieval time for popular web sites and reducing the amount of outgoing traffic. This means that the next time someone goes to that page, it doesn't have to load again from the web site. Instead it loads instantaneously from the proxy server.

Feature-Hungry Firewalls

Commercial firewalls are specialized computers running specialized operating systems with nice management interfaces. These devices are cleverly marketed and sold as complete one-stop solutions. In the process, the lines between firewalls, proxy servers, and other related security technologies have been blurred. In their race for one-upmanship, vendors advertise their firewalls as having capabilities such as virus scanning, web filtering, network performance enhancement, virtual private networking, and even intrusion detection. Different vendors provide different assortments of extra features. This makes it difficult to say exactly what a firewall does, since no two firewalls do exactly the same thing. The only sure thing is that a packet filtering router (usually with stateful inspection built-in) is at the core of every solution. Many of the other "features" are provided by integrated proxy servers and other software bundled with the firewall.

Nothing is particularly magical about an expensive commercial firewall. With a bit of knowledge, a very effective firewall can be built from scratch using old computer equipment.[1] All of the features of the most expensive firewalls would be available to this low-cost solution. However, commercial firewalls offer guarantees of reliability and service. A major vendor can make verifiable statements about the reliability of their products because millions of them are in use every day. It's not as easy to make the same statements about a homegrown solution, even if it is technically and functionally identical (or even superior).

Reliability is a significant concern when implementing a firewall solution. A firewall, whether hardware or software based, commercial or homemade, will ultimately fall into your gateway chain. This means that many, if not all, of the computers inside your network will be passing packets through the firewall. If the firewall should fail to function in any way the whole network can lose external connectivity. Things can get even more painful when firewalls used to protect internal network resources fail.

Some commercial firewalls are designed to perform automatic fail over when they are purchased in pairs. This is a feature that is more difficult to implement with homemade firewalls. This offers a big advantage to corporations who are willing to shell out big bucks for peace of mind.

Firewalls in Practice

The most direct use of a firewall is to protect a network from outside intruders. A typical company might have hundreds of computers that are networked together. Such a company will often have one or more connections to the Internet through some type of broadband connection. Without a firewall in place, all those computers are directly accessible to anyone on the Internet. Data packets can leave the inside to get to the outside and packets can come in freely. A malicious individual can probe those computers just as easily as those computers can access the Internet.

A firewall creates a doorway to the network, that can be used to prevent unwanted guests from entering the house. Firewalls are often placed between a broadband connection and an internal network. The firewall can then be used to enforce aspects of corporate security policies. For example, one of the security rules inside the company might be:

Out of the 20 servers and 450 workstations used at this company, only one of the servers can receive requests for web pages. This server will be permitted to act as a web server and people from the outside world can access web pages hosted on the server.

The firewall would enable this by blocking outside traffic to every machine, with the exception of the one web server. This firewall would only allow web requests to reach this machine, and no other types of traffic. Furthermore, only web responses would be sent back to the remote computer. In the event that server was compromised

---

[1]For most networks, firewalls do not require a large amount of computing power. A T1 Internet connection can be easily firewalled with just a Pentium II processor.

(through a web-based attack), it would not be possible to run other non-web services. All unauthorized communication would have to take place over the web channel.

Firewalls are also used to enforce network usage policies. They can control employee access to network resources and they can log activities in order to identify resource misuse. Companies commonly use firewalls to restrict access to the Web and other Internet resources. A firewall gives a company tremendous control over how people use the network.

What Firewalls Can and Can't Do

The key to understanding firewalls is knowing what they can actually accomplish. Thanks to confusing marketing techniques, many people think that a firewall is a complete security solution. Even the most advanced firewalls are only partial security solutions. To set things straight, we're going to look at some common security problems and the role that firewalls can (or can't) play in providing a solution.

Network vulnerabilities: Packet filters can help protect against attacks that exploit vulnerabilities in fundamental network protocols. For example, the path of a packet traveling through the Internet or any other network is usually chosen by the routers it meets along the way. However, the IP header of the packet can also specify the route that the packet should travel. This is called *source routing*. Hackers sometimes take advantage of source routing to make information travel through compromised systems. Normal applications almost never use this feature of the IP. Consequently, packet filters can prevent hackers from exploiting source routing by blocking packets that contain these spurious instructions.

Application vulnerabilities: Some programs have special features that allow for remote access. Others contain bugs that provide a backdoor or hidden access that provides some level of control of the program. Closely related to these are the next type of vulnerability.

Operating system vulnerabilities: Like applications, some operating systems have backdoors. Others provide remote access with insufficient security controls or have bugs that an experienced hacker can exploit.

Firewalls can be configured to block all unrequested external attempts at connecting to internal systems and servers. That's easy for a packet filter to accomplish. However, a packet filter can't tell the difference between a legitimate web page and an exploit page designed to crash the web browser and install a trojan. If a vulnerable application running on an internal system makes a connection to an external server, it's possible for that server to exploit the application.

Viruses: Probably the most well known threat is a computer virus. A virus is a small program that can embed itself in other programs and copy itself to other computers. Viruses can spread quickly from one system to the next via floppies, CDs, and the Internet. Most viruses come as attachments in email or through a file downloaded off the Web. Some spread by exploiting application vulnerabilities. Those that spread using network applications are called worms. Some viruses simply display harmless messages; others can erase all of your data.

Firewalls can usually stop some worms from spreading by preventing them from entering or leaving your network. Some commercial firewalls come bundled with virus scanners. These use integrated web, file transfer, and email proxies to intercept data that might contain viruses. But in general, packet filters can't tell the difference between healthy data and a virus. The only way a standard firewall can protect against viruses is by blocking off access to the Internet in general. Even this does nothing against viruses brought in via CDs and floppies.

Trojans: Trojans are closely related to viruses. The difference is that they provide some degree of remote access. Trojans often instigate connections to remote machines, opening the doorway from the inside and letting in the intruder.

Trojans disguise their communications to look like harmless traffic to packet filters. If stateful inspection is being used, the trojan will effectively create an open command channel. Packet filters alone are relatively useless against trojans. High-end firewalls with virus scanners and other features may provide somewhat better protection.

Spam: Typically harmless but always annoying, spam is the electronic equivalent of junk mail. Spam can be dangerous though. Quite often, it contains links to web sites. Be careful of clicking on these because you may accidentally install a Trojan that provides a backdoor to your computer.

Firewalls can do little about spam, other than blocking out email entirely. A better solution is to use a spam filtering solution. This is either a proxy that can be integrated with a firewall or an add-on for mail server software.

Denial of Service (DoS): You have probably heard this phrase used in news reports relating to major ISPs, such as AOL, and major web sites. What happens is that a hacker creates a situation where the target machine can no longer process information. There are many ways of instigating a DoS. A common technique is to flood the target system with requests. The target system becomes so overwhelmed by hacker requests that it can't process normal traffic. The attack effectively takes a system offline. Sophisticated DoS attacks leverage flaws in software and protocols to dramatically increase the impact of the attack. These attacks can crash servers and possibly do permanent damage.

Years ago, DoS attacks came from no more than a handful of systems. Firewall rules could be adjusted to block traffic from these networks until the attack subsided. Attacks that are more modern use thousands of "zombie" systems all over the world. These are machines that have been previously compromised for the purpose of launching distributed attacks. This type of attack is nearly impossible to counter. There's no way to block all the zombies—they're too numerous and they constantly change. Furthermore, the zombie systems aren't the actual hackers. Zombies are ordinary users who don't realize that their computer is participating in a massive attack.

One way to combat DoS attacks effectively is to work with ISPs and backbone providers to limit traffic in DoS conditions. They can use traffic shaping and proxy techniques to prioritize normal traffic, minimizing the effective impact of the attack.

Firewalls are not useful for stopping DoS attacks, but they can be used to prevent your network from unwillingly participating in one. A good firewall administrator will know how to create rules that will prevent abusive traffic from originating from within your network.

E-mail bombs: An e-mail bomb is a DoS attack usually focused on a specific person. Someone sends the same email hundreds or thousands of times until the recipient's email system cannot accept any more messages. A similar technique would be to fill up a person's voice mail system or answering machine. This technique could also be broadened to overload an entire company's mail server.

If the attack is coming from a single IP address, a temporary firewall rule can block email from the address until the problem is resolved. If the attack is distributed across thousands of IP addresses, packet filtering is not going to solve the problem. A proxy is needed that can filter out email messages that match a certain pattern or that are overly repetitive. Some spam prevention products can be configured to provide this protection. If the firewall includes spam filtering, it may be able to combat this attack. Otherwise, additional systems will be necessary.

Social Engineering: A hacker can always use social engineering techniques to bypass a firewall. He or she might convince somebody to install some software on an internal PC. This would get a virus or trojan past any firewall.

Dedicated Hackers: A highly skilled, dedicated hacker can find ways into almost any network. The compromises that a network administrator makes on a daily basis result in many opportunities for hackers to gain access. Good security philosophies and policies are the only effective defense against smart, dedicated hackers.

How Packet Filters Work

Packet filters work on a simple principle: they inspect packets and either accept (pass) or deny (block) them. This can be approached in one of two ways:

- Accept everything except for specific things that should be blocked, or
- Block everything except for certain things that should be accepted.

These are called *default policies* because they describe what happens if the firewall doesn't have explicit instructions matching a given packet. When a bouncer at a club is making sure patrons aren't wearing jeans or sneakers, he's using the former policy. When he's checking for invitations, he's using the latter. In general, the latter of the two is much more secure and usually more consistent with security philosophies.

The list of things to accept or block is called a *rule set*. Each individual instruction is a rule. Packet filter rules can make decisions based on the various information present in the TCP/IP header. This includes: source and destination IP address/domain name, source and destination port, protocol (TCP, UDP, ICMP, and so on), and specific protocol options (IP options, TCP state, ICMP message type). Quite a bit can be accomplished using those few levers. Let's look at these filtering choices more closely:

IP Addresses: Each machine on the Internet is assigned a unique address called an IP address. IP addresses are 32-bit numbers, normally expressed as four octets in a dotted decimal number. A typical IP address looks like this: 216.239.57.101. If a certain IP address outside the company is reading too many files from a server, the firewall can block all traffic to or from that IP address. The firewall can also block traffic coming from certain inside machines going outbound. Some companies actively block certain web sites (such as Hotmail, Yahoo mail, MSN mail, and so on) that they feel are inappropriate or harmful to the workplace environment. Others block everything and then allow access to a handful of pre-selected sites. The rest allow complete access to the Web.

Domain names: Most servers on the Internet have human-readable names known as domain names. These exist because it is hard to remember the string of numbers that make up an IP address, and IP addresses sometimes need to change. For example, it is easier for most of us to remember www.google.com than it is to remember 216.239.57.101. Furthermore, some of the larger domain names have large clusters of redundant servers, each with a different IP address. The domain name alternates between these IP addresses on a request-by-request basis to spread out the traffic load. When this happens, it's very difficult to block based on IP address alone. Most packet filters will allow filtering on the domain name, which gets around this problem. The only caveat is that looking up a domain name takes a little time and may noticeably slow down the firewall's performance.

Protocol: The protocol is the pre-defined way that someone who wants to use a service talks with that service. The "someone" could be a person, but more commonly is a computer program like a web browser. Protocols are often text, and simply describe how the client and server will have their conversation. *Hypertext Transfer Protocol* (HTTP) is the protocol of the World Wide Web. Firewalls can filter based on low-level protocols such as *Transmission Control Protocol* (TCP), *User Datagram Protocol* (UDP), and *Internet Control Message Protocol* (ICMP). Higher-level protocols can be filtered using ports.

Ports: Any server machine makes its services available to the Internet using numbered ports, one for each service that is available on the server. For example, if a server machine is running a web HTTP server and an FTP server, the web server would typically be available on port 80, and the FTP server would be available on port 21. A company might block port 21 access on all machines but one inside the company. No hard rule on ports exists though—a web server can run on port 21 and an FTP server can run on port 80. This usually doesn't happen because it's very confusing, but hackers might put their own server on port 80 in order to make a firewall think that it's a harmless web server.

Security Considerations

Firewalls are only as good their configuration, and configuring them can be quite difficult and time consuming. In fact, many firewalls in many companies are completely ineffective at protecting the networks on which they reside. Direct attacks against a network become possible when firewalls are poorly configured.

Misconfiguration isn't the only problem with firewalls. Changing business needs and changing technology infrastructures have minimized the value of firewalls in preventing certain types of attacks. Most of today's hackers attack services that few companies are willing to give up or block with a firewall. On average, these services center on web access and email access. The hackers have responded to the relentless installation of steel doors by going in through the window. web-based exploits, email viruses, trojans, and simple social engineering are all tricks used by modern hackers that can effectively bypass firewalls.

Commercial firewalls are highly marketed to consumers and businesses. They make you think that all you need to do is buy one, or maybe two, drop it on your network, and with a little help from their technician your network will be safe. This could not be further from the truth. Although commercial firewalls themselves will suffice as firewalls, they need to be configured based on a company's specific business and network needs. If you do not take a long hard look at all your network services, you can never configure a firewall properly.

To make an analogy, imagine if you wanted to pay a police officer to keep certain cars from entering a tunnel that you own. The cars you want to keep out are all makes and models that have a chassis of a certain weight or size. In order to know what kind of cars you want to prohibit entry, you need to know all about the possible cars that may come down the road. Well, many car manufacturers produce many different kinds of cars all over the world. Any car you do not explicitly tell the officer to flag down from the highway will be ignored and inevitably pass through the tunnel. Don't you think it would take a good deal of time to become knowledgeable about every car and its chassis specifications?

It takes even more knowledge for a network engineer or a security specialist to tighten down a firewall rule set. Even with the knowledge of every harmful packet of information, tuning a firewall still requires trial and error, experience, and a long-term strategy.

Even firewalls that appear to be configured properly can leave open doors that intelligent hackers can exploit. For example, stateful inspection is often used to allow outbound access while blocking all unrequested inbound traffic. The assumption is that all outbound requests are made by users, and therefore are safe. But a trojan server can establish a connection to a remote client by simply sending web packets out through the firewall. The server response will be allowed back in because the outbound connection looked just like a Web request. In reality, a hacker now can use this connection to directly control internal machines from outside the network, bypassing the firewall. A heavy steel door is useless if the door is held open for the intruder.

Making the Connection

Connecting Networks: Firewalls are often devices used to connect networks, or pieces of networks together. Core networking protocols are used by firewalls to perform basic networking services as well as advanced packet-filtering techniques.

Detecting Intrusions: Intrusions can occur at the network entrance point a firewall guards. Intrusion detection systems are often integrated into firewalls so blocking and watching can occur on the same physical device.

Best Practices

The level of security your philosophy and policies require will determine how many threats can be stopped by your firewall. The highest level of security would be to simply block all traffic. Obviously, that defeats the purpose of having a network connection. But a common practical rule of thumb is to block everything, and then select traffic to allow. This is a good rule for businesses that have an experienced network administrator who understands what the needs are and knows exactly what traffic to allow through. For most of us, it is probably better to work with the defaults provided by the firewall developer unless there is a specific reason to change it.

Transparent Bridging: Firewalls can be configured as bridging devices. This means that they can be transparently placed between two network devices. Traffic from one device is filtered and forwarded to the other device. If the traffic is allowed through, it happens as if the firewall isn't there. If it's not allowed through, the connection simply doesn't work.

The advantage to a bridging firewall is that it's invisible to devices on the network. There's no way to attack the firewall directly because it doesn't have an IP address. The only downside to transparent bridging firewalls is that they are more difficult to remotely manage.

Demilitarized Zone (DMZ): Sometimes you may want to have network segments that are deliberately insecure, or less secure than the private part of your network. These less secure segments are called demilitarized zones.

In the military, a DMZ is a buffer zone that sits in front of a well-defended, completely secure area. Although the firewall concept of a DMZ pulls its name from this example, it is not a very accurate analogy. A more accurate analogy would be a slightly less well known park in New York City called Gramercy. Gramercy Park is an anomaly that sits in the middle of a city full of anomalies.

Gramercy is a city-owned park that is semi-public and gated in as a result. To access it you need a key, but a copy of the Gramercy Park key is only given to residents who live in apartments surrounding the park! To the residents of Gramercy, the park is much like a DMZ on a network. It is an area that only residents have access to, but it is not as secure as their own private apartments. Although the residents prefer no

Advanced Hardening Techniques

Transparent Bridging Firewall

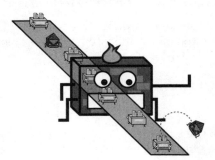

If a firewall has an IP address, a hacker can see it and attack it. Once a hacker controls a firewall, the rest of the network is often easier to compromise.

Configurating the firewall as a bridge device removes the IP address, making the firewall "invisible". It can still filter all of the traffic that passes through, but it becomes significantly more difficult to attack.

A change in network topology is *usually not necessary* when integrating transparent bridging firewalls within a network.

DeMilitarized Zone

The DeMilitarized Zone (DMZ) is a less secure part of a network. Risky software and services can be run in this zone without compromising the rest of the network.

A firewall can isolate the DMZ from other internal systems. Hackers that invade the DMZ cannot directly gain further access to the protected network.

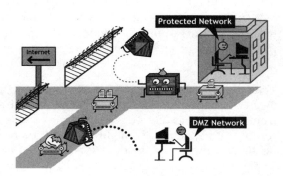

Honeypots

A honeypot is a system that appears to be really valuable to a hacker. Unable to resist, the hacker uses his tools to probe and compromise the honeypot.

What the hacker doesn't realize (until it's too late) is that the system's a sham. It's designed to expose the hacker's presence on the network.

By watching the action, an administrator can learn the hacker's techniques. This knowledge can be used to protect critical machines from similar attacks.

Concept by SageSecure (www.sagesecure.com) | ©2003 XPLANE.com®

Figure 18-2

one without a key enter their park, if someone does it will not affect the security of anyone's private domicile.

Many reasons exist for having a DMZ on your network. Some companies use them as party zones. They place computers in the DMZ that they expect to be compromised. They use these computers to browse the Internet in a wildly carefree manner. This can be useful for testing, or research, or just plain fun. In other cases, DMZ's are used to provide services to the public Internet. These servers will be offered some protection in a DMZ, but not the same level of total protection afforded to the private network.

Some examples of services that a business might place in a DMZ are as follows:

- Web proxy
- News server
- FTP download and upload area
- Unrestricted web browsing clients
- Unrestricted peer-to-peer file-sharing clients

Setting up a DMZ will involve a slightly different process on every network. Sometimes a DMZ is created in the area between the Internet connection and the firewall. This can be achieved by simply placing a computer or server in this unprotected network pocket. In other cases, the DMZ will sit behind the firewall, but in a different subnet, with a less restricted set of firewall rules. Most of the software firewalls available will allow you to designate a directory on the gateway computer as a DMZ.

Honey Pots: A honey pot is a tasty looking trap, set for the unwary, starving hacker. The goal is to keep malicious folk away from the truly important stuff by hanging a big juicy steak in front of them. A honey pot machine is one that looks important, but is not used for anything. It might be running a fake database, or a CRM system populated with fictitious data. A honey pot has no legitimate users or traffic; any activity on the system is the result of an intruder. This can alert a network administrator to the presence of an intruder on the network. The authorities can be contacted while the intruder is busy gnawing on the tasty looking treats.

It is common for people to confuse honey pots and DMZs. A DMZ is not meant to attract hackers; it is an isolated network zone for services that need less security. Honey pots, on the other hand, are *explicitly* designed to attract hackers. A network administrator needs to have some type of plan for capturing and neutralizing any hackers drawn to the honey.

A study by Global Integrity published in September 2000 found in the following:

- Honey pots are an excellent method of detecting insider attacks.
- Honey pots sidetrack attackers' efforts, causing them to devote their attention to activities that can cause neither harm nor loss.

- Honey pots allow security administrators to study exactly what attackers are doing without exposing systems or networks to additional risk that results from compromised systems.

- A moderate proportion of attackers refrain from attacking systems within networks in which they know that measures have been taken to capture and monitor their actions.

Final Thoughts

We apologize for the extensive section on firewalls. Wes previously wrote a book on building firewalls. Jay tried to keep this section short and relevant, but Wes kept adding more and more stuff. After hours of arguing, Wes was sent to his room and Jay finished the chapter and submitted it to the publisher. But then Wes hacked into McGraw-Hill's network and kept adding stuff. See, Jay doesn't know I'm writing this, man will he be mad when he sees the printed book!

Chapter 19
Hardening Networks: Network Address Translation

A technology that can convert Transmission
Control Protocol (TCP)/Internet Protocol (IP)
addresses from one subnet to another.

Technology Overview

Network Address Translation (NAT) has become a standard tool for both connecting and hardening networks. It is often implemented to solve certain network design issues, but it also offers significant security benefits as a hardening technique. To better understand and define NAT, it is important to know a little bit about the history and planning of the Internet itself.

For a computer to communicate with other computers and Web servers on the Internet, it must have an IP address. An IP address is a unique number that identifies the location of your computer on a network. All computers on networks that are set up using the TCP/IP protocol (the official protocol of the Internet) must have an IP address in order to communicate. For a more detailed explanation, please refer to Chapter 17, "Network Lingo."

Network Address Translation

Illustration by SageSecure

■ Figure 19-1

When IP addressing was conceived, it seemed as though there were plenty of addresses to go around. In theory, there could be as many as 4,294,967,296 (2^{32}) unique IP addresses. The number of available addresses is actually smaller (approximately 3.3 billion) because of the way that the addresses are separated into classes, and because some addresses are set aside for testing or other special uses.

With the explosion of the Internet and the increase in home and business networks, the number of available IP addresses was being rapidly consumed. Why? It turns out that IP addresses are wasted if you have many small networks (only a few IP addresses). Addresses need to be allocated in groups, and these groups can't be easily split across multiple networks. If one group has several unused addresses, they'll often just sit unallocated.

When the IP address system was created, its designers never thought millions of small networks requiring just a handful addresses would develop. They also didn't think that cell phones and other portable devices would eventually require individual IP addresses. When the growing demand for cell phones and computer systems throughout the world is taken into account, it's quite possible that there will be more than three billion systems requiring IP addresses in the near future. This would create a problem even if none were being wasted!

A long-term solution to the address crisis involves redesigning the address system to allow for more possible addresses. This new addressing scheme is called IPv6 (the current one is IPv4) and is supported by most new network systems. It will be several years before enough of the routers that make up the Internet's infrastructure are upgraded to support the new standard fully. Until then, most networks will continue to use IPv4.

Another more immediate addressing solution is network address translation, which allows a single device, such as a router, to act as an agent between the Internet (or any public network) and a local network. This means that only a single, unique, IP address is required to represent an entire group of computers. However, the shortage of IP addresses is only one reason to use NAT.

**Chapter 19
Hardening
Networks:
Network
Address
Translation**

NAT can be configured to create a one-way trapdoor between an internal network and outside networks such as the Internet. Essentially, a computer on an external network can only connect to an internal computer if the internal computer has initiated the contact. The internal system can browse the Internet and even download files, but somebody else cannot latch onto its IP address and connect without permission.

Many different network hardware devices are capable of performing NAT. Frequently NAT is implemented with a firewall or a router. In certain circumstances, specialized devices may also perform NAT depending on the complexity of the network.

NAT Terminology

It's difficult to describe NAT techniques without tripping over terminology. NAT can be implemented in a number of ways. Different ways of using IP addresses when dealing with NAT are also available. To keep everything straight, we're simply going to lay out some definitions.

Registered Versus Unregistered IP Address: Every publicly accessible IP address must be registered with the *Internet Assigned Numbers Authority* (IANA). This organization usually registers large blocks of IP addresses to *Internet Service Providers* (ISPs), who in turn allocate the addresses to customers. A number of unregistered address blocks have been permanently reserved for private internal use. These reserved unregistered addresses are only meant for *Local Area Networks* (LANs); they have no meaning on the Internet. Routers are supposed to discard these packets because they can never associate with a publicly accessible network.

Stub Network: In most cases where NAT is used, the internal network is usually a LAN, often referred to as a stub network. Most of the traffic in a stub network is local, so it doesn't travel outside the internal network. A stub network can include both registered and unregistered IP addresses. Computers in a stub network with unregistered IP addresses must use NAT to communicate with the rest of the world.

Global Addresses: These are registered IP addresses accessible to anyone on the Internet. The global addresses assigned to a network using NAT are called *inside global addresses*. Any packets sent to these IP addresses will be processed by the NAT router. All the other addresses on the Internet are called *outside global addresses*.

Local Addresses: These are the IP addresses accessible inside a network using NAT. They can be registered or unregistered addresses, but are almost always unregistered. A few of these addresses are used by the NAT routers and are called *out-*

side local addresses. The rest, known as *inside local addresses*, are used for all of the other computers on the stub network.

Static NAT: This associates an inside local address with an inside global address on a on-to-one basis. Also called inbound mapping, static NAT allows external devices to initiate connections to computers on the internal network. It's particularly useful when an internal server needs to be accessible from outside the network. For instance, static NAT allows an internal web server with an inside local address to be reachable via an inside global address.

Dynamic NAT: This maps an inside local address to an inside global address, drawing from a pool of available inside global addresses. This is usually used in combination with the following:

Overloading aka Port Address Translation (PAT): A form of dynamic NAT that maps multiple inside local addresses to one or more inside global addresses. This is also known as single address NAT or port-level multiplexed NAT. This is the most common usage of NAT and how it works is explained below.

Overlapping NAT: Sometimes the inside local addresses used on a network are registered IP addresses in use on another external network. In most cases, this is a networking mistake. But sometimes, a network needs to migrate from one set of registered addresses to another (perhaps because a larger block was needed). Without changing the IP addresses of the internal network, NAT can provide transparent migration by allowing machines to maintain their old addresses. When the internal machines communicate with the outside world, they will appear to have, and respond to, the appropriate new addresses.

How NAT Works

Many people have observed NAT in operation without being aware of it. That's because a typical office phone system employs many of the same concepts as NAT. A main public number connects to the office phone network. An extension (a private address that's only valid internally) can get to an individual's desk. This is how NAT is most commonly used—allowing many systems to be accessed via a single public address.

Sometimes people have direct-dial numbers. This is another NAT technique, where additional public addresses are directly mapped to private internal addresses. The device responsible for NAT performs this task automatically. In the case of the phone network it's either the telephone company or the office phone system that handles this. In the case of a network, usually a router or firewall performs the mapping.

How does NAT provide security? If no direct-dial is available, then every incoming connection has to go through the NAT device. In most cases, NAT acts like a receptionist that screens calls and only forwards the ones that are requested. For example, you have a large amount of work to do, so you've left instructions with the receptionist not to forward calls to your desk. Later you make a call to a client. She's not at her desk, so you leave a message. You tell the receptionist that you are ex-

**Chapter 19
Hardening
Networks:
Network
Address
Translation**

pecting a call from this client. The client calls the main number to your office, which is the only number the client knows. When the client tells the receptionist that she is looking for you, the receptionist checks a lookup table that matches your name with your extension. The receptionist knows that you requested this call, and forwards the caller to your extension.

Step-By-Step Dynamic NAT

NAT can be configured in various ways. In the following example, NAT is configured to translate unregistered IP addresses from the private (inside) network to registered public (outside) IP addresses. This happens whenever a device on the inside needs to communicate with the outside network.

An ISP has given a client's network a range of unique registered IP addresses for use with its broadband connection. There are not enough public IP addresses, however, to assign to each individual node on the internal network. Instead, a private internal network has been set up with IP addresses from one of the reserved unregistered address blocks. These addresses are non-routable since they are not unique. A NAT-enabled firewall is used to provide Dynamic NAT between the two address groups. Here is a play-by-play account of the life of a packet on this network:

1. A computer on the internal network attempts to connect to a computer on the Internet, such as a web server.

2. The firewall receives the packet from the computer on the internal network.

3. The firewall saves the computer's inside local address to an address translation table.

4. The firewall replaces the sending computer's inside local address with the first available inside global address out of the pool of available registered inside global IP addresses. The translation table now has a mapping of the computer's inside local address matched with the one of the unique inside global addresses.

5. When a packet comes back from the destination computer, the firewall checks the destination address on the packet. It then looks in the address translation table to see from which computer on the internal network the packet originated. It modifies the destination address to the one saved in the address translation table and sends it to that computer. If it doesn't find a match in the table, it drops the packet.

6. The internal computer receives the packet from the firewall. The process repeats as long as the computer is communicating with the external system.

The problem with this setup is that the number of available registered IP addresses limits the number of simultaneous outbound connections. If the network has only one external IP address, it can only have one outbound connection at a time. This doesn't help matters.

To solve this problem, *Port Address Translation* (PAT) is used. By doing a little extra manipulation trickery, the same single registered address can support thousands of simultaneous connections.

Given the same internal network, this is how PAT would work:

1. A computer on the internal network attempts to connect to a computer outside the network, such as a web server.
2. The firewall receives the packet from the computer on the internal network.
3. The firewall saves the inside local address *and originating port number* to an address translation table. It picks an unused port number from its own very large pool of available port numbers and stores this number in the table (we'll call this the *NAT-port*). The firewall then replaces the sending computer's inside local address with one of its inside global IP addresses. It also replaces the originating source port of the sending computer with the NAT-port. The translation table now has a mapping of the computer's inside local address and port number along with the newly assigned inside global IP address and NAT-port number. Many simultaneous connections can have the same inside global IP address, but the NAT-ports are unique for each connection. *The combination of IP address and NAT-port uniquely identifies each connection.*
4. When a packet comes back from the destination computer, the firewall checks the destination IP address (inside global) and port (NAT-port) on the incoming packet. The address translation table should connect the IP address and port number to an inside local address and port number. NAT then changes the destination address and destination port to those saved in the address translation table. It then sends the modified packet along through the internal network.
5. The internal computer receives the packet from the firewall. The process repeats as long as the computer is communicating with the external system.
6. Since the firewall performing the address translation now has the computer's source address and source port saved to the address translation table, it will continue to use that same port number for the duration of the connection. A timer is set whenever the firewall accesses an entry in the table. If the entry is not accessed again before the timer expires, the entry is removed from the table.

Security Considerations

Network Address Translation privatizes a network from the public Internet address space. It allows many computers to hide behind one little public IP address and this provides some security. It is a misconception, however, that this is all the security one should need.

**Chapter 19
Hardening
Networks:
Network
Address
Translation**

Gregarious Host: NAT only allows the return portion of a requested connection. But what if a computer inside the network *wants* to connect to the hacker's machine? Various tricks such as Web pop-ups and email viruses can be used to make the target machine actively contact the hacker's system, thus completely bypassing the protections offered by NAT. Sometimes it's even easier; commonly used applications such as Web browsers create many open ports on the NAT device. A smart hacker can forge packets that will be accepted by the NAT system and forwarded to the target computer. These packets can then exploit the waiting application.

Automated hacking tools can trivialize these rather complex attacks. In earlier years, when most hacking was manual, NAT may have been seen as adequate protection. Today it is just one part of a needed security strategy that includes solid network topology, effective firewalling, and constant planning.

NAT and VPNs: One of the essential technologies used with most *Virtual Private Networks* (VPNs) is the *IP Security Protocol* (IPSec). A fundamental conflict exists between the way in which the IPSec protocol and NAT operate. IPSec's *Authentication Header* (AH) protocol takes a snapshot of the entire IP packet, including invariant header fields such as source and destination IP address. In order to authenticate the packet the recipient uses this snapshot. If any field in the original IP packet is modified, authentication will fail and the recipient will discard the packet. The Authentication Header is intended to prevent unauthorized modification, source spoofing, and man-in-the-middle attacks. NAT, by definition, modifies IP packets; therefore, Authentication Header and NAT do not work together.

By far, the easiest way to combine IPSec and NAT is to avoid these problems by locating IPSec endpoints in public address space. That is: NAT after IPSec, your network's a wreck; IPSec after NAT, your network looks phat. This can be accomplished in two ways:

* Perform NAT on a device located behind your IPSec security gateway or
* Use an IPSec device that also performs NAT. These devices are set up to recognize and avoid this conflict.

Another option is to scrap the AH protocol. The *Encapsulating Security Payload* (ESP) protocol can offer many of the same benefits as the AH, and it cooperates better with NAT. This is discussed in more depth in the section on VPNs.

Making the Connection

Connecting Networks: Network Address Translation can be performed in many different places, especially on a large network. NAT commonly occurs on certain network devices such as routers and firewalls.

Best Practices

Some NAT routers provide for extensive filtering and traffic logging. Filtering allows your company to control what type of sites employees visit on the Web, preventing them from viewing questionable material. You can use traffic logging to create a log file of what sites are visited and generate various reports from it.

NAT is often confused with proxy servers, but definite differences between them exist. NAT manipulates header information. Proxy servers can manipulate the contents of packets. Proxy servers usually are slower, because inspecting application data is a more processor-intensive task than looking at limited header information.

Furthermore, NAT is transparent to the source and to destination computers. Neither one realizes that it is dealing with a third device. However, a proxy server is not generally transparent (although techniques exist that can be used to make it transparent). The lack of transparency means that every internal computer must be configured to make relevant requests to the proxy server.

A real benefit of NAT is apparent in network administration. For example, static NAT lets you move a Web or File Transfer Protocol (FTP) server to another host computer without having to worry about broken links. Simply change the inbound mapping at the NAT router to reflect the new host address. You can also make changes to your internal network easily because the only external IP address either belongs to the router or comes from a pool of global addresses. NAT and *Dynamic Host Configuration Protocol* (DHCP) are a natural fit. You can choose a range of unregistered IP addresses for your stub network and have the DHCP server dole them out as necessary. It also makes it much easier to scale up your network as your needs grow. You don't have to request more IP addresses from the IANA. Instead, you can simply increase the range of available IP addresses configured in DHCP to have room for additional computers on your network.

Final Thoughts

Got NAT? Network Address Translation is a powerful tool that has obvious benefits for large networks. What many smaller network owners often don't realize is that NAT can provide them with many of the same benefits. Even if only one machine is being protected by NAT, it still gains the benefits of having a private address that can't be reached directly from the outside world.

Activating NAT can result in an immediate increase in security on any network using registered IP addresses for internal machines. Most home/small office *Digital Subscriber Line* (DSL) and cable modem routers can be configured to use NAT. Almost every professional router can also perform NAT. So what are you waiting for? Get NAT today!

Chapter 20
Hardening Networks: Virtual Private Networks

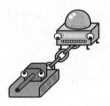

A virtual private network (VPN) uses encryption to
create a secure network connection between
two systems over an insecure network.

Technology Overview

The *virtual private network* (VPN) was originally a cost-saving solution for companies that had multiple offices connected via private and dedicated high-speed connections. If you think getting a T1 Internet connection in one office is expensive, try building your own internet using private lines connecting each branch office. It's an economic disaster.

Consequently, people started looking for ways to use the Internet to connect business locations. The Internet was relatively inexpensive, and most companies already had the necessary infrastructure in place. The only problem was the complete lack of security on the Internet. Sending critical business data across the Internet unprotected is like randomly handing out payroll and human resouces records to people on the street.

One solution is to use full-time encryption to secure all of the business traffic passing between two networks on the Internet. This creates the equivalent of a private network, but in the midst of the public Internet. The concept was creatively dubbed the "Virtual Private Network." Think of it as a giant outdoor tunnel going down the middle of an Interstate expressway (or an Autobahn). Only authorized cars can get into the tunnel, and from the outside nobody can see in, so it's impossible to see what type of cars are traveling and how many passengers are in each car.

The virtual private network is a simple, yet powerful, extension of encryption. Instead of merely encrypting a single message or a data file, every piece of data exchanged between two systems or networks is encrypted. Any system with the right authentication information can become part of a VPN. Ultimately, every VPN is created between two machines. It doesn't matter what type of machine; laptops can connect to firewalls and servers can connect to other servers. Any two machines that encrypt communications over a network are creating a VPN.

This broad definition of a virtual private network includes some things that we don't immediately associate with VPNs. Secure web connections are essentially a temporary VPN between a web browser and a web server. Likewise, secure remote access connections such as *Secure Shell* (SSH) create a limited VPN between two machines. SSH is actually a special case, since it can actually be used as a VPN solution on its own. We will provide more information on that topic later.

In the commercial world, the definition of a VPN is a little bit narrower. Commercial VPN solutions try to ensure that all business data travels across an encrypted link. A commercial VPN is usually a black box that sits on the network near the firewall. Each branch office has its own VPN box. The VPN device encrypts all of the traffic between the two or more locations. Individual remote users present a more difficult scenario—most will be connecting via modem or broadband connection. These individuals are not going to be able to install a black box, so instead they use software that mimics the VPN device, encrypting communications between the remote users and the corporate network.

How VPNs Work

There are two basic components to a VPN connection: authentication and encryption. The first step is for both endpoints to prove their identities to one another. Once the two points have exchanged authentication information, the communication can be encrypted.

Authentication is usually performed via certificate exchange and validation. The "digital certificates" section of the determining identity chapter goes into detail on how certificates can be used to verify identity.

The communications are encrypted using either a shared-secret encryption scheme or a public/private key system. In the shared-secret scenario, every VPN endpoint needs to know the secret key. This is relatively easy to set up, but has some serious drawbacks. If an endpoint is compromised, the entire network is compromised and the key needs to be changed. A public/private key infrastructure offers much better security and flexibility, but requires more thorough network planning.

Virtual Private Networking

**Chapter 20
Hardening
Networks:
Virtual
Private
Networks**

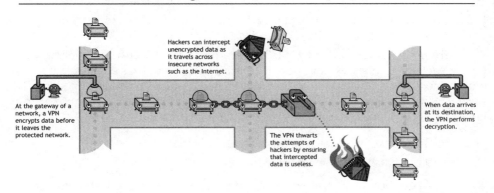

Hackers can intercept
unencrypted data as
it travels across
insecure networks
such as the Internet.

At the gateway of a
network, a VPN
encrypts data before
it leaves the
protected network.

When data arrives
at its destination,
the VPN performs
decryption.

The VPN thwarts
the attempts of
hackers by ensuring
that intercepted
data is useless.

Concept by SageSecure (www.sagesecure.com) | ©2003 XPLANE.com®

Figure 20-1

The section on "Cryptography" in the hiding information chapter explains both shared-secret encryption and public/private key encryption.

Security at the Protocol Level: IPSec

One of the essential technologies used to create VPNs is an enhancement to the *Internet Protocol* (IP) called the *IP Security Protocol* (IPSec). IPSec adds support for authentication and encryption at the protocol level. It was created because the original Internet Protocol does not provide mechanisms for protecting data or guaranteeing the identity of the sender. IPSec remedies this situation by providing the following additional security:

Confidentiality: Using encryption, IPSec can ensure that only the receiver can see what data has been communicated. For example, IPSec can be used transparently to ensure the confidentiality of Internet email messages sent between branch offices and the central office.

Integrity: IPSec can also guarantee that data has not been altered while in transit. This is critical for financial transactions and any other Internet-based data exchange where reliability is a concern.

Authenticity: The ability to sign data cryptographically can be used to prevent forgeries. Only the original sender can generate the proper signature. Hackers can't inject forged data into the network if each packet is digitally signed.

Replay protection: Hackers can create mischief even without creating a forged packet. By resending captured packets later, hackers can fool a system into doing something twice. For example, imagine if a hacker could capture and replay the code that authorizes an ATM to dispense cash. IPSec can be used to track the processing of packets, ensuring that the same packet is never processed twice.

IPSec Protocols

IPSec provides these services using two new protocols: *Authentication Header* (AH) and *Encapsulating Security Payload* (ESP).

Authentication Header (AH): This protocol provides an authenticity guarantee for packets. It takes a snapshot of the entire packet and uses a secret key to generate a unique identifier. The identifier is attached to the packet and sent along. The receiving system uses the same secret key and snapshot process to generate its own identifier. If the identifiers match, two things are guaranteed:

1. The packet was not generated by an impersonator.
2. The packet was not modified in transit.

The AH protocol doesn't encrypt the data portion of the packet, therefore it cannot guarantee confidentiality.

Encapsulated Security Payload (ESP): This protocol provides a confidentiality guarantee for packets by encrypting the content. If you receive a packet with ESP and successfully decrypt it, you can be sure that the packet was meaningless to any computer that may have captured it en-route.

ESP can ultimately provide nearly all of the benefits of the AH and more. In most cases, only ESP is used to provide confidentiality, integrity, authentication, and replay protection. AH is used only in special circumstances and is generally not needed for common VPN operations.

IPSec in Operation

Technically, ESP doesn't provide any protection or guarantees about the IP header of the packet. This gives hackers a potential point of weakness. In practice, a technique called tunneling solves this by wrapping the packet inside of itself. Say what? It's not as complicated as it sounds. If this is a normal packet:

[IP header] [TCP header] [data . . .]

then the straightforward use of ESP (called "Transport Mode") would generate the following packet:

[IP header] [ESP header] [*TCP header*] [*data . . .*]

The parts shown in italics are being protected by ESP.

If we apply ESP in tunnel mode to the original packet, we get:

[IP header] [ESP header] [*IP header*] [*TCP header*] [*data . . .*]

The IP header is used twice—once in the beginning of the packet (so that it can get where it's going) and once inside the packet. The inner IP header is protected from tampering. This can be used to provide a number of different benefits. On a basic level, the two IP headers can be compared. If different, the packet may have been tampered with. Another use is to hide the true source and destination addresses of

**Chapter 20
Hardening
Networks:
Virtual
Private
Networks**

a packet. The only visible addresses are the IPSec endpoint systems (the VPN device). The IPSec endpoint then uses the internal, encrypted IP header to route the packet to the correct machine on the internal network.

Security Considerations

In an ideal world (from the VPN solution provider's point of view) every supplier, distributor, and customer would connect to the corporate network using VPN hardware and/or software. In the real world, this is very expensive and very risky. The risk comes from insecure endpoints and compromises due to network incompatibilities.

Too much access: In many cases, remote users don't need full access to the network. Proxies, FTP servers, web mail systems, and other software tools can provide limited access in a much more secure manner. The "best practices" section has some suggestions along these lines.

Overhyped: VPNs are often sold as security panaceas. The vendors focus on the fact that hackers can't see the traffic in transit. So what? It offers no protection if the hacker can access either end point. In fact, it makes it harder to detect a hacker, because they appear to be legitimate users and their hacks are encrypted. External intrusion detection systems will not be able to identify hacks conducted over the VPN.

Multihoming: Many companies use VPNs or dedicated lines to provide access to the corporate network from employees' home computers. For normal Internet access, the home computer often has a separate modem or broadband connection. If both connections are active simultaneously, hackers can gain access to the VPN through the Internet connection through the back door. This defeats the entire purpose of the VPN.

Insecure end point: In most cases a remote end-user's machine can't be trusted. If it has any access to the Internet or if software can be installed upon it, then its security may have been compromised. A trojan could allow a hacker to gain access to the VPN by stealing the connection settings (passwords, encryption keys, and so on). Or it could queue commands for the next session on the VPN, establishing back doors on the corporate network.

Making the Connection

Cryptography: These two concepts are inseparable. Really, VPNs are an extension of cryptography. Think of VPNs as taking cryptography to the evolved plane of full time, streaming data encryption.

Digital Certificates: These are frequently used to authenticate each endpoint on a VPN. If your network has several users connecting via VPNs, you'll probably want to centrally manage the authentication certificates using a *Public Key Infrastructure* (PKI).

Remote Access: SSH is a remote access tool that has overlapping functionality with VPNs. All other remote access tools should be run either over SSH or a full VPN.

Best Practices

Providing suppliers and distributors with access to a database or file server is a common need in business today. Using a VPN to allow direct access to the resource is not always the best bet. If a remote system that is granted VPN access is compromised by hackers, the VPN will give them direct access to your network. This is too risky, as you have no control over the network security of your business associates.

One solution is to create or purchase an application that can provide intermediary access. The application can limit the scope of what each remote system can accomplish. For example, a web-based pricing and availability system might make predetermined queries to the central product database. The most a hacker on a remote system can do is access these predetermined queries. Compare this to a VPN, through which a hacker could create arbitrary queries and attack the database directly.

Remote users also have limited needs. Frequently, the only thing a remote user needs to do is check email and scheduling information. Occasionally files are also needed. In this case, a web-mail application could provide secure external access. Files could be obtained through a web-based intermediary. The file application could check for viruses and perform other tasks that might not happen if the user is sharing files over a direct VPN connection.

Network administrators should be particularly careful when connecting remotely. Home connections should be on a dedicated machine with proper VPN hardware. The machine should have no direct connection to the Internet other than through the VPN. This machine should be thought of, and secured as, an essential corporate network resource. If a laptop is being used for remote network administration, consider the following:

- Run a highly secured UNIX-based operating system.
- Run firewall software to block external attacks.
- Encrypt the hard drive, protecting any critical information if the laptop is lost or stolen.
- Keep all encryption keys on non-identifiable removable media (flash cards, smart cards, and so on). Take this media out whenever it's not in use. Do not store it with the laptop (carry it in a different pocket, and so on).
- If the laptop has built-in Wi-Fi, disable it. If you want to use Wi-Fi or otherwise access the Internet, install a second operating system on the machine. Make sure the second operating system cannot access the files of the primary operating system.

SSH can help: The SSH program is a very powerful tool. Although it was primarily designed for Unix, it can also work on Windows. Its best feature is the ability to tun-

**Chapter 20
Hardening
Networks:
Virtual
Private
Networks**

nel other *Transmission Control Protocol* (TCP) connections. Here's how it works: let's say you want to access a web server on a remote machine securely. You make an SSH connection to that machine. You then tunnel your local machine's TCP port 80 (the port used for web connections) across the SSH connection to the remote machine's TCP port 80. If you tell your web browser to look at your own machine, SSH will automatically take the web request and send it over its encrypted connection. On the other side, it will pass it to the web server that's waiting on port 80. The web server's response will likewise travel across the encrypted channel, back to your web browser. The entire communication gets encrypted, as if you had a VPN between the two machines. This can be done with any TCP service. It requires a little more effort to set up (although the process can be automated), but works using freely available software versus $10,000 devices.

Focus on end-user security: Do not allow multi-homing. Give the user a separate machine for accessing the VPN that has no other Internet access. Use remote auditing tools periodically to check the configuration and security of the remote machine. Have policies in place to reprimand users for installing unauthorized software or connecting the machine to the Internet in an insecure manner. Require anti-virus software to be running at all times. Make sure that the virus software is constantly updated. Encrypt the hard drive, preventing unauthorized access to the VPN if the machine is lost or stolen, or if a visitor starts playing around.

Use public/private key authentication: The major advantage to public/private key authentication is that any particular user's key can be removed from the authorized list if their system has been compromised. A new system and key can be issued and the rest of the network doesn't have to be reconfigured.

Final Thoughts

Virtual Private Networking has become very popular in recent years. In fact, it has gone so far as to become a trendy item in business. "Oh, you don't have a VPN yet? Everyone is getting them." The VPN, like so many others, is just a network security tool and nothing more. It seems as though this point has been made repeatedly throughout the chapters of this book, but the authors cannot stress this enough. There is no single solution to the problem of network security. While a VPN can benefit certain situations, if the entire private environment cannot be controlled, security can never be guaranteed. Consider very carefully the access needs of the situation before deciding if a VPN is really a more secure option.

Chapter 21
Hardening Networks: Traffic Shaping

Traffic shaping is a system for controlling
access to bandwidth in order to improve
efficiency and data security.

Technology Overview

Picture this: you're going to the beach and you want to pack all your beach toys into
your beach pail. If you just throw your oddly shaped toys into the pail, you'll only be
able to fit a few of them in before the pail is full. Actually, the pail is still mostly
empty; you've just wasted space between the toys. But if you take your time, you can
carefully pack the pail and everything will fit. You can even put the lid on, and then
your friends won't be able to guess what you have in the pail.

What people think: We need more bandwidth; our connection to the
Internet does not feel fast enough.

What we think: Traffic shaping can help maintain a tight control over
network bandwidth. You may not be using your bandwidth efficiently.

Traffic shaping is a way to use bandwidth more efficiently, packing a greater amount of data into a fixed amount of space. Ever notice that your network clogs up during certain points of the day? Adding a traffic-shaping device to your network can eliminate those clogs. It can also provide security by making it harder for hackers to analyze encrypted traffic.

Network applications tend to be greedy. This isn't a statement of prejudice; it's a statement of fact. Due to inherent design, most applications will use as much bandwidth as possible at any given moment. Often, this ends up being more than the application needs to reasonably function. For example, when downloading large email attachments or files from the Web, software will try to download the data as fast as possible. This will result in a poor balance of network resources, as there may not be enough bandwidth left over for other users during the download. Ideally, an application's download should only use bandwidth if it is available. By using less bandwidth at any given time, other users will notice less degradation in performance. This way, any application can still have good bandwidth access and others can use the network effectively during that time.

One solution is to separate traffic based on the type of application and user. Some applications are more important, and should always get priority. With traffic shaping, mission critical communications (*Virtual Private Network* [VPN], email, database) can be given guaranteed access to bandwidth. Likewise, some users are more important. For example, it is possible to make sure that the CEO always has a lightning fast connection (although it might be more advantageous to do the opposite).

The Shape of Traffic

Why the word "shaping"? If you look at the distribution of bandwidth broken down by application and user, you'll probably notice that some users and applications consume the vast majority of the bandwidth, while others take up very little. You'll also notice that when the bandwidth hogs are active, all of the other services suffer. On a graph, this would appear as a series of big bumps (representing the hogs). The goal of traffic shaping is to lessen the bumps—changing the shape of the traffic.

How Traffic Shaping Works

Several different systems for optimizing bandwidth utilization are available. Most rely on the concept of a "queue." As an example, picture a bag of colorful chocolate candies (the type designed to melt only after they've entered your mouth). Now imagine a couple of long, narrow tubes, just wide enough for a single stack of these candies. As candies are pulled out of the bag, all of the red candies are placed in one tube, greens in another, browns in the third, and so on.

While pulling the candies out of the bag, there is no control over what color comes out next. Once they're in the tubes, any color can be released. If only reds are desired then all reds can be had. This is the advantage of queues; once data is sorted, fine-grained control over how it's used exists.

Now imagine five tubes (red, green, yellow, brown, and black) that can each hold 10 candies. Every second, 10 candies are pulled out of the bag and added to the tubes. There's someone at the bottom of the tubes who can eat five candies a second.

This is certainly a problematic situation. Within 11 seconds, one or more of the tubes is going to overflow. It might even happen very quickly if one color is pulled out of the bag more frequently than the others. This is what happens when a network is congested. The overflowing candies are like lost packets.

Mind Your Poms and Queues

If you've ever been to England, you've probably noticed that Brits "queue up" whenever they get a chance. It's so ingrained into their culture that it's become instinctual. Consequently, you can really mess with them by getting a few friends to form a line in a crowded place. Within minutes, dozens of Brits will have joined the line, even though they have no idea what the line is for, or why they're in the line. After a little bit, you can leave the line and watch from a distance. It can sometimes take hours before people realize that the line had no purpose.

Of course, every candy lover knows that the red candies are much better than all the others, and the brown ones are the worst. Our candy eater is very particular, and has prioritized the tastiness of the colors as follows: red first, green second, yellow and black tied for third, and brown last. What this means is that he will eat up to four red candies if they're available, followed by up to three green, two yellow or black, and one brown.

The result of this candy eating system is that the red candies will rarely overflow. On average, the tube will have two candies per second, meaning that he'll get a chance to eat some greens, and possibly some yellows and blacks. Browns will rarely be eaten, if ever. If the pace of tube filling slows, the green, yellow and black tubes will also rarely overflow. Only the undesirable brown tube will end up overflowing.

Instead of candies, imagine the colors were different types of network services (fileserver, database, web, email, or peer-to-peer file sharing). Packets associated with each service would go into a separate queue. Priorities would be given to each service. If the total amount of traffic were to exceed the network's capacity, the higher priority services would continue to operate smoothly. Only the lower priority services would experience congestion. Figure 21-1 illustrates the basic principles of shaping.

Queues can be created based on many different factors. As mentioned, different types of network applications can have different queues. Traffic pertaining to individuals or groups of users can also be assigned to queues. Even the destination network can be used as a parameter for queuing. Furthermore, some systems allow queues to be hierarchical. This means that complex conditions can be created, such as "give high priority to developers requesting database connections to the Boston network."

Traffic Shaping

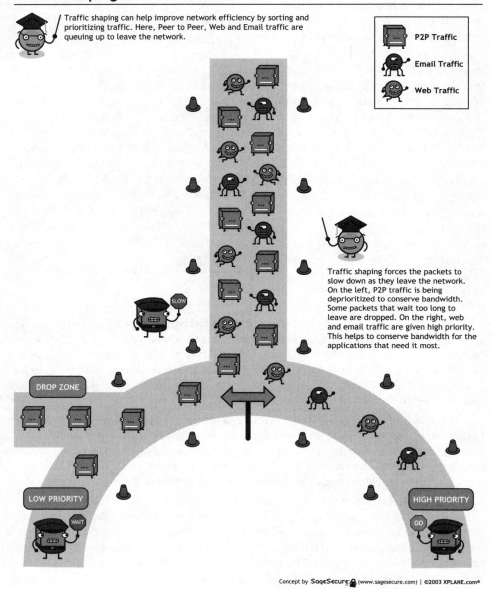

Traffic shaping can help improve network efficiency by sorting and prioritizing traffic. Here, Peer to Peer, Web and Email traffic are queuing up to leave the network.

P2P Traffic

Email Traffic

Web Traffic

Traffic shaping forces the packets to slow down as they leave the network. On the left, P2P traffic is being deprioritized to conserve bandwidth. Some packets that wait too long to leave are dropped. On the right, web and email traffic are given high priority. This helps to conserve bandwidth for the applications that need it most.

SLOW

DROP ZONE

LOW PRIORITY

WAIT

HIGH PRIORITY

GO

Concept by SageSecure (www.sagesecure.com) | ©2003 XPLANE.com®

Figure 21-1

A number of related technologies help in determining when to start dropping packets. Some traffic shaping systems will start dropping a few packets before the system gets totally overloaded. This is called *Random Early Detection* (RED) and can help delay congestion or minimize its overall effect on the network. Another technology called *Explicit Congestion Notification* (ECN) can notify applications that congestion conditions are about to occur. Smart applications will then use their own techniques for reducing bandwidth usage, which is preferable to simply losing data with no warning.

Traffic shaping does have a significant conceptual limitation: it can only effectively control outbound traffic. If you think about it, this makes sense. You can control what goes out your door, but you can't stop people from walking up to your door and ringing the bell. The best you can do at that point is not to let them in. This is called *rate limiting*.

Functionally, rate limiting is built into the same software and hardware that performs traffic shaping. It simply drops excess inbound packets. It's up to the sending application to figure out how to handle the situation. Smart applications will reduce the speed at which they send information. Less intelligent applications might have problems, ranging from inefficiencies to lost information.

Applications based upon *User Datagram Protocol* (UDP) or any other stateless protocol will simply lose data. Applications that are based upon *Transmission Control Protocol* (TCP) will find that the TCP views a rate limited connection as network congestion. It will slow the transmission rate until the amount of packet loss is minimized. That's good, but TCP also assumes that network congestion will eventually clear up. It will keep trying to increase the transmission rate again, resulting in more lost packets. This means a rate limited TCP connection will be somewhat inefficient if the application isn't smart enough to compensate.

Security Considerations

Traffic shaping is a positive thing. It is uncommon for security issues to be associated with it at least on the level of the common business network. Traffic shaping as it pertains to other types of networks (ATM, Frame Relay) and core Internet routing is a different story, one that is out of the scope of this book.

A few functional issues exist, however, that could affect network security and integrity at the same time. One such issue is hardware selection for traffic shaping. Since traffic shaping controls bandwidth, it tends to require a significant amount of processor power. To make matters potentially worse, traffic shaping is often coupled with firewall hardware. One black box is capable of providing firewall protection and traffic shaping simultaneously. It's important to ensure that this device has sufficient processor power to handle both firewalling and traffic shaping of the entire network load. If traffic shaping hardware becomes overloaded, it could begin to fail intermittently or cease to work entirely.

Traffic shaping devices can become single points of failure on a network. While there are benefits to combining firewall and traffic shaping services on one piece of hardware, the possibility of failure is at least doubled. If the black box device in this scenario fails, not only does traffic shaping shutdown, but also so do firewall services, and network connectivity in general. But don't fret yet, a number of high-availability solutions are available that can be used to remedy this problem described in Part XI, "Ensuring Availability."

Making the Connection

Connecting Networks: Traffic shaping begins where network design ends. Tightening down a network with traffic shaping techniques should not be attempted until a network has been properly designed and redesigned if necessary.

Best Practices

As with any technology, a poorly configured system can wreak havoc upon your network. However, assuming you have a traffic-shaping device properly configured, it can only help to improve the security of your organization's network.

Minimizing Inefficiencies: Needless inefficiency equals insecurity. If applications are inefficiently using bandwidth, it becomes easy for hackers to hide in the noise. If your network operates efficiently and suddenly slows down, it's a good indication that something is wrong and needs to be investigated. Furthermore, traffic analysis may reveal successful hacking activities that would otherwise go undetected.

Preventing denial of service: It takes very little effort to create a serious traffic problem. A single PC can use up multiple T1 lines worth of bandwidth. Some viruses and Trojan programs are capable of this as well, making the network completely unusable until the offending machine is found and repaired. This is called a *Denial of Service* (DoS) attack.

Traffic shaping will constrain the problem to either a single class of user or a single class of application. For example, if the offending computer is sending out a flood of web requests, then the queue for web traffic will fill up, but other traffic will still be able to pass through. An even more advanced queuing strategy will create queues for different workgroups. Now only the offending workgroup's queue will fill, leaving the rest of the network unharmed. With a proper analysis and monitoring system, the flooding of a queue will trigger an alarm, immediately indicating the existence of a problem.

Rate limiting and traffic shaping can also be used by *Internet Service Providers* (ISPs) to minimize the impact of DoS attacks directed at their customers' networks. The further upstream (closer to the backbone) the rate limiting occurs, the more effective the protection. Most major ISPs have become good at using these and other techniques to handle DoS attacks.

Traffic shaping can be used very effectively to control people's access to networked data. Take, for example, a web server that contains heavily accessed content. The web server is connected to the Internet by six T1 lines of bandwidth. With traffic shaping, the webmaster can ensure that no individual downloading content

Part VII Hardening Networks **251**

Chapter 21
Hardening
Networks:
Traffic
Shaping

from the web server can eat up more than a specified amount of bandwidth. This way, the bandwidth to the web server can be split up between hundreds of surfers to ensure everyone gets a fair shot at decent download times.

Firewall Integration: Many traffic-shaping systems are tightly integrated with firewall systems. This allows for some incredibly powerful, fine-grained network tuning. Firewalls provide a simple rule-based approach to analyzing traffic. These rules can also be used to place different types of packets into appropriate queues. Connections to trusted systems and networks could be given priority over unknown networks. Parameters such as the source and destination address, protocol, and type of application can be used to create highly effective traffic shaping policies based on security needs as well as business needs.

Another advantage to firewall integration is that it forces strong security on the shaping device. If a hacker gets control of the traffic shaper, they have control of the network, similar to what would happen if they obtained control of the firewall. It makes sense for these two devices to be the same from a security position. However, if your traffic-shaping device is independent of your firewall, you had better make sure it's as secure.

Preventing traffic analysis: Even if all of your data is encrypted using a very strong encryption system, it may not be safe. Sophisticated hackers can analyze the patterns in your bandwidth usage to make educated guesses about the contents of the encrypted data packets.

How does this work? Certain common applications like web browsers and email programs send very predictable requests across the network on a regular basis. This creates patterns in the traffic flow. Hackers look for patterns that correspond to simple requests in which the entire content of the data can be guessed ahead of time. For example, when connecting to an email server, an email program will send the word "HELO." When a hacker sees an email pattern, he or she can be sure that the first part of the data contains the "HELO" command, and can make some educated guesses about the rest.

If a hacker knows what the message contains, breaking the encryption key becomes much easier. By capturing enough patterns, a sophisticated hacker can reverse-engineer an encryption key in a relatively short amount of time.

Final Thoughts

For years, traffic shaping was considered a technology for the big boys. Currently, even the small boys have a decent amount of bandwidth to share with many people. The time to use the best tools available is now at hand, and traffic shaping should be a consideration. Many network administrators would be surprised by how many of their day-to-day network problems would be solved by traffic shaping. Essentially, traffic shaping is a macroscopic, proactive approach to smoothing out network troubles.

VIII
Storing Information

Summary

Once information worth protecting is created or obtained, it needs to be stored somewhere. Different types of storage systems come with different security risks. The following chapters cover the various technologies available for securely storing information.

Key Points

- Mankind creates and collects by nature, so mankind needs a place to put all of its stuff. People collect stuff they might need again at some point so they store it for later.
- Security is critical to storage. The information stored may or may not be valuable to you at the moment, but it may be valuable to someone else at any time.
- Data Storage systems have weaknesses that are independent of the systems they run on, the applications that access them, and the specific data they contain.
- Many important digital storage systems were not designed with security in mind.

Connecting the Chapters

Several effective methods exist for storing information. When databases and traditional flat file storage are combined with network file systems data can be stored and retrieved quickly over great distances. The following chapters explore how data is stored both locally and over networks:

- **Chapter 22, "Storage Media,"** examines the physical devices that hold information.
- **Chapter 23, "Local Filesystems,"** describe structured environments established on a hard drive that enable it to store files.
- **Chapter 24, "Network Filesystems,"** shows how a central storage system that can be accessed over a network is convenient and efficient, but also creates a single point of failure.
- **Chapter 25, "Databases,"** looks at systems that organize a collection of data so it can be easily accessed, queried, and updated.

Introduction to Storing Information

Collecting stuff is part of human nature. Many people spend their whole lives accumulating things, and over the course of a lifetime that can mean a lot of stuff. Once stuff is acquired, it needs to be put in a place. Storage space often becomes a critical element in most peoples' lives. Sometimes people even need to change their living quarters just to accommodate the volume of stuff they own.

The digital world is much the same. Millions upon millions of 0s and 1s make up the digital items businesses and people want. Whether it's software, digital photos, spreadsheet data, or whole databases, those bits need to be stored somewhere. The desire for more space in the physical world is mirrored in the digital world. Eventually hard drives fill up and people find themselves squeezing their digital possessions into nooks and crannies.

As computer applications evolve, they seem to be getting larger and larger. A word processor fifteen years ago was less than 400 kilobytes in size. Today, a word processor requires over 100 megabytes of hard drive space. Sometimes we wonder if today's word processor is really any better than the ten-year-old version that was $1/250$ the size, but we digress.

Whatever the reason, storage demands have grown exponentially and it appears the trend will continue. To meet storage demands, storage technologies have advanced in leaps and bounds. Fifteen years ago, an entire room of equipment would have been needed to store the same amount of information that a tiny chip can hold today. From punch cards to flash cards, storage systems have come a long way.

Don't Leave Me Unprotected

The push for more storage space may never end, especially with peoples' tendency to save everything. Storing data means that tons of information will be sitting in a repository, waiting to be accessed. In many cases, infrequently accessed information will be taken offline. The offline storage unit (floppy, CD, or tape) may be placed in a filing cabinet or taken off site. Frequently, data is archived in this manner and then forgotten. Why is the poor data left all alone in a dark room? Because the information stored may have limited value in the present, but extraordinary value later.

The value of stored data is a matter of perception. Usually, the data is of little value to whomever stored it, until it's needed again. However, it may hold great value to an outside party at any point in time. What has been stored and forgotten could be worth stealing.

One person's garbage is another person's gold: This is the preeminent security issue with the storage of data. Putting something away is not enough; it needs to be highly secured. With the proper security comes a guarantee of the data's integrity when it does, once again, become important to those who stored it in the first place.

A good example of storage versus priority is the tax return. Tax returns are very important when they are being filed. The accountants want to get the numbers right and corporations and individuals do not want to pay more than necessary. Once the return is filed, it is stored away. In fact, accounting firms are legally obligated to store seven years of returns. That takes up a significant amount of space, both physically and digitally (as many firms use a combination of both).

Old tax returns are often long forgotten, until an audit comes along. All of a sudden, the aging tax returns are worth their weight in gold. It would not be pleasant for those being audited to learn that their old tax returns are missing or damaged. Even if the old returns are intact, unauthorized individuals may still have viewed them. Someone with malicious intent may perceive those dusty returns as highly valuable. The information contained in just one individual's tax return is enough to give the ability to commit identity theft.

Physical vs. Virtual Security

It's dangerous to draw analogies between physical storage and data storage. In the physical world, when something is stolen, it's gone. In the digital world, information can be stolen from storage yet still be there. Often, people don't realize that they've had digital information stolen; after all, how can they tell?

Digital valuables do have a few advantages over their physical counterparts. Data that is stolen or destroyed can always be recovered from a backup. Corruption can also be easily detected. Compare this to the invisible deterioration that might be damaging a valuable physical object or the permanent loss if it's stolen or destroyed.

Treating all stored information with equal care is a critical aspect of a solid security strategy. A good rule of thumb is not to differentiate between active and archived data. Both types of data are subject to the same dangers of theft and destruction. Data that is stored and archived may be considered yesterday's news, but is often just as valuable to an outsider as actively used data.

Storage Caveats

Sometimes, modern storage technology appears too good to be true. It is fast, stable, reliable, comes with huge capacity, and best of all it's cheap. Storage vendors have been releasing a variety of newer technologies that take storage options even further. Devices are available that can store large amounts of data, yet fit in a pocket. For example, keychain USB devices are available that store data for easy transfer to other computers. Tiny flash cards enable cameras and other digital devices to exchange information with PC computers and one another.

Each type of storage device brings with it new conveniences and new problems. The truth is that storage systems have weaknesses that are independent of the systems they run on, the applications that access them, and the specific data they contain. This means that regardless of the precautions taken on the application level, the hardware holding critical data can and eventually will fail to do its job. This aspect of storage leaves administrators with the need to ensure reliability despite inherent and unavoidable flaws in the physical storage systems.

Databases, for example, are great at storing large amounts of information while allowing hyper-fast accessibility. They often run on independent servers that other applications hook into when retrieving data. Unfortunately, databases frequently corrupt the data stored within their tables. This can happen for a myriad of reasons, including too much use or not enough maintenance. When tables become corrupted, it becomes difficult or impossible to access critical data.

Database replication is one solution to the problems of database storage failure. In short, this takes all the data from one database and duplicates it in real time to another database server. Replication can be done on or off site, but always entails the use of separate hardware. If one database fails for any reason, the other database can remain unaffected and provide continuous service to its users.

Tape and floppy media have been around for a long time and are still in wide use today. A problem that has always plagued this form of storage is exposure to magnetic fields. All magnetic media (including hard drives) can be severely damaged when placed near a strong magnetic field. The slightest brush with a magnet can result in the corruption or deletion of part of the data stored on such a device.

Ostensibly, tape and floppy media have a shelf life. If they are left for more than a few years, background magnetic radiation can corrupt the data, or the media itself may simply degrade. This is one reason many people have transferred their old floppy data to CD-ROM. CD-ROMs also can degrade, but their shelf life is at least 30 to 50 years.

Old-fashioned hardware failure is one of the biggest problems plaguing storage devices today. Even the highest quality hard drives will fail over time. Hard drives are mechanical devices and mechanical parts eventually wear down. Another problem is that manufacturers focus on building storage devices that can hold the largest amount of data for the least amount of money. This is, after all, the primary demand of the consumer. The result is a certain loss of quality control, which translates into hard drives that simply stop working. Sometimes a whole line of hard drives end up in recall. The race to be the first to market with the largest, fastest, and cheapest drive puts great pressure on the manufacturers.

Storing Securely

Most storage systems are not designed with security in mind. Storage devices in use today rely on the security of the applications or methodologies used to access the data they contain. Nothing is inherently secure about a hard drive, a flash memory card, a tape drive, or any other storage media. For example, a tape from a server backup may be sitting on a desk at someone's home. If the home of that person is robbed, the tape may be stolen. If the data on the back up tape was not encrypted, then it will be completely accessible to any third party that places it in a tape drive. There is no security system built into the tape media itself.

Network, operating system, and application level security systems usually dictate access to storage devices. This means that it's the user or administrator's responsibility to ensure that information is stored securely. That said, advanced storage systems such as network files systems and databases can directly provide data security if properly configured.

Summary

The desire for secure storage is only in its infancy. In time, security will be integrated into storage devices and storage media. Already, some of the newer memory cards have built-in security systems. This may help secure data, especially in circumstances of remote storage. In the future, a backup that has the financial data of a company might not be viewable in the wrong person's hands, regardless of whether the backup system used encryption.

Chapter 22
Storing Information: Storage Media

A discussion of the actual devices
that can hold information.

Technology Overview

Storage media have come a long way since floppies. A few years ago, the word terabyte was a mystical concept—a thousand gigabytes. Only serious data centers had a terabyte of storage. The average desktop PC today comes with over 100 gigabytes on a single hard drive. Putting a terabyte worth of storage into a desktop PC has not only become possible, but it can be done for less than a thousand dollars.

Simultaneously, removable storage is both increasing in capacity and decreasing in size. The latest flash memory technology can store a gigabyte on a device no bigger than a postage stamp. Removable media could be made even smaller, but people might have a hard time holding it in their hands.

There wouldn't be a need for larger storage systems if there weren't demand for more space. The demand comes from high-resolution audio and video media, general file bloat, and applications that now require gigabytes of storage to install. Developers can count on continuously increasing storage and processor capacity. As a result, they design systems for flexibility, not efficiency. *Extensible Markup*

Language (XML) is a perfect example. It's essentially a database, but in a format that is easy for people to read. Naturally, this is incredibly inefficient; the files are huge, but the storage space is there, the bandwidth is cheap, and text-based files don't faze powerful processors.

The distinction between storage media and computing devices may become a gray area. Small devices are starting to have significant storage capacity. New, portable MP3 players can hold many gigabytes of data. These devices are not necessarily limited to storing music data. Cell phones are beginning to have significant storage capacity as well. Eventually, a cell phone may be used as a portable hard drive to carry files from work to home. It will also be possible to send files to other cell phones or directly to email accounts.

Large storage systems are also now being sold as independent devices. Instead of buying hard drives and a file server, *network connectable storage systems* can now be purchased. These are plug-and-go black boxes that automatically provide a large amount of highly reliable storage. In reality they are complex computer systems.

Security is a concern whenever storage media come packaged with a functional computer. The storage system may have unique security vulnerabilities, exposing data to risks that would not have been otherwise present.

How Storage Media Works

Storage Media

Illustration by SageSecure

■ **Figure 22-1**

Security Considerations

You might think that securing storage media simply means sliding the "write protect" tab into place. In fact, there are a few non-obvious security features and pitfalls in most modern storage media.

Lifespan: There is an ongoing debate in the authors' office as to which has a longer shelf life, a CD or a Twinkie. Wes insists it's a CD, but Jason claims he has a Twinkie in his house that is over 20 years old and still looks tasty! Whichever one lasts longer, one thing is certain: neither will last forever. The optical surface of a compact disk will deteriorate over time. Eventually, a CD may not be readable; of course "eventually" might be over 30 years from now. Likewise, eventually Jason will get hungry enough to eat his ancient-yet-somehow-still-moist Twinkie.

Frankly, in addition to old Twinkies, the authors have floppies that are still readable even after 15 years of use. Nonetheless, it's a good idea to copy all long-term archival data to new media every few years. This also avoids the problem of being unable to find current hardware capable of reading older forms of media. You don't think CD players will go away? Try to find a record player today. Even finding a decent cassette deck is tricky.

Built-in Protection: Floppy disks always used a write protect tab for preventing users from accidentally deleting their files. Newer media go well beyond write protection and have built-in encryption systems. This can be used to provide added protection if the tiny storage device is lost.

Walkabout: As removable storage gets smaller in size and larger in capacity, critical data can leave the home office on a key chain. New devices that are smaller than a thumb can connect directly to a PC and carry hundreds of megabytes of data. These types of removable storage systems can be hooked up to USB and other ports. Floppy disk adapters can allow any PC with a floppy drive to write to flash cards— which can hold gigabytes of data. Perhaps you thought that it would be too difficult to get any significant amount of data out of the office via a floppy? Think again.

It is not a good idea to have floppy drives or CD-R drives on machines that have access to critical data. Physically securing access to the workstations in general can prevent many problems, including theft and unauthorized equipment modification.

Policy Enforcement: Removable storage can lead to situations where security policies become hard to enforce. If PCs have CD drives and floppy drives, users can bring in software and install it on their systems. In the process, they may bring in viruses and Trojans inadvertently.

Policy may also require storing all files on a central server for revision control, management, or auditing purposes. Removable storage can provide an alternative that may prove to be more convenient (it lets people easily take work home or move it from one machine to another) yet is less secure and makes tracking the data that much more difficult.

Unauthorized Duplication of Licensed Media: Keep data that has value locked away. Inexpensive and versatile storage media make duplication a breeze. Software that is licensed to a business can easily be copied and spread to others for free. An investigation might trace pirated software back to an organization that was lax in securing its software, which could result in a lawsuit.

Damage From Handling of Media: Most system backups are sent to tape media. Unlike other types of media, tapes are quite fragile. They need to be rotated often to prevent overuse or abuse. Wear and tear will ultimately cause a media meltdown. A backup tape will have no value if it cannot perform during a critical restore job.

Throwing Away Old, Broken Media: There's more than meets the eye, or the disk drive, when it comes to data retrieval. Professional data forensics experts can get data off a drive that has been long since erased. Broken hard drives, damaged tapes, failed burns of CDs—these should NOT be thrown in the regular trash if they ever contained sensitive information. Before junking or selling PCs, an eraser program should be used to properly wipe the hard disk clean. Even after erasing a drive, traces of the old magnetic alignment still exist. Sensitive equipment can read these traces and retrieve "old" data. Proper erasing software eliminates any chance of this by writing meaningless noise to the entire disk repeatedly. Eventually the noise will weaken the old magnetic pattern to the point of illegibility. Then 0s can be written, blanking out the disk.

Chapter 23
Storing Information: Local File Systems

A local file system is a structured environment
established on a hard drive to enable it to store files.

Technology Overview

Computers see data as nothing but 0s and 1s. A blank hard drive is a giant sea of 0s,
ready to have 1s strategically placed like buoys in a busy harbor. But how should the
computer organize the data on the hard drive? That's a tricky question. Every operating system deals with this question in a different way. These organizational strategies are called file systems. The most common file systems have names like FAT16,
FAT32, NTFS, JFS, FFS, UFS, VFS, and ext2/3.

Early file systems were just responsible for getting information on and off a storage device. The operating systems were responsible for controlling the way in which
the information was used. More recent systems have direct support for access control, error recovery, and data security.

The majority of users and organizations today employ two basic types of operating systems, Unix or Windows. For this reason, the file systems used by these operating systems are covered in the greatest depth.

How File Systems Work

One of the earliest Microsoft file systems was called (*File Allocation Table* 16 [FAT16]). It integrated with an operating system called *Disk Operating System* (DOS). It worked by breaking the hard drive up into regions. Each region was given an address, which was a number between 0 and 65,535 (this is 2 to the 16th power, thus the 16 part of FAT16). When a file is stored, the data starts at the beginning of a region. If the file is larger than the region, it keeps flowing into additional regions. A lookup table links the filename with the starting addresses of each region used. Any unused space in a region is lost. These address regions are often referred to as clusters or blocks.

The size of the address region has an impact on the overall efficiency of a file system. At a basic level, the number of total addresses multiplied by the region size can't be smaller than the drive; otherwise the remaining space is wasted. For example, on a 2-gigabyte drive formatted with FAT16 each address region needs to be 32 kilobytes. This is ok when storing a small number of large files. However, when saving many small files a large amount of space is going to be wasted, possibly more than a gigabyte. This space is wasted because no matter how small the data actually is, it will take up 32 kilobytes worth of space.

This problem prompted Microsoft to increase the address range of their file system and resulted in FAT32 (released with Win98). Under FAT32, over 4 million addresses are possible. This allows very large hard drives to use relatively small region sizes, which can minimize wasted space to under 10 percent. Why not use even more addresses? The larger the address space, the longer it takes to find and retrieve files on the storage system. The goal is to strike a balance between the performance and efficiency of space allocation.

In between FAT16 and FAT32, Microsoft developed a next-generation file system to go with their *New Technology* (NT) line of server operating systems. In an unparalleled burst of creativity, they called it NTFS. We'll let you figure out the acronym. NTFS was their first file system that provided more than basic load/save functionality. It interacts with the operating system to provide users with file and directory access control. This means that users can protect their information from other users, or choose to share information with a limited selection of users. NTFS also prevents users from directly undeleting information removed from the file system. Most importantly, NTFS implements systems for improving the reliability of the storage process. It is very difficult for an application to write data to NTFS in a way that results in a corrupted file or directory, even if the application or operating system crashes midway through the process. The system automatically will attempt to repair any errors in the background, another useful benefit.

With all those features, NTFS is still not very secure. If the hard drive is accessed from another operating system, all the data becomes available without security restrictions. This is relatively easy to do with the right bootable floppy disk.

As a result, Microsoft created an enhanced version of NTFS for their Windows 2000 operating systems. The enhancements focused on security improvements and scalability. Security was improved by implementing direct support for encrypting the entire file system and all user files. This would prevent the "bootdisk" attack described above. Scalability was improved by moving to a 64-bit address table, enabling up to 18,000,000,000,000,000,000 possible drive region addresses. We had to print out the number because we just don't know the name for something that big.

While this was going on, the Unix world was busy creating its own file systems. Two fundamental differences between the Unix world and the Windows world were apparent. First, Unix was designed as a multiuser environment from the beginning. This meant that user-level security was an early concern. This led to an early adoption of security and reliability features only found in the more recent versions of NTFS. Second, many different vendors were creating competing versions of Unix. As a result, a number of different and incompatible file systems were in use. The result was the *Virtual File System* (VFS), a generic approach to dealing with arbitrary file systems. The VFS is a powerful concept. As far as the operating system is concerned, only one type of file system is available. This means that the development of file system code can be totally separated from the development of the operating system. Any data source that provides the right access commands can be treated as a VFS. For example, in many Unix systems, the kernel (main processing code) can appear as a file system. No actual "files" exist—instead, various kernel code and parameters can be viewed and modified in a directory structure.

Thanks to the Virtual File System, it is easy to access foreign file systems (many Unix systems can read NTFS and FAT16/32 systems with ease). It's also possible to create network-level file systems. These systems don't interact directly with storage devices, but treat the entire local file system as a storage device. *Network File System* (NFS) and *Andrew File System* (AFS) are two examples of file systems designed to operate over a network, which we'll discuss later in this chapter.

The most current Unix-world file systems support fault tolerance and prevention, as well as the automatic recovery of information. These systems are known as "journaling" file systems. Also, certain file systems are capable of providing file system and user-file encryption. At the moment, both Unix and Windows systems are fairly equal when considering the potential security level of the file system.

Security Considerations

Permissions: Not every file system truly supports permissions. Even if the file system does support them, they only work if they've been configured correctly. The user should never be expected to set permissions—all of his data should be given the most security by default. Permissions can also be negotiated around. Flaws in software can allow users to access files they shouldn't be able to reach. If a user has physical access to the machine, he can boot the system into an alternate operating system that is capable of bypassing permission. Trojans and other devices can also be implemented to bypass permissions. These programs would either run as an administrator, or with the same privileges as a particular user.

Ghost Data: When you write on a chalkboard and erase the chalk, you can often see the faint outline of the previously written message. It's not until the board is washed that these faint outlines disappear. The same is true of magnetic and optical media such as hard drives and floppies. Previously written data might be "erased," but faint traces can still be detected with the right tools. Therefore, it's actually possible to restore files that have been erased and "zeroed."

Temporary Files: Swap files, spool files, AutoSave, cache, and other temporary files are sometimes hard to find and can contain copies of the data that you're trying to protect. Sometimes unauthorized users can easily read these files. If an encryption system is in use, make sure that temporary unencrypted files are thoroughly deleted (no ghosts).

Undelete: Ever notice that it only takes a second to move or delete a giant file, but copying the file to another drive takes forever? That's because the actual data is not being moved or deleted on the hard drive; instead, only the file system table is altered. This means that deleted data is still actually on the hard drive—it simply doesn't have a file "handle" associated with it, so the file system doesn't have any way of locating the data. Eventually, new files will be written over the old data.

Plenty of tools can "undelete" data by simply restoring the file handles. This means that one user can obtain information that another user thought they had deleted. The only way to properly delete something is to write "0s" throughout the entire region of the disk on which the data resides. Now, nothing is available to recover . . . well, almost nothing. The hard drive can still be haunted by the "ghosts of data past."

Malicious Denial of Service: The size of the address region is important. A number of file systems can't support large drives (100 gig-terabytes) without increasing the block size to at least 16k. If a malicious user creates a large number of very small files (1 byte) on a drive with 16k clusters, each file would be stored in a separate address region, wasting essentially 16k per file. 250 million such files would consume 4 terabytes of space, or 16 terabytes with a block size of 64k.

Although 250 million files might seem like a lot, often no limit is placed on the number of files a user can create. It is possible, however, to limit the amount of disk space each user can have by enabling quotas. Unix file systems will look at the amount of actual space used on disk and can therefore solve this problem. Under NTFS, the quota system adds up the size of the data (250 megabytes), not the size of the space used on disk. This means the quota system won't solve this problem under NT, since even a 50 megabyte quota (small for the files generated by today's applications) could be used to consume nearly a terabyte of disk space.

Three possible solutions exist for NTFS systems. The first is to not use address regions over 4k. Under NTFS, this means that a drive system can size up to 2 terabytes. If you need more storage space, you can use multiple storage systems. Another option is to use third party software that can calculate quotas based on the actual disk space used. Finally, this particular problem is due to limitations that

Microsoft has imposed on the current versions of NTFS. The file system can theoretically handle up to 16 million terabytes using 1K clusters. If they were to allow the system to function as it's supposed to, this problem would go away.

Making the Connection

Accessing information: Local file systems allow local users with the appropriate permissions and groups to access data. This information is retrieved using methods and technologies covered in this part of the book.

Connecting Networks: Network design is heavily dependent on the selection of workstation operating systems and their local file systems. Local file systems will determine client/server file system compatibility over the network.

Best Practices

File systems have vulnerabilities that make them susceptible to many different types of malicious attacks. This does not mean that file systems cannot do anything to protect the often-valuable data they store. One weapon that some file systems have in their arsenal to offer added protection to data is encryption. The following are the four approaches to encrypting data on local file systems:

File-by-file encryption: There are many software packages that can encrypt individual files or directories. This gives users plenty of control, but is time consuming and highly susceptible to user error.

Encryption of the entire hard drive: When the computer first boots up, a password or token is necessary to decrypt the drive. This prevents someone from removing the hard drive or using a special boot disk to get information off of the machine. It's used with machines that are hard to physically secure, such as laptops. Once the system has loaded, any user with the right access can get at any particular file.

File system level encryption: This process is transparent to the user since the files look like they're readily available. But in reality, the data is encrypted on the disk drive. When a user goes to access a file, the system decrypts the data in the background. The administrator can chose directories to automatically encrypt. When a user places files into these directories, they can only be read by the creator or by other explicitly specified users (see Figure 23-1).

The basis of most encrypted file systems is public/private key encryption. Each user has a public and private key that is used to encrypt their files. These keys are normally stored on the file system. This, however, is an insecure approach toward security since access to the key gives an intruder access to all of the user's files. A better option is to store the keys on a floppy or on a smart card. Some systems give the administrator a key that can be used during system repair. This key can unlock every user's files. It is very important to protect this key and to use it as rarely as possible.

File Encryption Options

Illustration by SageSecure

1 Encrypt the entire hard drive. This protects against physical access, such as a lost laptop or an office break-in.

2 Encrypt files individually using software such as PGP. This gives users the ability to directly protect the information that's most important to them.

3 An Encrypted File System locks information automatically and transparently. Files appear as normal to an authorized user, but are unavailable and protected from others.

Figure 23-1

Numerous commercial and noncommercial *encrypted file systems* (EFS) are on the market. In the Windows environment, NTFS now includes an EFS. It has come bundled with Microsoft network operating systems since Windows 2000. Third-party encryption software can be used to supplement or replace the built-in system as well. A number of tools that will encrypt the entire hard drive are available, and others will monitor user directories and provide transparent encryption.

In the Unix world, two major encrypting file system initiatives exist: the *Cryptographic File System* (CFS) and the closely related *Transparent Cryptographic File System* (TCFS). A number of methods for encrypting the entire hard disk as well as many steganographic file system initiatives are also obtainable; these will be discussed in the next part of the book.

Encrypted File System Key Management: Normally, public/private key encryption uses two authentication factors: possession of an encrypted private key and a password for decrypting the private key. However, few users want to type a password every time they access a file. As a result, encrypted file systems either cache the decrypted key/password or just use a "passwordless" key in the first place. This lets the system operate transparently to the user, but at the same time removes one of the authentication factors. The result is that possession of the private key is the only thing necessary to decrypt the files. Storing the private key on the file system defeats the entire purpose of an encrypted file system.

This is where smart cards come in to play. The private key can be placed on a smart card—only available to the system when the card is in the reader. If a processor-based card is used, the key can remain on the card because decryption will occur on the card itself. The only way for a hacker to get the key would be to hack into the smart card. This requires techniques that are currently beyond those of the average hacker. If someone wants to get at specific data they'll probably find many other ways that are far more direct.

Of course, an encrypted file system is useless if a hacker can capture passwords and decrypted data through a Trojan. This is one of the reasons for hardware-level encryption at the processor and memory level. Further discussion of this is in the chapter on hardening systems.

Final Thoughts

Most users, system administrators, or organizations seldom think about local file systems. In fact, most user-based operating systems are pre-installed on computers by the PC manufacturer. Ironically, by taking control of an organization's computer systems on every level, security philosophies can be carried out more consistently. Local file systems should be well understood as they embody the beginning and end-points of where all data travels.

Chapter 24
Storing Information: Network File Systems

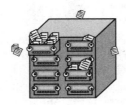

Putting data in a central place that can be accessed over a network
is convenient and efficient, but also creates a single point of failure.

Technology Overview

If you don't like acronyms, you should avoid this chapter. It's full of acronyms of the worst kind—the three-lettered devils. NFS, SMB and AFS are three different versions of the same concept each with its own uniquely annoying acronym. What they all have in common is the fact that they are network-driven file systems. Maybe that point alone was enough to provide an educated guess as to what the FS stands for? Let's fill in the other blanks as well.

One of the most commercially successful and widely available remote-file system protocols is the *Network File System* (NFS), designed by Sun Microsystems. NFS is the most widely used file system found on network servers. It currently serves more data in volume than any other network file system in the world.

Two components are important to the success of NFS. First, Sun placed the protocol specification for NFS in the public domain. Second, Sun sells that implementation to all people who want it, for less than the cost of implementing it themselves.

As a result, many vendors chose to buy the Sun implementation. They are willing to buy from Sun because they know that they can always legally write their own implementation if the price of the Sun implementation ever rises to an unreasonable level.

The *Server Message Block* (SMB) protocol is a protocol created by Microsoft for sharing files, printers, serial ports, and communications abstractions between Windows-based computers. It's a relatively simple system with a design focused on ease of information exchange rather than security or administrative control. SMB is the most commonly used file system, and is used by the more computers than any other file system in the world.

The *Andrew File System* (AFS) is a distributed file system that enables cooperating hosts (clients and servers) to efficiently share file system resources across both local area and wide area networks. AFS was originally developed at Carnegie-Mellon University, but is now marketed, maintained, and extended by the Transarc Corporation. AFS is mostly used in academic circles and rarely seen in modern operating environments.

How NFS and SMB Work

NFS: The NFS protocol can run over any available stream or datagram-oriented protocol. Common choices are *Transmission Control Protocol* (TCP) and *User Datagram Protocol* (UDP). Each NFS message may need to be broken into multiple packets to be sent across the network. A big performance problem for NFS running under UDP on an Ethernet network is that the message may be broken into as many as six packets; if any of these packets are lost, the entire message is lost and must be resent. When running under TCP on an Ethernet, the message may also be broken into as many as to six packets; however, individual lost packets, rather than the entire message, can be retransmitted.

The NFS protocol is *stateless.* Being stateless means that the server does not need to maintain any information about which clients it is serving or about the files that they currently have open. Table 24-1 illustrates a typical step-by-step data exchange between a client and server using NFS.

In practice, the server caches recently accessed file data. However, if there is enough activity to push the file out of the cache, the file handle provides the server with adequate information to reopen the file.

The benefit of the stateless protocol is that state recovery is not necessary after a client or server has crashed and rebooted, or after the network has been partitioned and reconnected. Instead, the server can simply begin servicing requests as soon as it begins running; it does not need to know which files its clients have open. Indeed, it does not even need to know which clients are currently using it as a server.

SMB: The SMB protocol is a client server, request-response protocol. Servers make file systems and other resources available to clients on the network. Client computers may have their own hard disks, but they also want access to the shared file systems and printers on the servers. Clients usually connect to servers using TCP/IP (*Internet Protocol*) (specifically NetBIOS over TCP/IP) NetBEUI or *Internetwork*

Table 24-1. Step-by-Step Data Exchange

Client Side—Requesting	Server Side—Providing
1) A read request is sent to the server. It will include the credential of the user that is issuing the request.	1) This initial information allows the server to open the file.
2) The file handle on which the read is to be done is sent to the server.	2) This information allows the server to verify that the user has permission to read.
3) The offset in the file to begin the read is sent to the server.	3) This information tells the server to seek to the appropriate point in the data.
4) And the number of bytes to be read is sent to the server.	4) This information is used by the server to read the specific contents.
5) At last the process is finished and the server closes the file.	

Packet Exchange (IPX)/*Sequenced Packet Exchange* (SPX). Once they have established a connection, clients can then send commands to the server that allow them to access shares, open files, read and write files, and generally do the things that clients do with a file system. However, in the case of SMB, these things are done over the network.

Samba is a Linux version of Microsoft's SMBprotocol. This enables Windows NT servers and 95/98 workstations to share files with Linux machines. As far as the Windows based client is concerned, it thinks it's talking to another Windows machine. This is an excellent way to expand many network resources, including printers, which are supported by Samba. In fact, Samba can even act as a *Primary Domain Controller* (PDC) for the Windows clients on a network. Samba can perform *Windows Internet Naming Service* (WINS) resolution and act as a WINS proxy as well. This can speed up browsing or even fix problems across slow *Wide Area Network* (WAN) connections without the cost of licensing a Windows NT or Windows 2000 server.

Security Considerations

Access Control: Frequently, network file systems are set up with very few restrictions. In a workgroup environment it's common to see directories and entire hard drives available via the network without any authentication. This is convenient for the users, but eliminates any hope for security. Even if proper access control is used, hackers can easily bypass it by compromising a user account or operating system.

Data Interception: Sometimes a hacker doesn't even need to compromise the networked file system. When a client requests data from a server the information is often sent across the network without encryption. The hacker can simply capture the file in transit. It's never a good idea to use networked file systems across insecure networks.

Protocol Vulnerabilities: Network file systems exchange information using protocols that may have inherent vulnerabilities. These protocols are layered upon, and inherit the security issues of, TCP, UDP, and IP.

Server Vulnerabilities: Even if the protocol is secure, the implementation may not be. Many file servers have had numerous security vulnerabilities discovered over the years. Hackers can exploit these vulnerabilities to gain unauthorized access or deny service to a file server.

Cache Manipulation: Caching causes the client and server to experience periods of desynchronization between cache updates. There are many exploits that a skilled hacker might be able to perform by keeping the client and server out of sync.

Reliability: File servers place a high degree of wear and tear on their underlying storage hardware. Data corruption and equipment breakdown can render an entire network useless. High availability solutions (described in part 11) can minimize the impact of a file system failure.

Denial of Service (DoS): Many parts of a network file system can fail. A hacker simply has to cause one part to fail in order to deny system service to users. Perhaps the hacker could thrash the hard drive by rapidly reading and writing a lot of information. Server vulnerability might enable a remote system crash. DoS techniques can be used to exploit other vulnerabilities or create larger network troubles.

Making the Connection

Accessing Information: Network file systems provide access to data across networks. This information is retrieved using methods and technologies covered in this part of the book.

Connecting Networks: The hardware covered in this part is what makes network file systems necessary. Networking hardware connects networks together and opens endless pathways for data to travel across. Networking protocols are used to bring data between clients and servers that use network file systems to store and retrieve files.

Best Practices

Network file systems integrate with various security protocols to ensure a secure exchange of data across the network. AFS, SMB, and NFS all have unique and overlapping methods with which they handle security.

AFS integrates with Kerberos to improve security. Kerberos uses the idea of a trusted third party to prove identification. This is a bit like using a letter of introduction or quoting a referee who will vouch for you. When a user authenticates at time of login, the user is prompted for a password. If the password is accepted, the *Kerberos Authentication Server* (KAS) provides the user with an encrypted token. This token contains a "ticket-granting ticket". From that point on, it is the encrypted token that is used to prove the user's identity. These tokens have a limited lifetime (typically a day) and are useless once they expire.

Kerberos improves on network security because a user's password only needs to be used once, at the time of the initial login prompt. AFS uses Kerberos to do complex mutual authentication, which means that both the service requester and the service provider have to prove their identities before a service is granted. This level of security integration that comes with AFS is a big win for the users and the system administrators.

The SMB model of network file sharing integrates security in a different manner. The SMB model defines two levels of security:

Share level: Protection is applied at the share level on a server. Each share can have a password and a client only needs that password to access all files under that share. This was the first security model that SMB had implemented. Windows for Workgroups' vserver.exe implements share level security by default, as does Windows 95.

User level: Protection is applied to individual files in each share and is based on user access rights. Each user (client) must log in to the server and be authenticated by the server. When it is authenticated, the client is given a *user ID* (UID) that it must present on all subsequent accesses to the server.

NFS is not secure because the protocol was not designed with security in mind. Despite several attempts to fix security problems, NFS security is still limited. Encryption is needed to build a secure protocol, but robust encryption cannot be exported from the United States. So, even if building a secure protocol were possible, doing so would be pointless, because all the file data are sent around the Net in clear text. It makes no difference if a hacker is unable to break into an NFS server to retrieve a sensitive file. Instead, they can just wait until a legitimate user accesses the server and then grab the unencrypted file as it travels over the Net.

Final Thoughts

Network file systems have two inherent characteristics: they are complex and taken for granted. As it turns out, both of these features support each other a little too nicely. Most users do not think about how their data is stored or retrieved. Those that do open a rabbit hole that goes deeper than expected.

Network file systems rely on highly technical, fundamental information technology concepts to operate. Sometimes these are the pieces of knowledge that get brushed over and forgotten, or simply avoided. Unfortunately, taking network file systems for granted limits the extent of the good network design. For example, network file systems rely on network protocols to bring them data. Knowing how data is sent to network file systems can help determine what ports can be closed on a firewall that connects separate network nodes. With a firm understanding of network file systems, a network can be designed with much greater efficiency.

Chapter 25
Storing Information: Databases

Databases organize a collection of data so it can be
easily accessed, queried, and updated.

Technology Overview

Much like messy papers on a desk, data needs to be organized. An important piece
of paper serves no purpose if it cannot be found when it is needed. Likewise, if data
is not organized, it holds little value. Important data, whether analog or digital, needs
to be highly accessible.

A database is an advanced method of storing and organizing data so it can be
easily retrieved. Databases have been a standard in computing since the 1970s. The
original databases, called *flat file systems* (FFS), were little more than a consistent
way of storing records in a digital file.

As needs for data handling expanded, more complex database systems were developed. *Relation Database Management Systems* (RDBMS) hit the market and
their popularity exploded. These systems worked by enabling vast amounts of data
to be organized and stored in tables. The data could be rapidly manipulated by

creating relationships between different tables. Relational database systems became the standard in database technology for years, but standards eventually change.

In the late 90's the major database vendors released a plethora of new products. The new products are still geared toward handling large volumes of complex data, but now some of the products are *middleware* oriented. Enabling these new products is an extended version of relational database technology called an *object relational database management system* (ORDBMS).

Object-oriented databases take the concept of relational databases to a more advanced level. Unlike relational databases, object databases take the focus away from tables and place it on object-oriented programming instead. This is an attempt to make the interaction with large-scale databases less specialized and more straightforward for the average programmer.

Databases are now widely used and have become a commodity. As a result, many traditional database vendors are moving away from selling database engines as their primary product. Vendors are now exploring other areas of business that surround data storage and retrieval. This includes multimedia types (text, image, audio, and video), or any data type a user may wish to define. These are extensions from the very limited, simple, traditional data supported in the mainstream relational database products.

Relational databases have been employed to automate most of the obvious back-office and, more recently, front-office applications for today's enterprises. Any competitive advantages derived from that automation activity are diminishing. To find other information technologies to leverage for competitive advantage, organizations are turning to the Internet/intranet and to a richer set of data types.

To keep pace with their customers' needs, almost all relational database vendors are scrambling to extend the capabilities of their product lines to support Internet-enabled applications and the multimedia data types typically found on the Web. The World Wide Web promises global access from a "universal client." Why not then a universal database or server? Well, this dream realized would certainly make Ellison a happier and even wealthier man. I guess you could say it would have the same effect as landing a big right hook on Bill Gate's face.

Applications are now more frequently implemented in object-oriented or object-based architectures. As a result, application developers have high-performance storage mechanisms that are fully compatible with the entire object-oriented model. This forces the need for object database management systems as they can provide efficient storage for object-oriented applications. In short, the evolution of software development is being traced by the evolution of database systems. As these worlds continue to merge, good data security will rely on a working knowledge of the underlying database systems.

How Databases Work

In the beginning, all databases were flat. This means that the data types contained within the databases were completely unable to relate to one another. It also means that the information was stored as a simple delimited text file. Delimited simply means that data segments are separated by specialized character such as a pipe or vertical bar. A popular delimiter is the comma; many applications recognize *Comma Separated Values* (CSV) files as a simple, flat database file. The following diagram illustrates what a typical delimited text file looks like, using the | character as a delimiter:

```
Firstname, Lastname, Age, Height, Weight|Robert, Johnson, 42, 6'2,

195|Sarah, Clementine, 34, 5'6, 135|Timothy, Sanders, 23, 6'1,184|Kenny,

Thompson, 66, 5'11, 176|Peter, Roth, 15, 5'5, 128
```

The diagram makes it clear that data stored in flat file databases is fairly difficult to search through. This is because a search must look sequentially at the data to find a result. For example, if a search for Peter Roth's weight were performed on the above data, the search would have to look through every name, age, height, and weight until it reached the end of the data stream. This is an extremely slow and clumsy method to retrieve data.

As opposed to flat file, a relational database management system stores data in a database consisting of one or more tables of rows and columns. The rows correspond to a record; the columns correspond to attributes (fields in the record). Each column has a data type. Some data types include character, string, time, date, numbers (fixed and floating point), and currency. Any attribute of a record can store only a single value. Here's an example:

FNName	LNName	Age	Height	Weight
Robert	Johnson	42	6'2	195
Sarah	Clementine	34	5'6	135
Timothy	Sanders	23	6'1	184
Kenny	Thompson	66	5'11	176
Peter	Roth	15	5'5	128

The simplified table in the diagram illustrates how a relational database stores data. The columns represent the data fields and the rows represent the actual records. Running a search through a relational database table is much faster and more efficient than a flat file database. The table allows the sorting of any field. In addition reports can be generated that contain only certain fields from each record. For example, the relational database table illustrated here can quickly compare the heights and weights of all the records shown.

In relational databases, relationships are not explicit, but rather implied by values in specific fields. This is implemented through the use of keys. A key in one table matches records in a second table to signify that a relationship exists. Many-to-many relationships typically require an intermediate table of nothing but keys. This table of keys only contains data on relationships and their definitions. This is how database structures begin to get rather complex for large organizations.

The *Structured Query Language* (SQL) is used to define, manage, access, and retrieve data from a relational database system. With SQL, data is retrieved based on the value in a certain field in a record. The types of queries supported run the gamut from simple single-table queries to very complex multitable queries that link tables based on complex parameters and calculations.

Relational databases provide a simple, easy-to-learn user interface via their row-and-column metaphor. However, it is important to note that very few users interact with relational databases directly via SQL. The relational database vendors and their partners have provided a myriad of tools that hide the guts of SQL from the user by automatically generating appropriate statements for common tasks. Currently, user-oriented tools are not as common with object-oriented database products, but this will change over time.

Object databases are very compatible with organizations that regularly use object-oriented programming. There is a direct, one-to-one correspondence between the application data object and the stored data object. In other words, the application doesn't have to worry about converting the object data to a table format. The objects can be directly stored "as-is" in the object database. This makes the development process very efficient and also simplifies maintenance. Most importantly, object databases enable a more consistent approach to securing both the application and the stored data.

Security Considerations

There are many security concerns that are unique to database systems. Database systems are unusual in that they do not just store data, but usually link stored data to multiple, active applications that rely on the data to function. As a result, unauthorized access to a database can result in several problems such as downtime, invasion of privacy, theft of intellectual property, and access to other secured systems.

Databases of all types have what is considered to be "built-in" security. Generally speaking this is little more than user permissions control. The administrator of any database system has the ability to grant access to data based on user or groups of users. This secures data from users who are trying to gain access to databases through the front door. However, security systems built into database systems offer no protection against downtime or the corruption of data.

To prevent application downtime, databases need to be constantly available to the applications that query them. If one database is hacked and disabled, another version of the database needs to be ready to take its place. Of course the data in databases is constantly being modified and updated by users and applications. This means if one database stops functioning, the database that replaces it needs to be up to date with the most recent data. How is this possible?

Distribution, Replication, and Federated Databases

Three major concepts are used in providing database redundancy and high availability.

- A *distributed database* transparently stores its data across multiple volumes and even different locations.

- A *replicated database* has all or portions of its data replicated at one or more different sites. Replicated databases periodically synchronize the contents of the replicated data. Data replication is the foundation for data warehousing.

- A *federated database* integrates several isolated, heterogeneous databases into a single virtual database system for use by applications such as transaction processing.

Replication is the common thread between all of the above redundancy techniques. Database replication can be used for:

Efficient Data Access: Accessing a local database is more efficient than accessing a database over a *Wide Area Network* (WAN) such as the Internet or through a *Virtual Private Network* (VPN). With replication, only the databases need to talk over the network. Every other user and application can communicate with a local database. This cuts down on network traffic and latency.

Disconnected Use: Remote users may not always be connected to the database via the network. In many cases, offline database access is useful even if the data isn't completely current. Laptop users might be able to access a customer database on the road, for example. The replication system will bring the remote user up to date whenever the user is connected to the network. This can greatly extend the practical uses for many types of database applications.

Load Balancing: Replicated databases can share user loads. This puts less stress on each database system, speeding up access times and queries and reducing network traffic. The less a database is stressed, the less chance there is for data corruption or downtime.

Backing up Databases: Some types of databases need to be taken offline in order to perform a full backup. This might be difficult to do if the database is mission critical and highly active. Replication can solve this problem. Users and applications can still access a database while it is being replicated. The backup is constantly being updated in real time while users are modifying the database. This has many security advantages. If one database is corrupted or taken down, the replicated system will kick in and there will be no loss in data service. There is also no need to restore a downed database because the replicated backup can be accessed in real time from the moment the original database goes down. This is known as real time fail over.

Relational databases can support some level of replication, especially for read-only replicates. Object oriented databases are capable of a much more complex form of replication and distribution. Object databases are inherently designed to integrate with applications with complete transparency, over multiple database servers. In addition, they have better support for federated database structures.

Making the Connection

Ensuring Availability: Databases need to stay up for applications that rely on them to work. Techniques covered in this part can be combined to work with database systems to maximize uptime and minimize recovery.

Replication Copies the Good and the Bad, Which Can Get Ugly

When databases replicate, they don't usually know if what they're replicating is good or bad. Corrupted data can get replicated to another database. This is particularly problematic if replication is being used for backup. All of the backup databases might end up with corrupted data.

The solution is to replicate to a read-only database that is never accessed by users or applications. Periodically, this database can be taken down and backed up using traditional offline backup techniques. When it comes back up, it will catch up on any updates that happened when it was down. The main database never goes down, so users don't experience any service interruptions.

Best Practices

The easier a database is to manage, the easier it is to properly secure. Which types of databases are easier to manage? That depends on the organization's management style and the applications it uses for routine business operations. It's not easy to determine the best database system for a particular task. Certain facts within the database industry can provide insight into making this difficult choice.

Vendors have optimized their databases to best serve their target markets. High-end vendors such as Oracle have systems that can handle extreme situations in which other databases might fail. Midrange vendors such as Microsoft have products that function well for many types of common enterprise applications. Low cost or free databases tend to be optimized for lightweight applications and academic computing.

Object-oriented databases are better than relational databases for certain applications, but they are not always the best choice as they are still relatively new. Relational database vendors have been around for a longer time, are very large, and can offer better support. It is also quite likely that these vendors and their products will be around for a long time to come.

Maturity: Relational database products have been used much longer than object database products. Relational databases are simply more mature products. As a result, they have been fine-tuned for optimized performance and provide a very rich set of functionality, including support of advanced features like parallel processing, replication, high availability, security, and distribution.

Compatibility: The RDBMS model allows the stored data to maintain independence from the applications that use the data. With SQL as a query language, any application can access and use data in an independent fashion. A wide variety of tools and applications that support the relational databases and work with SQL are available. The object-oriented databases should be able to take advantage of this support because they are based on relational database systems. However, relational database systems are built around the concept of tables. Object databases have been built with new ways to manage recovery, indexing, and caching. As a result, traditional RDBMS tools are frequently incompatible with ORDBMS systems.

Tradition: The other advantage that RDBMSs and the SQL-based ORDBMSs have is the availability of experienced developers and the plethora of SQL-based developer tools, books, and consultants. SQL is the most universal database language. As a result of the investments made into the SQL platform over the years, most developers are familiar with SQL and own the development tools needed to maintain the systems.

The relation database model of tables with simple data is easy to use, but only if it maps well to the application's data structures. If the application's structures are complex, mapping them to tables is like forcing a circular peg into a square hole. In addition, this traditional approach has created a need for specialized database programmers. Most relational database programmers need expertise in the following:

- Translating data back and forth from tables to application structures
- A comprehensive understanding of SQL
- A knowledge of SQL tools for testing and development
- Designing table structures to match complex data relationships
- Optimizing SQL queries to best run on the chosen database engine.

Conversely, object-oriented database programmers find it simpler to directly use objects without having to force them into tables. All programmers today are being trained in object programming, which opens up the use of database technology to a much broader base of programmers. It has been said many times before that traditions are made to be broken. If object oriented-databases continue to increase in popularity a new standard may be born.

Final Thoughts

The ease with which a database system integrates into a specific organization depends on:

- Staff knowledge of database management
- Application specific database requirements
- Network topology
- Requirements for data across multiple offices.

Security needs for databases boil down to availability, control, privacy, and access. If one platform integrates with your organization in such a way that these elements of security are easier to achieve, then that's the platform to choose.

IX
Hiding Information

Summary

A handful of techniques for keeping critical information away from wandering eyes are available, such as cryptography and steganography. Not only can these methods aid in privacy (if used correctly), but they also continue to protect information even if the data has been intercepted or stolen.

Key Points

- Hiding something effectively is difficult.
- You can hide information by covering it up (obfuscation), disguising it (steganography), or putting it somewhere safe (cryptography).
- You can't use something that's hidden. It is not possible to effectively hide something you need to access frequently.
- No matter how well something is hidden, its location can be revealed when the hider accesses the item. So, the better hidden something is, the less convenient it is to access. Hiding something very well might be better for peace of mind than it is for practical security.

Connecting the Chapters

The most effective method for hiding information is cryptography. It's also possible to hide information in more subtle ways, such as with steganography. When steganography is combined with cryptography, the result is an extremely powerful data hiding technique. The following chapters explore how digital data is hidden, and how it can be found:

- **Chapter 26, "Cryptography,"** explores the science and art of scrambling messages to keep the contents secret.
- **Chapter 27, "Cryptanalysis,"** covers the science and art of code breaking.
- **Chapter 28, "Steganography,"** looks at techniques for effectively hiding one piece of information.

Introduction to Hiding Information

The desire to hide stuff may be instinctual. Dogs hide bones, squirrels hide acorns, and many species of animals hide their eggs. People hide valuables. Why? Hiding is a means of protecting things that can't be constantly guarded.

Information, although less tangible, can also be hidden. It is common for individuals and businesses to make an effort to protect:

- Information that could be damaging, misunderstood, or embarrassing if found by the wrong hands.
- Personal and organizational information that a business has an obligation or competitive need to protect.

Unfortunately, *hiding things effectively is hard.* Dogs leave visible mounds of torn-up earth after hiding their bones. Birds often make nests, providing evidence of where their eggs are hidden. Squirrels are much better at hiding acorns, but when winter comes they sometimes forget where they put the acorns and end up digging everywhere.

People tend to have the same problems. Either we hide things poorly (in the underwear drawer, for example) or we forget where we put things, in effect hiding them from ourselves. Anyone who has spent an hour in the morning looking for keys is aware of this. When we find our keys, we usually relearn a fundamental truth about hiding things:

> *Putting something in plain view, but where it isn't expected, can be a very effective hiding technique.*

The desire to find hidden things is also probably instinctual. When growing up, we play games like "hide and seek" and "search for buried treasure." Some people never grow out of it—spies, journalists, and tabloid writers (to name a few professions) spend their lives looking for juicy, hidden secrets.

It's usually not too difficult to find things that have been hidden because most people are bad at hiding. The irony is that most people think that *they* are not "most people." Research does show that the average individual is a creature of habit and convenience. People with similar backgrounds will react similarly when placed in similar situations. In the case of hiding this means people will identify the same handful of hiding spots when confronted with a particular room.

Many items are hidden when they should be destroyed or placed under monitored security. A suburban burglar simply puts himself in the shoes of a "hider" and says, "Gee, what would a person living here consider a good hiding spot?" Thieves probably observe the same hiding spots being re-used from home to home. Nonetheless, people are still surprised and/or embarrassed when a thief walks off with their hidden loot or a steamy adulterous letter makes the front page of the neighborhood gossip column.

How Things Can Be Hidden

Hiding information can be done in three simple ways:

- Cover it up
- Disguise it
- Put it somewhere safe

Of the three methods, covering something up is the most obvious and instinctual response. It can also be quick and effective in many circumstances. For example, let's say someone is sitting in a room, examining a letter they're not supposed to read. Suddenly, they hear footsteps approaching the door. What is the response? To quickly throw a newspaper over the letter and casually answer the door as if nothing were amiss. If the person at the door has no reason to suspect that something's amiss, he or she won't look twice at the newspaper. The "cover up" will be effective.

The problem with covering something up or ducking it away comes when somebody suspects that something is being hidden. Perhaps it is the sight of a flushed faced, or the sound of hastily shuffled papers. Regardless, now the other individual is suspicious. The oddly positioned newspaper on the table might be noticed, prompting a casual straightening out and, "Oh my, what do we have here?" Or perhaps a thorough search later on will find other things hidden in folders or drawers.

In security terms, this is known as *obfuscating* something. As many security experts will attest, obfuscation does not provide any real security. With enough effort, most obfuscation is transparent. For example, giving a critical computer file a misleading name and putting it in an obscure directory does little to actually secure the information. A hacker can quickly search the entire hard drive for interesting information.

A better system for hiding things involves using a disguise. We've all seen the mystery TV shows where a switch for a secret passage is disguised as a candlestick or a bookcase is really a doorway. Perhaps a secret message could be written on the inside of a lampshade, only to be revealed when the right type of light bulb is placed in the lamp.

A disguised object can be effectively hidden in plain view. This is because people tend to ignore the obvious things in front of them when they go looking for something. This is the lost keys phenomenon. Every location where the keys may have been placed is searched. In reality, they're sitting on the stairs, right where they were dropped on the way to bed. The keys aren't found because the assumption is that they have been put in a safe place. Nobody is going to look closely at the common objects that are lying around the room. People are too busy looking for crafty hiding places.

Data can be disguised using a technique called *steganography*. This is a process that takes important data and hides it inside more common data. For example, a secret message can be easily hidden in a digital picture or music file. Looking at the picture or listening to the music would give no indication that a secret message exists. As a result, the combined file can be left in plain view on a computer system.

Disguises don't just make things hard to find,they make it hard to tell that something valuable exists in the first place. When a thief looks in a window and sees a safe, he can guess that something valuable is inside. But if the safe is hidden behind a painting, the thief has no idea if valuables are kept on the premises or somewhere else.

That said, a disguise may not be necessary if the safe is strong and secure enough. Even a safe in plain view is effective at protecting its contents from prying eyes. It also ensures that only authorized people (those who know the combination or have the key) can get at the valuables inside. A good safe will deter all but the most skilled of criminals.

For protecting information, *cryptography* (encryption) provides the equivalent of a digital safe. To an unauthorized observer, encrypted information looks like a jumbled mess. Extracting the message without the proper key is as difficult as breaking into a bank vault, if not more difficult.

Encryption and steganography can both be used to hide information, but the approach is different. Figure IX-1 highlights the key differences.

Cryptography vs. Steganography

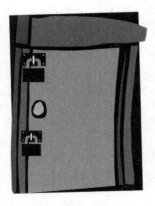

Padlocked Door (cryptography)

A padlocked door can hide the details of what is in the next room. Even though someone sees the door and may assume it is hiding valuables, they cannot gain access to the room.

Secret Door (steganography)

The secret door shown above hides the fact that another room exists. If no one knows that a door exists than they cannot gain access to the protected room. The hidden door can even be locked, adding an additional layer of security.

■ **Figure IX-1**

Looking at the illustration, one might wonder, "Why not lock the secret door?" In fact, that's exactly what people do. It's common practice to first cryptographically protect a message before disguising it with steganography. The result is a very well hidden message that is also protected in the event that it is discovered.

How Hidden Things Are Found

> When you go looking for something specific, your chances of finding it are very bad. Because of all the things in the world, you're only looking for one of them. When you go looking for anything at all, your chances of finding it are very good. Because of all the things in the world, you're sure to find some of them.
>
> Daryl Zero, *Zero Effect*

It's hard to use something and keep it hidden at the same time. When you're reading a hidden message, another person can read it over your shoulder. When something is put into a hiding place, somebody can watch and observe where it has been hidden.

Observation and manipulation are the most effective weapons in a thief's arsenal. A smart thief won't just try breaking into a safe using brute-force tactics. They might try to get the combination from observation, or try to trick somebody into giving it up. Likewise, observation and trickery can be used to obtain keys needed to access encrypted information.

In order to effectively hide something, *whom it's being hidden from* needs to be considered. The nature of the opponent can influence the choice of technique:

Law Enforcement: In most cases, a court will issue a subpoena requiring all materials to be made available to an investigation team. If information is encrypted, the court will require you to produce the necessary keys. In this situation, steganography provides the best protection, because the investigators might not be able to find the hidden data. A number of techniques exist for using steganography to achieve "plausible deniability." Not that we're endorsing illegal activities, but if you happen to have files containing the dates and times of contraband shipments, you might want to use steganography. You might also want to plan a fast escape route, just in case.

Hackers: Casual hackers can be thwarted by hidden information, but if a focused hacker gains significant access to a network, hiding information is generally not going to help. A hacker will simply wait until someone needs to use the hidden information. Then he'll watch as the information is retrieved and record any passwords that are used. As the file is accessed, the hacker will also have access to the file.

Your Employees: They may just be curious, but curiosity killed the cat. Hiding information from employees is difficult. They know if and where information might be hidden. They have the best opportunities for manipulating others into revealing information. Ultimately, hiding information from employees is like putting a lock on a window. It will thwart the casually curious, but a determined intruder will break the glass. If an employee becomes an internal hacker, the best hope is that someone might notice him poking around and report the suspicious activity.

Notice a few things in common? Hiding information can offer protection against casual discovery, but anybody determined enough could get at the information some other way. Even law enforcement can use hacker techniques to find information that is hidden via steganography. Plausible deniability doesn't work very well if there is evidence (such as usage logs) showing that a file once existed.

Final Thoughts

The problem with hiding information is that it can be very inconvenient. Think about it: how does a secret hiding spot stay secret? Nobody can be watching when it's used. Every time you want to hide something, you need to make sure the environment is secure. That's tough to do without making the entire operating environment very unfriendly.

If a hiding system is implemented poorly, it can be worse than having none at all. It gives a false sense of security and may call attention to information that otherwise would have been ignored. The hassle of dealing with the hiding system may affect productivity without actually providing any real security benefits.

The question is: does a hiding system provide adequate protection against casual hacking/snooping? Sometimes hiding information is critical to ensuring the privacy and security of information. In other situations, hiding information effectively is very difficult and may be impractical. The deciding factors will be the nature of the business and the nature of the information that needs to be protected.

There is one absolute: if hiding information is necessary, then it needs to be done correctly. The rest of the section will look at the two major techniques for hiding information: cryptography and steganography. The chapters explore how the technologies work, as well as how they can be circumvented.

Chapter 26
Hiding Information: Cryptography

Cryptography is the science and art of scrambling
messages to keep the contents secret.

Technology Overview

For most of its modern history, cryptography has been shrouded in secrecy. Political and military organizations have been using cryptography to communicate sensitive information securely. They employ the most brilliant mathematicians and information theorists in the hopes that they will create unbreakable codes. At the same time, these cryptologists are put to the task of breaking the codes created by the enemy's brilliant theorists. The successes and failures of cryptologists have been known to profoundly influence the outcome of wars.

For most of the history of cryptography, advanced techniques for creating secret codes and ciphered transmissions were the exclusive domain of government organizations. Code makers and code breakers were kept far away from the public and academic spotlights. Any unauthorized research into cryptography was strongly discouraged. The existing literature on cryptography went little further than the theory needed to solve the cryptogram puzzles in a newspaper. Truly effective cryptography techniques require an advanced understanding of mathematics, and the few people with these skills were often hired by government agencies. As a result, the general public knew very little about cryptography.

This all changed in the mid-1970s as computers became a viable tool for academic research. A number of mathematicians started exploring cryptography and realized that it would be a powerful tool for protecting the communications of individuals. Over the next few years, the public's understanding of cryptography would dramatically advance. By the end of the '70s, cryptography would become a viable tool for securing personal and business communications.

The most straightforward use of cryptography is for secure communication. Encrypted messages are sent between two parties to ensure that the message, if intercepted by a third party, cannot be read. Thanks to the development of public key systems, secure communication has now become commonplace.

Identification and authentication is another area where cryptography is commonly used. An example of encrypted authentication and identification is observed when withdrawing money from a bank. The automatic teller machine obtains the data on your bankcard and your secret pin code. Cryptography protects this information when it is sent to your bank for verification.

Electronic commerce has become the rage of the past decade. Millions of people worldwide make e-commerce transactions over the Internet. Included in the category of electronic commerce is online banking, online brokerage accounts, shopping, renting of cars, hotels, and online reservation systems among others. All of these transactions require the sending and receiving of confidential information between two parties, the vendor and the consumer. Simply sending information of this nature over the Internet provides the opportunity for it to fall into a third party's hands. Cryptography allows the vendor and customer to communicate securely over an untrusted network.

We've used the terms cryptography and encryption, but we haven't precisely defined what they mean. Before we go any further, let's look at some of the terms that we're going to use over the next few pages.

Cryptography: To most people, cryptography is the study of how to keep communications private. Cryptographic devices are those that enable secure communication between two parties. Literally, cryptography means "hidden writing."

Cryptanalysis: Sometimes people intercept private messages they were not supposed to receive. When this occurs, the interceptor's attempt to decode the message without the proper cryptographic device is known as cryptanalysis. This process is commonly referred to as code breaking.

Cryptology: No, this has nothing to do with the constellations or your personal lifeline. It's the field of study that encompasses both cryptography and cryptanalysis.

Encryption: This is the process of taking information and modifying its form to disguise its actual content. Unencrypted information is called *plaintext;* encrypted information is known as *ciphertext*. Don't let the word "text" fool you—any type of data can be encrypted, including images, sounds, and computer code.

Decryption: In all cases of encryption, the intended recipient(s) of the message possess a decryption instrument that is specifically capable of modifying the content of the message to its original form.

Encryption Algorithm: The mathematical process used to scramble the information is called an algorithm. Some algorithms are simple: substitute B for A, C for B, and so on. Others rely on a secret dictionary of substitutions where, "The eagle has landed," might mean, "Jimmy will make the delivery tonight." The problem with these systems is that anyone who knows the algorithm can easily decrypt messages.

The most advanced encryption algorithms use complex mathematical equations that combine a *secret key* with the plaintext in order to create the ciphertext. Knowing the algorithm doesn't help—without the key it's mathematically "impossible" to reverse the equation and obtain the original plaintext. We put impossible in quotes because it's impossible to prove that something is impossible. Impossible essentially means, "not computationally feasible in a reasonable time based on current mathematical techniques." Most modern cryptanalysis techniques (Chapter 27, "Cryptanalysis") attempt to find new mathematical techniques that can extract the plaintext or the secret key from a message encrypted with these highly advanced algorithms.

Today's encryption algorithms are classified into two categories. Those that use the same key for encryption and decryption are called "symmetrical" (shared secret) systems. "Asymmetrical" systems are the opposite; they use different keys for the encryption and decryption process. Both are used extensively and, in many cases, both techniques are used at the same time.

Symmetrical Encryption and Key Length

Symmetrical algorithms are called "shared secret" systems because both parties need to agree on (share) a secret key ahead of time. This key is usually a number. The longer the number, the more difficult it is to guess. For instance, if 9 digit numbers are used, there are a billion possible keys (0 to 999,999,999). A hacker would need to try each key in order to find the right one. Imagine having to go through a billion physical keys to open a door lock. These keys, stacked against each other, would stretch between New York and Washington, D.C.

It's important to realize that only the exact right key will work. A key that's only one digit off will be no more effective than one that's completely wrong. In other words, if 12,345,678 is used to encrypt a message, then only this number will decrypt it correctly. Using 12,345,677 or 12,345,679 will produce unintelligible plaintext. Each decryption attempt with an incorrect key will produce a different piece of gibberish, and will reveal no clue as to how close the guess is to the correct key.

In the encryption world, the length of a key is measured in bits, not digits. A bit is a digital "on and off" switch—its value is either 0 (off) or 1 (on). Eight "bits" can represent all of the numbers from 0 up to 255 (there are 256 possible combinations of eight on/off switches). Thirty-two bits can represent all of the numbers up to 4 billion, and then some. Today's most commonly used key length is one hundred and twenty eight bits, which gives approximately 340,282,366,900,000,000,000,000,000,000,000,000,000 possible keys.

The fastest supercomputer that we know about at the time of writing can perform 41 trillion math operations a second (Japan's Earth Simulator super computer). Assuming only one operation is needed to test a key (in reality it requires many more), it would take over 267,000 trillion years to search through all of the possible keys using 128 bit keys. We are talking about hundreds of thousands of trillions of years, which is still a long time to some people.

Asymmetrical (Public Key) Cryptography

For many years, a single key was used to both encrypt a message and to decrypt the resulting ciphertext. Two people who wanted to communicate securely both had to have a copy of the same key. The problem was, how to transfer the key securely? The only secure solution was to place the key on a floppy and physically transfer it. This was inconvenient, as it required a face-to-face meeting or some other means of exchange.

In 1976, Whitfield Diffie developed a new approach called *Public Key Cryptography*. In this approach each person has two keys. One—(the private key) is kept secret. The other —(the public key) can be given freely to anyone. The sender uses the recipient's public key to encrypt the message. The recipient uses her private key to decrypt the message. This means that someone can send a message without the need to get a secret key to the recipient first. As long as the sender can get the recipient's public key, a secure message can be sent.

The concept of public and private keys utterly changed the usefulness of cryptography. Previously, physical couriers were needed to transport the single keys to both ends of an anticipated communication path. No electronic path could be trusted with the key. Public key cryptography, when properly implemented and used, enables people to easily communicate in secrecy without any prearrangement.

```
-----BEGIN PGP PUBLIC KEY BLOCK-----
mQGiBDxiVVARBACHsgq167F+woSZd/r4g657I3Jrc6Lrz7Q3V/OMqOeuKhMO01EZ
UvjfKnCrTef8x4FDsu1gJdgH/6KTYZ+tVX4M+UNjSua2q47sGAyKtFnGMa1RsVTO
IMTFkw/gL9fwl5p6uHKTs9uKb3F36zdLFPGpPZlNChUb9zryDhtoem0UIwCgkKL6
rijYU1Cmm/TqtxHOdagP0S0EAINNzM5FTX3avUfDiq66UnTQ4Rusa8DrtLYa+GhI
/Y5g1naDRhPrOZe3Yobrj+11X7pjL/lNCcdickNpybAXESoDkxh2W/YoZ7VxitQe
D/y5NRg4BPWbu+l4vG9AdBst63GAEaDQXOt3yWWXapYYbVp9DEwcD9LVnh1XFMMf
gVo+A/9Vs6QEO9fR01C7gQ4LhUqcAz24WpWnAuhmqtrBDxmTd1QUz+T/filHV4HC
adTNfeLsMmK5+EjRjY44mpaHNmT1IgyiouGhygssUh4fAP3+CD0267Z63ywmh94w
DAJWb4jwW7RnsQffRiPIpT/h98t9ShyxRWMpJnDJez3kd0d05rQlV2VzIFNvbm5l
bnJlaWWNoIDx3ZXNNAc29ubmVucmVppY2guY29tPohXBBMRAgAXBQI8YlVQBQsHCgME
AxUDAgMWAgECF4AACgkQOga48ERXxNZ63QCfWL0XSqYrSU2Lxvxif+tZDPUOLy4A
n1OO7xKI64NOvJu1f7Iu8gg7sOfJiEYEEBECAAYFAjxi9QMACgkQb/iQJ6CyDOwG
zwCgirDS/Dq1xIHwuhqI1Fl0aZ3xlowAnAsb5kA61rz3dyUUmCliV2pTvnhKuQEN
BDxiVVQQBACWXGF8xfSffPO/0tABP5Sn2DT0zhw+4jLyOk64VgoP8GiuJ8Xu+Fvu
```

Kzx++gveIOY4rk40Z92TILFunxd8bWYu35X1+cuSRq6yxRCuDvZkhmPpn7Jplzk7
hF/y2P4fJnwLhj9isLR8guMkl9YMiZw7AWs/AjWHBpRIntqBR+sRewADBQP+PwtB
y79R2Q0dG2oN8+fo62dNQsPK/+/3gzepH2VSTl7rYMxQsW24VZGOz0sIFWmP8NfF
TK3HX5yPrGBhrZOHeq4fnkTX3h8K/4O8N9TjuBHxQa//ppOxoI0g8AObT5LAMmvR
BPJUtb+MGIyeQWCivRaKh4UFbonXG/8RMJ/fWweIRgQYEQIABgUCPGJVVAAKCRA6
BrjwRFfE1g3dAJ95e30nhBUnAvwAcAT1rPELfkoiUACfcqpyfGFhVm/H7QtIQeoD
th6esU0=
=EKGL
-----END PGP PUBLIC KEY BLOCK-----

This is a sample 1024-bit public key (Wes's key, in fact). It doesn't look like a number, but looks can be deceiving. Every 4 characters actually represent 24 bits of data using a special encoding system designed to be email and Web-friendly. The astute observer will note that this means there's far more than 1024 bits of data here. It turns out that these types of public key blocks contain additional information besides the public key, such as certificates from others who have verified the authenticity of the key. Jason's digital signature is part of this key block, but its relatively useless because he loses his keys once a year on average.

How Cryptography Works

We're not going to get into the mathematics of cryptography. For that, you can read Bruce Schneier's excellent book, *Applied Cryptography*. We will, however, point out a few of the basic principles.

The keys used for public key cryptography are not as simple as a standard symmetrical key—which can have any value at all. Every public key cryptosystem relies on what is known as a "one way trapdoor function." The concept is pretty simple. Some math equations can be reversed. Knowing the result makes it easy to find the original values. With one-way functions, this doesn't work. For example: $x + 1 = y$ is a simple equation that can be easily reversed. If somebody tells you that y is 3, you know that x was 2. But $x + y = z$ is not so easy. If someone tells you that z is 3, you have no idea what x and y might have been (2 and 1? -345 and 348?).

The real equations used for cryptography are much more complex, but the principle is the same. A message gets encrypted using the recipient's public key. There's no way to reverse the process and get the original message from the ciphertext and the public key. The private key is the "trapdoor" that lets the intended recipient reverse the equation.

The most famous public-private key system is known as the RSA cryptosystem, named after its developers (Rivest, Shamir, and Adleman). The public key and private key are both derived from two large prime numbers. The product of multiplying those two large prime numbers together is known as the modulus. The modulus plays a critical part in both public key and the private key cryptosystems.

Technically, if you could factor (split) the modulus into the original two, very large, prime numbers, you could figure out the private key. Therefore, anyone who could factor the modulus could decrypt messages. It turns out that factoring a large modulus is a very difficult task. This is known as a "hard problem" because current mathematical techniques for solving the problem are very slow and inefficient. The relative security of the RSA algorithm is dependent on two elements: that factoring remains a hard problem and that no other method of attack is possible.

Security Considerations

Cryptographic systems that have stood the test of time are considered "strong" systems. For users of cryptography, the word "strong" has a practical meaning. It means that there is little to no chance of a secret message being decoded by anyone who does not possess the private key.

For cryptographers, "strong" has a much more specific meaning. It is not a word that is thrown around lightly. "Strong" means that no known methods exist to break the cryptosystem unless anything less than astronomical time scales or expenditures are permitted. It is never true to say that a cryptosystem is uncrackable because given a few billion years, the key to decrypt a message can certainly be found. So, "strong" generally means "To all intents and purposes uncrackable—with existing technology and knowledge."

When deciding if a cryptosystem is strong, cryptographers look at what's necessary in order to break the system. A cryptosystem can fail in three basic ways:

Brute force: New techniques in super computing and distributed computing have allowed unprecedented processing power to be harnessed and focused on key cracking. It was actually possible to figure out a 512-bit RSA key using this technique, but it took 5 years and hundreds of thousands of computer users. That said, encryption systems that were considered adequate a number of years ago (56-bits) can now be cracked in less than a day using a single computer system. This was not entirely unanticipated. On page 153 of *Applied Cryptography*, Bruce Schneier estimated that a system to find a 56-bit DES (*Data Encryption Standard*) key in 3.5 hours could be built for $1M in 1995. So, using a strong system and a big enough key is an essential defense against brute force.

The "hard problem" becomes "easy": Somebody might manage to figure out how to reverse a one-way function quickly using some nifty new math trick. This requires a mathematical or computer science breakthrough, and these are few and far between. Strong systems rely on hard problems that will hopefully remain hard for many years, if not forever.

The implementation is poor: This is the bane of most cryptography systems. Some logical oversight or a programming bug creates a fatal weakness in the system. For example, Netscape's 128-bit symmetrical key, used to provide security in their U.S.-released browser, was cracked in a far shorter time than expected. This was due to a poor key generation technique that was only capable of producing a tiny subset of the possible key range. By limiting the brute force search to the possible subset, a successful cryptanalysis was possible within a reasonable time period. Netscape quickly changed the implementation of their key generation technique.

Communication with "strong symmetric" cryptography can never be as secure as with "strong public key" cryptography for the simple reason that with symmetric cryptography, the decryption key must leave the site of the person who encrypted the communication so that the receiver can decrypt it. Once that copy has left that person's possession, he or she can never be absolutely sure what has become of it. With public key cryptography, the decryption key is the private key of the person who receives the communication. This never needs to leave that person's possession; its security and the secrecy of the communication can be, for all practical purposes, absolute. That is, of course, unless the cracker is an ESP master.

Making the Connection

Digital Rights Management: Encryption is the core technology needed to protect rights when information is outside of your direct control.

Determining Identity: Critical identification factors must be encrypted, otherwise hackers can use the information to impersonate valid users.

Preserving Privacy: Sensitive and personal information should always be encrypted, allowing only authorized users to access the data.

Virtual Private Networks: Encryption is used to protect information as it travels through an insecure network.

Storing Information: Some storage systems provide direct support for encrypting information.

Accessing Information: When transferring or accessing critical data, the entire communication process can be encrypted.

Backup Systems: If backups are not encrypted, the backup media can become a major security risk if lost or stolen.

Best Practices

The ideal form of cryptography is "strong public key" cryptography because it does not require the exchange of secret keys. Symmetric cryptography, whether for encryption or signatures, has no advantage over public key cryptography or similar security except for the two following considerations:

The keys used by public key cryptography need to be somewhat (6 to 30 times) longer than the equivalent symmetrical key for a given level of security. For instance (according to *Applied Cryptography* page 166) a 128-bit key for symmetrical cryptography is as secure as a 2304 bit key for public key cryptography.

Public key cryptographic algorithms are much more complicated than symmetrical algorithms. On the same computer, a public key algorithm will be much slower than a symmetric algorithm.

From a storage and network performance point of view, long keys are no problem. After all, even a 10,000-bit key takes up less space than an average email. As data, long keys are still pretty tiny.

The performance issue is a much more serious matter. In fact, for encryption of communications and data storage, pure public key cryptography is almost never used on its own. Encryption and decryption operations on relatively short messages would take an excessive amount of time. Instead, a digital envelope is used.

The concept behind a digital envelope is simple. The main message is encrypted using a randomly generated, one-use symmetric key (called a session key). Only the session key is encrypted using a public key cryptosystem. When the recipient gets the message, his private key will open the envelope, revealing the shared secret session key necessary to decrypt the rest of the message. In this way, public key encryption solves the problem of securely transmitting a shared secret key.

A digital envelope is a big win if the message being encrypted is larger than the size of the symmetric key. In almost every instance this will be true because most documents contain more data than a 1024-bit symmetric key. The slow-to-process public key system only has to handle a few bits. The faster symmetric system takes care of the heavy labor.

Final Thoughts

Encryption would be ideal if it only used a key-pair once. This would be feasible if a two-way real-time communication channel between the two computers existed. As more and more homes and offices are connected to the Internet via broadband, two-way real-time communications may replace the current method of message exchange. The "encrypt, send, and pray" model may soon be a thing of the past.

Imagine two computers, A and B, each with the ability to generate a key-pair and send the other their public keys. Now, communications can take place securely with these key-pairs and they can be erased at the end of a session. If the private keys can truly be deleted, then anyone intercepting the communications must find the key. This would involve a *very* lengthy search process. This scenario presents a major security improvement but is not the ultimate in cryptographic security. What if the goon squad breaks down the doors while the aforementioned communication session is in progress? Even a couple of goons could conceivably obtain the public key from the computer and decrypt the whole session up to that point.

To take security even further, imagine two radios, each being equipped with a single chip, which performs the encryption, decryption, and key-pair generation functions. Each chip in each radio is linked by an insecure two-way data connection. These chips are built in such a way that it is physically impossible to extract the private key from them.

This is quite an assuming example. It assumes that the chips use a cryptographically secure algorithm, that their key length is sufficiently long, that the key-pairs are generated from genuinely random numbers, and that the private key cannot possibly escape the chip. If all of this is true then two such chips in communication form a perfectly secure channel.

Today encryption is used transparently in many common digital transactions. Most online and offline financial exchanges use encryption at some level. From web-based e-commerce to the ATM in the corner grocery store, encryption thwarts thieves who hope to profit from tapping phone lines or network connections.

Chapter 27
Hiding Information: Cryptanalysis

The art and science of code breaking.

Technology Overview

Cryptanalysis is the exact opposite of cryptography. Specifically, it is the science of code breaking, decoding secrets, cracking authentication schemes, and destroying cryptographic protocols. You might be tempted into thinking that cryptanalysis is practiced exclusively by criminals and government code breakers. The truth is that cryptography has a tremendous need for cryptanalysis.

Cryptanalysis is a tool that all cryptographers use to test and prove their latest and greatest encryption schemes. The only way to determine the strength of a particular method of encryption is to try to break the key and decode the message. Companies that sell encryption systems, such as RSA, actively challenge cryptanalysts to crack their algorithms. The results can either help them fix a broken system, or prove the effectiveness of a good system.

The techniques that are used in cryptanalysis are called attacks. There are many different types of attacks; some specific in nature and others aimed at more general methods of code breaking.

How Cryptanalysis Works

Cryptanalytic attacks are broken down into six groups, organized by the type of information that the cryptanalyst has in his or her arsenal. The following categories of attack have been placed in order based on the quality of information available to the cracker, from least to most:

- A *ciphertext-only* attack is one in which the cryptanalyst obtains a sample of ciphertext without the plaintext associated with it. This data is relatively easy to obtain in many scenarios, but a successful ciphertext-only attack is generally difficult, and requires a very large ciphertext sample.

- A *known-plaintext* attack is one in which the cryptanalyst obtains a sample of ciphertext and the corresponding plaintext as well.

- A *chosen-plaintext* attack is one in which the cryptanalyst is able to choose a quantity of plaintext and then obtain the corresponding encrypted ciphertext.

- An *adaptive-chosen-plaintext* attack is a special case of chosen-plaintext attack in which the cryptanalyst is able to choose plaintext samples dynamically, and alter his or her choices based on the results of previous encryptions.

- A *chosen-ciphertext* attack is one in which a cryptanalyst may choose a piece of ciphertext and attempt to obtain the corresponding decrypted plaintext. This type of attack is generally most applicable to public-key cryptosystems.

- An *adaptive-chosen-ciphertext* is the adaptive version of the above attack. A cryptanalyst can mount an attack of this type in a scenario in which he has free use of a piece of decryption hardware, but is unable to extract the decryption key from it.

Using the arsenal of weapons listed above a cryptanalyst attempts to carry out his or her objective. The objective in all cases of cryptanalysis is the ability to decrypt new pieces of ciphertext using only the limited information available. The ultimate achievement for any cryptanalyst is to come up with a process for quickly and consistently extracting a secret key from the encryption system being attacked. If this is accomplished, the process can forever be used to decrypt any communications that use the cracked encryption algorithm.

Security Considerations

Cryptanalysis doesn't have any real problems associated with it, as it is a method used to solve problems. It is, however, important to discern between cryptanalysis, security in general, and social engineering. For example, if an intruder were determined to crack a cipher to gain access to a bank account, she might have several options available to her. One option is to purchase the amount of equipment necessary

Cryptanalytic Attacks

Illustration by SageSecure

1) I like cold **+** 🔑 **=** 0mxkj89qh377rgf7j

The first equation illustrates the basic process of encryption: a message is combined with a secret encryption key, resulting in ciphertext.

The goal of cryptanalysis is to figure out the secret key based on the information available.

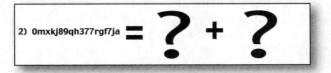

2) 0mxkj89qh377rgf7ja **= ? + ?**

This shows a ciphertext–only attack. Just the encrypted cipher is available to the cryptanalyst.

Successful cryptanalysis usually requires a large amount of ciphertext, a lot of time, or an exploitable weakness in the encryption system.

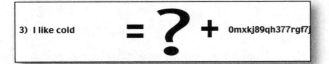

3) I like cold **= ? +** 0mxkj89qh377rgf7j

Line three depicts a known plaintext attack. This attack can be used when a cryptanlyst has the encrypted cipher and the original plain text message.

There are variations on these attacks where the cryptanalyst can control the choice of plaintext

■ **Figure 27-1**

to perform cryptanalysis on the cipher. This equipment may cost millions of dollars and then still take several months or years to perform a successful cryptanalysis.

If the bank account that is under attack only contains a million dollars, it is clearly not worth the money that it would take to crack the encryption. The key could, however, be successfully obtained by bribing an individual or committing extortion. In this case, the difficulty of cracking the encryption is irrelevant; the intruder has found another way around the problem. This is why cryptography should never be confused with security. Thorough security is not obtained through cryptography or any of its counterparts; it requires a comprehensive strategy that protects the "doors" and the "windows" of your network.

Best Practices

Are you good at the cryptograms in the Sunday paper? Do you want to be a top-secret government code breaker? Sorry, bucko. You're playing stickball and they're rigging the World Series. Did you celebrate your sixteenth birthday by defending your doctoral dissertation on field theory and quantum computing? Then men in black will be baking your seventeenth birthday cake, because you'll never be seeing your family again. Not that they were your natural parents anyhow—anyone that smart is the result of secret government genetics experiments.

Seriously though, cryptology is the domain of mathematics and physics PhDs. It requires a massive amount of computer and number theory to understand how most modern cryptosystems work. Actually analyzing these systems for weaknesses is a completely different level of pain and only a handful of minds are up to the task.

Final Thoughts

Cryptographers are among the smartest people in the world. Be glad that a few of them are actually in the public spotlight. If it weren't for the likes of Diffie, Schneier, and Rivest, there'd be no concept of individual encryption. Without their efforts at creating and analyzing cryptosystems, many aspects of networks and communications would be impossible to secure.

Chapter 28
Hiding Information:
Steganography

Hiding one piece of information inside of another.

Technology Overview

Sometimes, the fact that a communication is taking place needs to be as secret as the contents of the communication. What if the CEOs of two companies suddenly begin frequent communications? In the business world, this might be enough to verify rumors of a merger. Under certain circumstances scrambling a message isn't enough: the existence as well as the content of the communication needs to be hidden.

Many ways to communicate invisibly exist. Cloak and dagger movies have illustrated this concept with their spy characters relying upon "dead drops." A dead drop is a pre-arranged point where information is left behind. After a drop is made, a signal is given. A typical signal could be an ad in the paper or a mark made on a mailbox. With a positive signal, the intended recipient can pick up the dropped information. Using a dead drop, the involved parties do not need to meet and it will be difficult to know that a communication has occurred.

The problem with dead drops is that they're very inconvenient, time consuming, and require physical access to the drop zone. A better solution is for both parties to "anonymously communicate" using pseudonyms and a public forum. Classified ads, bulletin boards and online forums can provide this type of anonymity.

Even if the sender and recipient remain anonymous, messages discussing merger details may attract attention in a public forum. It might even be possible to determine the senders' identities by the contents of the messages. Encrypting the messages could keep the communication details secret, but anybody watching will know that something important has been said. Once again, attention has been attracted. With some effort, the anonymity could ultimately be compromised.

One solution is to hide the encrypted communication in an otherwise innocent message. This concept is known as steganography. Literally, steganography means "covered writing." It refers to any technique used to hide one message inside of another where the "innocent" message is referred to as the "cover." There are three basic categories of steganographic techniques: media-specific, pattern, and signal/noise (yes, we made these names up, for lack of something official).

Steganography and Cryptography

Steganography is often confused with cryptography. Steganography protects a message by hiding it; cryptography protects a message by scrambling it. The two technologies are complementary, however; often used in combination for added secrecy—a cryptographically scrambled message can be hidden using steganographic techniques the same way a normal message can.

How Steganography Works

Media-specific steganography relies on physical properties of a communications media. A good example of this is invisible ink. Ever write with invisible ink when you were a kid? If so, you were a budding steganographer! The ink remains hidden unless heated, chemically treated, or viewed under special lighting. Some other interesting media-specific techniques are the following:

- **Painting over:** This techique entails covering one painting with another. A message could be written on the canvas and then painted over. Upon reaching its destination, the paint could be removed. An early variation on this concept was used in ancient Greece. From the Web site of Neil Johnson, a prominent steganography researcher: "One of the first documents describing steganography is from the Histories of Herodotus. In ancient Greece, text was written on wax covered tablets. In one story Demeratus wanted to notify Sparta that Xerxes intended to invade Greece. To avoid capture, he scraped the wax off of the tablets and wrote a message on the underlying wood. He then covered the tablets with wax again. The tablets appeared to be blank and unused so they passed inspection by sentries without question." An even more bizarre variant: A messenger would have a secret message written on his scalp. When his hair grew back, the message would be invisible. Upon reaching his destination, he'd shave his head again, revealing the message. A little bit time consuming, but it gets points for creativity (Bruce Norman, *Secret Warfare: The Battle of Codes and Ciphers*).

- **Microprinting:** Shrinking the text of a message such that it appears to be a speck, or period at the end of a sentence. A normal looking letter could contain many different microprinted messages. Note that retyping, photocopying, or otherwise manipulating the original sheet would easily destroy the message. U.S. currency uses microprinting as a security technique because of this very property.

- **DNA:** The building blocks of DNA are the four proteins: Adenine, Cytosine, Thymine, and Guanine, which are represented by the letters A, C, T, and G. This four-letter alphabet can be used to encode a message in the same way that 0s and 1s can represent data. An actual chain of DNA can be assembled from the message using currently available biotech tools. The microscopic DNA strand would be easy to send in an envelope with an innocuous letter. It's a bit James Bondish, but very possible. Likewise, the same principle could be applied to other types of molecules, encoding messages in the design of the molecule.

These media-specific techniques have a similar problem to the dead drop technique—something needs to be physically moved from one side to the other. These techniques also require a significant amount of effort and special equipment to produce.

Another style of message hiding known as *pattern steganography* encodes the secret message by creating patterns in the cover media. Often, this involves significantly adjusting, or completely creating the cover media based on the message. For example, the following message was intercepted from a Nazi spy during the Second World War:

Apparently neutral's protest is thoroughly discounted and ignored. Isman hard hit. Blockade issue affects pretext for embargo on by-products, ejecting suets and vegetable oils.

If we eliminate all but the second letter of each word, we get

pershing sails from ny june 1 (the 1 is the "i" in "oil")

Pattern steganography is closely related to code book cryptography, where a phrase such as "the eagle flies at midnight" might mean "attack tomorrow." The difference is that hiding a message in text is more flexible, you can say things that are not pre-arranged. The downside is that it is very difficult to create meaningful text surrounding the message that doesn't look incredibly forced or strange. Some interesting techniques for generating secret keys include the following:

- **Text Generators:** These are programs that can hide messages in cover text that is automatically generated. The text might be random, or it might look like poetry, art criticisms, baseball commentary, and etcetera. Peter Wayner's book *Disappearing Cryptography* gives an explanation of how these programs work, along with a sample program.

- **TCP/IP Sequence Steganography:** uses properties of *Transmission Control Protocol/Internet Protocol* (TCP/IP) packets (sequence numbers, acknowledgements) to transmit messages. A normal-looking web browser/web server session might actually be a cover for transmitting a secret message. This requires highly customized client and server software.

- **Blank Spaces:** Text files can contain blank spaces at the end of each line. These spaces are not visible to the casual observer. A message can be encoded by varying the number of spaces at each line ending. Many current email systems do this arbitrarily when adjusting the text for readability.

- **Textual Rephrasing:** Possibly the most interesting of all, this technique takes normal text (a typical email message between friends) and hides a message by subtly adjusting the phrasing of the cover. The result looks completely natural—the secret message is potentially impossible to detect without comparing many samples of the author's writing style to the manipulated message.

The most active area of steganography research is signal/noise steganography. Noise is a natural byproduct of electronically transmitting visual or aural information. Sound and light travel in infinitely complex waves. Noise is produced when these waves are captured and reproduced. The properties and limitations of the capturing device influence the way in which the waves are recorded. These influences distort the signal, creating noise. Signal/noise steganography attempts to hide information in this noise.

Steganography software looks for noise in digital media and replaces it with useful information. A digital media file is nothing more than a large list of 0s and 1s. The software determines which of these 0s and 1s correspond to redundant or irrelevant details. For example, the software might identify details in an image that are too fine for the human eye to see and flag the corresponding 0s and 1s as irrelevant noise. Later the flagged 0s and 1s can be replaced by the secret message.

The advantage to signal/noise steganography is that it can be simply performed on a modern computer using easily available software. Images, music, and video files are all excellent noise sources. Messages of significant length can be hidden in a series of images or a single audio file. The resulting image or song is perceived as identical to the original— no visible or audible distortion is present.

For example, the CEOs from the previous example could encrypt their merger discussions and hide the data in photographic images of dogs. They could then set up anonymous accounts in an online forum about dog grooming and post their photographs. To the casual observer, the messages are just the proud boastings of dog fanatics. Nobody would even think about analyzing the message or pictures closely.

The problem with signal/noise steganography is that a slight manipulation of the data could result in the destruction of hidden messages. For example, slicing up an image or compressing the image would change the pattern of noise, and destroy

Steganography

Illustration by SageSecure

A secret image is hidden in the original image

The change in noise pattern indicates the possibility of hidden information.

original image noise

noise with hidden image

Figure 28-1

most hidden messages. It is possible to hide messages in a robust manner that can withstand manipulation, but this requires a small message and a lot of data to hide it in or visible/audible changes to the cover data.

Security Considerations

Steganography seems like the sort of tool only spies would use. Recent terrorism events have caused some to conclude that terrorists are using steganography to coordinate efforts over public forums. This hasn't helped build credibility as a viable business tool. The following are some other potential reasons for not using steganography:

Other Equally Effective Techniques: In reality, many ways to communicate in a covert manner without resorting to steganography are possible. Robert J. Bagnall wrote an excellent article for the *System Administration Networking Security* (SANS) Institute (www.sans.org/rr/steg/myth.php) on why terrorists are probably not using steganography as a primary communication tool. This article illustrates dozens of simple techniques that are easily available to anyone with a laptop. These include secure wireless communications, communication over free mail/forum/chat/web systems, anonymous Wi-Fi access, hiding data on mp3 players, data sharing through peer-to-peer networks, and passing information on tiny flash cards that can be concealed in a pack of gum. Many of these techniques are as undetectable as steganography and require less technical knowledge.

Easy to break: Most steganographic techniques are sensitive to manipulation of the cover. In some cases breaking potential steganographic messages might be desirable. For example, a major public service like Ebay might wish to prevent people from creating fake auctions as covers for steganographic communications (embedded in pictures of the alleged sale item). A number of basic techniques can be used to reliably eliminate steganographic messages:

- **Visible manipulation:** Simply resizing an image would be enough to break most types of underlying steganographic messages. Also, slicing up an image into four or eight pieces and then displaying the pieces side-by-side would force the recipient to go through an extra step of reassembling the image. This in itself might destroy the message due to subtle changes in the alignment. On the other hand, automated programs designed to detect steganographic messages might be foiled by this approach, whereas a human would know to download the pieces and re-assemble them later.

- **Overwriting the message:** Adding a digital watermark to a file can change the noise pattern of the cover, essentially overwriting the first message with the watermark message.

- **Compression:** Many audio, video, and image files are stored in a compressed format. Uncompressing and recompressing the file will distort the noise in which a message might be hidden.

Steganalysis: Some people just love learning secrets. Although most secret lovers are in primary school, there are a few who work in various three-letter U.S. government agencies. These agencies have spent time and effort learning how to intercept and decode communications. This also includes detecting communications hidden by steganography. Steganalysis is an entire field of research devoted to creating automated tools for detecting the usage of steganography. Most steganalysis researchers are working with a government agency, although some are simply in it for the challenge.

The difficulty of steganalysis depends on how much you know ahead of time. There are a few things you need to know, or assume, in order to detect a hidden message:

- **That a hidden message exists**   Obviously at least one hidden message is out there, but what about the one you're looking for?
- **The technique/media used to hide the message**   Is it hidden in digital noise, a letter, or on the scalp of the hairy guy in seat 3A?
- **Where to look for the message**   Is it on a web page, sent in email, over instant messaging, in snail mail, or in a newspaper?

Without this basic information, finding a hidden message is shooting in the dark. With all three bits of information, there may be a decent chance of success. The success rate ultimately depends on the ability to test every possible cover. For many techniques and message locations this is either impractical or impossible. Imagine the look on the airline security agent's face if he or she were asked to check every passenger's scalp for hidden messages.

Most steganalysis research focuses on digital data as a cover medium. A few years ago, researchers showed that existing steganography tools left "signatures"—detectable patterns visible in the combined data. Some automated tools were developed that could scan large numbers of images for these signatures. This was a big breakthrough because it proved the possibility of automatic steganography detection. It was bittersweet though, because slightly modifying any of the steganography tools would change the resulting signature and thus evade the automated steganalysis systems.

The new goal is to be able to achieve "blind steganalysis," which is the ability to recognize steganography in digital data without knowing anything about the software tool used to encode the message. A successful blind steganalysis tool would not rely upon signatures. Instead, it might use statistical analysis to spot anomalies that indicate the presence of a hidden message. The current state of blind steganalysis is not well known, as much of the research is confidential.

Making the Connection

Cryptography: As described earlier in the sidebar of this section, encryption and steganography can be used together for added security.

Storing Information: Quite simply, steganography is a subset of information storage. Steganography is used in combination with many types of storage technologies and techniques to maximize its effectiveness.

Best Practices

One of the major commercial uses for signal/noise steganography is digital watermarking. With digital watermarking, the message being embedded is usually very small. The purpose is not to communicate secretly; rather it's to secretly mark commercially valuable media to detect theft. As a result, plenty of effort is put into making the watermark as robust as possible, capable of surviving extreme manipulation. Tools for automatically detecting the watermark are also necessary. There is more information on this topic in the digital watermarking section of the book.

Another area of development for signal/noise steganography is steganographic file systems. The idea is to fill up a disk with cover files. The real data is then steganographically stored in the cover files. This requires a large amount of disk space, as only a portion of each cover file is noisy enough for hiding data. Another variation is to fill the disk drive up with random data (noise) and then to write the files onto the drive in a random-seeming manner. One file might be spread out across the entire drive. Only the file's owner has the code necessary to reconstruct the original file.

What's the point of a steganographic file system? Why not just encrypt files for protection? In the words of the creators of StegFS, a steganographic file system for Linux (taken from their web site):

> *"Assuming correctly implemented encryption software is used as designed, and cryptanalysis remains infeasible, an attacker can still chose among various tactics to enforce access to encrypted file systems. Brief physical access to a computer is, for instance, sufficient to install additional software or hardware that allow an attacker to reconstruct encryption keys at a distance. (UHF burst transmitters that can be installed by non-experts inside any PC keyboard within 10-12 minutes are now commercially available, as are eavesdropping drivers that will covertly transmit keystrokes and secret keys via network links.) An entirely different class of tactics focuses on the key holders, who can be threatened with sanctions as long as there remains undecryptable ciphertext on their storage device . . .*
>
> *. . . Steganographic file systems are designed to give a high degree of protection against compulsion to disclose their contents. A user who knows the password for a set of files can access it. Attackers without this knowledge cannot gain any information as to whether the file exists or not, even if they have full access to the hardware and software.*
>
> *They aim to provide a secure file system where the risk of users being forced to reveal their keys or other private data is diminished by allowing the users to deny believably that any further encrypted data is located on the disk."*

Even more interesting is the combination of steganographic file systems and peer-to-peer networking. Instead of hiding data on a single hard drive, the vast repository of audio and video data available on various peer-to-peer networks can be used as a source of cover files.

Final Thoughts

Steganography is an interesting technology that has many potential uses. Unfortunately it has garnered a bad reputation in the last five years because of its association with evildoers. The truth is that steganography is a security tool like any other. Myriads of legitimate situations exist where it is used to add an extra layer of privacy to communications. Conversely, many people who are hiding information are doing so because they are breaking the law, or have malintent. In either case, steganography is an extremely effective way to communicate privately and ensure that no third parties are even aware a communication took place.

X

Accessing Information

Summary

We can hide and store information, but how do we actually use it? This chapter talks about tools for remotely connecting to and/or managing information sources and services in a secure manner. Secure storage is great, but what happens when you need to access that data, or move it from one machine to another? How do you do this securely?

Key Points

- The ability to access information from anywhere is one of the most powerful benefits of network computing, but the lack of universal standards for computing makes it hard to capitalize on the opportunity.
- Different operating system and networking architectures support different approaches to accessing information. Visually focused operating systems such as Windows and MacOS emphasize usability over flexibility. More complex systems like Linux use a textual interface for efficiency.
- It's important to match the right method of accessing information to a particular task. Some tasks are best accomplished within a visual environment (creative design) while others benefit from the flexibility of a textual interface (servers). A mismatch can lead to security problems.

Connecting the Chapters

These chapters describe the major technologies for data access across networks:

- **Chapter 29, "Client-Server Architecture,"** examines the networking concept that allows workstations and servers to work together to produce an efficient environment for running applications.
- **Chapter 30, "Internet Services,"** looks at the most commonly used systems for accessing data over the Internet.
- **Chapter 31, "Remote Access,"** explains how to securely get at applications and data on remote systems.
- **Chapter 32, "Peer-to-Peer,"** covers the world of connecting individual, unrelated, computers directly as if they were on the same network.

Introduction to Accessing Information

Information is a funny thing. It is not physically tangible, yet it can be extremely valuable. When information is unwanted it sits around idly, and untouched. But when a piece of information is sought after, getting at it quickly and efficiently is the only thing that matters.

Accessing information has been the most profound argument for computerizing the world. With computers, people can access needed information more quickly and easily than ever before. Computers contain potentially endless storage capacity, software tools to catalogue and organize information, and high-powered search engines that speed through infinite bytes of data to retrieve specific requests.

Despite the advancements made in information technology, a lack of standardization has undermined the full potential computers bring to accessing information. Different combinations of computer systems within organizations make accessing data surprisingly difficult. Systems that aren't innately compatible require other, additional systems to translate information and integrate resources. Extra systems create additional points of failure and aren't always cost-effective to implement or maintain. This reduces the efficiency of access to information in many real-world situations.

Computers may, in time, evolve to a more consistent standard, but for now there are still many choices and no consensus as to which platform is the best all around. Computers are a result of the combination of hardware and software; the type of hardware chosen will affect the software that can be used, and ultimately what function the computer will have. In the business world most organizations perceive their computing needs as very specialized and unique. As a result the choices are seemingly infinite and very confusing.

The Burden of Choice

Choosing a computing platform means choosing an operating system. But which system is right? The easy answer is to go with the flow and use Microsoft Windows. But there's been a lot of press on using Linux or other forms of UNIX as an alternative to Windows. And every office has some guy or girl who won't stop raving about their Mac—about how the world would be a better place with peace and love for everyone if more people would just use Macs.

Zealotry and evangelism aside, the differences between operating systems ultimately comes down to different approaches to accessing information. Most operating systems fall into one of three predominant models of information access:

Textual: Information is represented in hierarchical lists or tables. This can be very powerful and flexible, but it's completely unintuitive. UNIX-based systems (such as Linux, SunOS/Solaris, and BSD), DOS and various specialized systems use this approach. Even when a UNIX system offers a graphical interface, it's almost always less powerful and therefore secondary to the textual interface.

Visual: Information is represented using visual metaphors. Windows can be used to group related items or controls. Trees can represent hierarchical information. Icons provide mnemonic cues when looking for a particular piece of data or application. Applications are entirely controlled using visual representations. This is very intuitive, but the amount of control over the information is limited by the boundaries of the visual metaphor. The Apple Mac system (prior to OS X) epitomized this approach. The latest versions of Windows are now centered on visual access.

Mixed: Textual and visual systems were originally mixed together as a result of evolution. UNIX and DOS vendors realized the value of a visual metaphor and attempted to layer a visual interface over their existing textual system. Windows 3.x was one example of an early attempt. It had less flexibility than DOS, but advanced users could always revert back to a textual mode when necessary. Later versions of Windows have maintained this duality, although there are now many aspects of a Windows system that can only be controlled by visual tools. Ironically, it is the Apple Mac that has been the first to properly mix a visual and textual interface with the OS X system. They have designed a hybrid system (based partly on UNIX, partly on the old Mac OS) that tries to ensure that both approaches are equally usable and powerful.

Figuring out which information access approach works for your business is tough. Chances are, different approaches are needed in different parts of the organization. Creative people tend to find highly visual systems to be the most effective —thus the popularity of Macs among artists, designers and academics. Technical people tend to prefer the power and flexibility of textual systems. UNIX-based systems are often used in research, financial environments, and as a back-end for data intensive applications. Mixed systems such as Windows are used everywhere else, allowing some degree of choice in balancing power and ease of use.

Of course, with the rapid pace of technology these generalizations are constantly being challenged. The newest UNIX systems, including Mac OS X and Linux, are showing surprising aptitude in functioning as flexible business platforms. The reputation of UNIX is extreme power, scalability, and stability. This is why it is often associated with servers, instead of workstations. But many people today have adapted versions of UNIX to behave more "normally" and function as a workstation. Graphical interfaces have been laid on top of the traditional command line environment. This has made different "flavors" of UNIX more reasonable to use for the average Windows enthusiast. Even so, with the exception of OS X, the look and feel of a windowed environment on the UNIX platform is little more than sheet metal on a tractor-trailer. It may look "purty," but it's the diesel engine under the hood that gets the job done. And like any good truck driver knows, life on the road is easier when you're a good mechanic.

Textual vs. Visual Access: UNIX and Windows

The first time an avid Windows or Mac user encounters a traditional UNIX-based system, frustration is the inevitable outcome.

There is no mouse, no graphics, and a bunch white text on a black background. The interface between the user and the operating system occurs through a program known as a shell. The shell allows the user to issue commands to the operating system by typing them in at a prompt. Shell prompts are the core interface to any UNIX operating system.

To complicate matters, there are many different types of shells. A UNIX shell, much like the desktop in Windows, provides a direct interface with the operating system. Each shell has certain unique qualities. No shell is "perfect" or "intuitive" to a novice user, mostly because UNIX is not perfect or even remotely intuitive. Over time, most users learn to prefer one shell in particular and will go to great lengths to avoid using systems that don't support their "shell-of-choice."

UNIX forces the user to become familiar with the intricate details of a computer system since UNIX shells require the user to know many commands to accomplish tasks. Furthermore, access controls infiltrate every aspect of the operating system, even when a network connection is not established. Applications cannot be run, and files cannot be accessed without the correct user privileges. There are no convenient menus with command choices to click on with a mouse. There is no "dragging and dropping" when it is time to copy a file. Instead, the copy command needs to be typed, along with the path of the source file and the path of the destination. Specific knowledge requirements like these create a learning curve steep enough to make glacier climbing look like geriatrics physical therapy. A lot of studying and practice is needed to become proficient in a UNIX environment.

Graphical operating systems like Windows eliminated the need for a strong background in computer science. This was accomplished by implementing a complete and fairly intuitive graphical, menu-driven interface. Through a series of "windows," files and applications present themselves in an easy to understand format. This allows beginners to quickly move along the learning curve, becoming proficient enough to accomplish a variety of tasks on a computer.

With the goal of ease of use in mind, Windows has been heavily marketed to the home and casual business user. Until recently, it was never seen as a platform of choice for serious computing needs such as serving files, applications or databases. When network administrators needed to set up extremely secure and robust environments UNIX was generally the system of choice. After all, these were the people that really understood how computers worked and were able to take advantage all UNIX could offer.

As the demand for computer technology began to skyrocket, so did the demand for robust server side applications that would be as easy to manage as a Windows desktop. As a result, everything from web servers to databases were redesigned so "anybody" could use them. The people had spoken, and they said, "We want to run and manage scalable server applications too!"

Initially, the release of Windows-based high-end server products was not taken seriously by computing's elite. What soon took the cynics by surprise was that these new GUI-driven server products were actually being purchased. Instant customers were found in small to mid-sized businesses that couldn't afford to properly administrate UNIXsystems but still had significant data access needs. Suddenly, the ability to serve and access information was available to everybody… perhaps a bit too soon.

The advantage to the steep learning curve of UNIX is that the average UNIX administrator has a very solid understanding of networking and information security. That sort of background is developed as a side effect of the training necessary to become proficient in the UNIX environment. As a result, UNIX administrators understand the risks and issues involved in providing any type of network service.

Windows administrators generally don't have this depth of background. It's not that they're incapable of learning it—it's just that many of the underlying details have been hidden by the pretty graphical interface. Few Windows administrators have ever needed to peruse a program's source code to find out how a critical application works. It is uncommon for Windows application vendors to make the source code available at all. On the other hand, most skilled UNIX administrators reference source code all the time In fact, UNIX applications frequently are distributed as source code. The administrator needs to compile (build) the program before it can be installed on the system.

What does this all mean? It means that most small to medium-sized businesses are relying upon technology that appears easy to use, but really isn't. They lack the in-house expertise to make the correct decisions about how to best provide the necessary services, or how to ensure that the information they need will be constantly available. It's true that what used to require a large, climate controlled room and a dozen skilled technicians can now sit in a closet under a desk and be activated with the push of a button. You can also start a helicopter by pushing a button, but nobody would dare fly a helicopter without extensive training. People should feel the same way about running network servers, but they don't (probably because a server failure doesn't send the administrator plummeting to a fiery death).

Access Bold As Love

UNIX was designed to work in scaled networked environments; as a result it was also designed with security and remote access in mind. UNIX servers and workstations can all be managed from a remote location as easily as from the actual keyboard attached to the system. This is accomplished not only through remote access tools, but also through the simple power of the shell interface. Since most UNIX applications have been designed to be managed through a command line shell, any basic terminal can provide the necessary interface to redesign a website, or directly manipulate information in a database. All an administrator needs is the remote machine's correct IP address and the proper login credentials. With UNIX, an administrator can jump from system to system, making changes without leaving the comfort of their remote access terminal. Systems can even be rebooted or shutdown from halfway across the world, simply by typing the right command at the shell prompt.

To remotely manage applications that rely on a complex visual interface, that interface must be viewable from a remote location. This is far more complicated a task than providing command-line access. Graphical remote access systems make trade-offs between flexibility, performance, complexity and resource usage. Very flexible systems tend to require a lot of bandwidth to ensure decent performance. More complex software can minimize the use of bandwidth and processing resources, but the added complexity increases the potential for security problems. Flaws in complex remote access tools have long provided hackers with an easy entry into targeted systems.

Final Thoughts

New techniques for accessing information across disparate systems are constantly evolving. Much of the Internet's value has come from those tools that enable the exchange of data in a platform-independent manner (email, the web, file transfers, newsgroups). The issue of a visual versus textual interface is rapidly becoming moot for the average user.

Unfortunately, the servers that enable information access are still complex machines. Accessing and controlling these machines is difficult, and best done through powerful textual interfaces. Eventually the visual interfaces for servers may offer this level of power, but until then, the skilled administrator will find a friend in UNIX.

Chapter 29
Accessing Information: Client-Server Architecture

A current method for networking systems that allows workstations and servers to work together, producing an efficient environment for running powerful applications.

Technology Overview

Most people who work in any modernized business environment have used a client-server application. They may have even heard the term "client-server" thrown around by the MIS team. But few people actually understand how they get email, or how their web browser works. The best part of client-server is that they don't need to know.

A server is simply a computer like the one sitting on your desk. Generally, however, a server has hardware that is a bit more specialized. It is designed with the following goals in mind:

- Staying powered on for long periods of time
- Storing massive amounts of data on large hard drives
- Serving (providing) that data to many users at the same time

- Communicating with other servers as well as many clients
- Making plenty of seemingly unnecessary noise.

To achieve these goals a server uses advanced versions of the same hardware found on a desktop computer. More importantly, a server uses specialized software, commonly referred to as server applications or services. This software, much like server hardware, is designed with the following goals:

- Staying up without crashing for long periods of time (stability)
- Organizing and storing massive amounts of data
- Retrieving stored data quickly and accurately and delivering it to users on the network
- Communicating with client versions of the software and other server applications
- Being extremely expensive, more so than one could imagine.

The terminology becomes tricky when people refer to servers. A common line one may hear is, "The server is down." Well, what does that really mean? Is the physical server itself malfunctioning or has the server software crashed? It can also be confusing that one physical server may run a multitude of server applications (services). The services running on any individual server machine may not be related in any way. It's possible for a single server application to crash without affecting any of the other services running on a server machine. Awareness of these points can help an individual user to ask the right questions when the hear, "The server is down."

A client is any computer that connects to a server or any software program that connects to a server application. Think of it as a window into the world of the server. Internet Explorer is an example of a client that lets somebody connect to a web server and retrieve data. Microsoft Outlook is a client that connects to a mail server.

The Rise and Fall of the Titans

In the early days of computing, if a business needed computing power it used a mainframe system. Mainframe architecture works quite differently than client server architecture. A mainframe system has a very powerful back end computer that does all of the work. These systems were behemoths that often filled one or more rooms with flashing lights, whirring tape storage systems, and techies who physically loaded tape reels onto the mainframe.

Access to a mainframe system is provided through a terminal, which is little more than a monitor with a special cable that connects it to the mainframe computer. It does nothing except provide a window, or looking glass, to the mainframe computer. Not only does the mainframe process the data, but the application which the user interfaces with is run off of the mainframe's processor and hard drive. Mainframe terminals don't provide any resources whatsoever, which is why they were affectionately called "dumb terminals." As more people use dumb terminals simultaneously, the overall performance of the entire system degrades. Everyone is pulling resources from the same resource pool.

When the Personal Computer hit the scene in the mid-1980's, it had enough power to do some of its own processing. Mainframes could let the PC handle the application processing work. As a result, mainframes evolved into application and file servers. The PC loaded the entire application and data from the mainframe and then ran the application using its own processor. This dramatically lessened the processing load on mainframes and allowed a single mainframe to handle many more end-users simultaneously. However, the mainframe architecture was ultimately unable to support the use of graphical user interfaces. Eventually, organizations began to phase out mainframes from networked environments (see Figure 29-1).

Chapter 29
Accessing
Information:
Client-Server
Architecture

It was not long before stand-alone PCs were networked directly to each other. The original PC networks were based on file sharing architectures, where the server downloaded files from the shared location to the PC's desktop environment. This basic file sharing concept works if shared resource usage remains low and the volume of data being transferred is minimal. In the 1990s, the number of users connected to the average network was growing beyond the capacity of the file sharing model. At the same time, *graphical user interfaces* (GUIs) became popular. These required significantly more overhead and were further taxing the file-sharing system. This was the beginning of the end of the mainframe architecture. As a result of the limitations of file sharing architectures, the client/server architecture emerged.

The Evolution of Client/Server

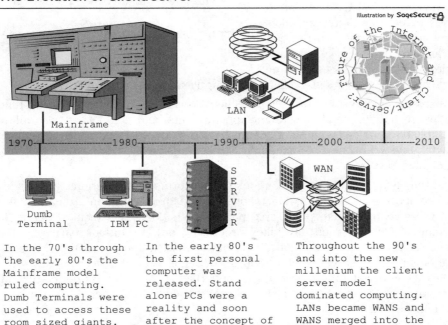

Illustration by SageSecure

LAN

1970 — 1980 — 1990 — 2000 — 2010

Mainframe

WAN

Dumb Terminal   IBM PC   SERVER

In the 70's through the early 80's the Mainframe model ruled computing. Dumb Terminals were used to access these room sized giants.

In the early 80's the first personal computer was released. Stand alone PCs were a reality and soon after the concept of the server was born.

Throughout the 90's and into the new millenium the client server model dominated computing. LANs became WANS and WANS merged into the Internet.

■ Figure 29-1

The Client Server Application Model

Client/server applications split up the work that needs to be handled in a very unique way. These applications have been designed to take advantage of the fact that workstations (client PCs) are very capable machines. In fact, the capability of these machines may, in certain ways, supersede the abilities of the server itself.

Today's client/server applications are designed to take advantage of the enhanced processing power of the workstations. Instead of letting the server do all the work, the applications are designed to let the workstations do most of the work. Of course, this requires the client side of the application to be installed on the workstation. This installation process takes time, and also takes up resources on the workstation such as hard drive space, memory, and processing power. Let's examine a very commonly implemented client/server application as an example.

Shouldn't the Server be Faster?

When people think of servers that operate in client/server environments they assume that they must be some pretty powerful machines. It is ironic that in fact, clients may have more gusto than the servers themselves. *Central processing units*, or CPUs, have become so inexpensive and have increased in speed so rapidly that client workstations are often replaced with great frequency. In addition, it is much easier to replace a client workstation n which one user relies than it is to replace a server on which hundreds or even thousands of users may rely. As a result, it is not uncommon to see an office with clients that have more processing power than the servers can provide.

Microsoft's Outlook is a prototypical client/server application. Its primary functions are to handle email, scheduling/appointments, and contacts. It also allows group collaboration on these functions so all users in an office can see the "company contacts" or view anyone else's schedule. The server side of the application is known as Exchange Server.

On the client side, Microsoft Outlook can perform many functions. It will allow you to compose email, check contacts, or schedule an appointment, to name a few. All of these features take processing power when they are executed. Even the launching of Outlook each day takes processing power. Outlook harnesses this power from the processor and hard drive of the workstationon which it resides. In fact, it only uses the server when it requests or sends a specific piece of data. When an email is composed, for example, it isn't until the "send" button is pressed that any interaction with the server exists. Only at that moment will the server be contacted and the email sent.

An industry term for a client that is robust enough to run applications that requires a lot of workstation power, is a *Fat Client*. Fat client software relies heavily on the specification of the workstation, and only pulls occasionally from the resources of the server. As a result, the server is able to handle many requests from many users, each one for a small bit of resources. This entire process within the realm of client/server is known as two-tier architecture.

The two-tiered application model introduced the *relational database management system* (RDBMS) as an alternative to the file server model of accessing data. Databases enable a user's data requests to be answered directly with greater speed and efficiency. The database response contains just the information the user needs. This generates significantly less network traffic. When compared with the previous method of sending an entire data file, databases consume significantly less network resources.

How Client/Server Applications Work

One of the most powerful concepts in networking is the "layer model." The idea is that different types of

The Thick and Thin, Application Model Wars

Thick client/server applications have clearly been the leading contenders in the business world for the last decade. Mainframes are dead, but some of the powers that be would like to see a shift back to that computing model. While the mainframe architecture may have no place in the modern world of client/server, its data storage concept is still very relevant. Specifically, some very large database companies believe that their massively scalable database engines are capable of serving whole applications. Instead of a workstation needing an installed application, all it would use is a thin client, such as Web browser, to access a complete application. The database, in combination with a web server, would provide all the functionality, and the user would never need to install any large application files on his or her machine. The question remains whether this is an improvement on the model for everyone, or just for the people whose stock price would rise as a result.

network tasks are split into layers. Each layer relies on the layers beneath them, the lowest of which deals with the physical connections and the raw data (such as Ethernet). The next levels up make sure data gets to where it needs to go (such as TCP/IP). The application layer sits on the top and just assumes that data will safely get to the desired destination.

An application sees a network as a socket; an apt metaphor as the application treats network data much the same way a lamp treats electricity. Once an application plugs into a socket, it either listens for data or sends data.

Sockets are at the heart of the client/server model. When a client wants to connect to a server, it opens up a network socket to the server. Usually the only information the client needs is the IP address and the port number; the operating system's network software takes care of the rest. Once the socket is open the client can send commands across to the server. If the server has a response, it can communicate using the client's existing socket or open a new socket back to the client. Some systems keep sockets open (such as chat tools); others open and close sockets for each communication (such as the Web).

Many different companies try to market their own client and server applications, all which server the same purpose. For example, the World Wide Web is a client/server system. Many different types of web clients (browsers) and many different types of web servers are available. A user with one vendor's browser might be trying to connect to a web server created by a different vendor. This exchange will work if all the clients and servers for a particular type of application use the same commands when communicating. The set of commands a server will accept from a client is called a *protocol*.

Protocols are often created or managed by international standards groups. The *Internet Engineering Task Force* (IETF) is the group responsible for keeping track of client-server protocols, in addition to many other networking standards.

Security Considerations

Many companies would cite their client/server applications as the top reason for having a *local area network* (LAN). Often these applications become integral to daily workflow and to any business success. Their ease of use and flexibility does not come without its costs.

Client/server applications have a tendency to touch every part of your network. When they become infected, or compromised due to a lack of security planning, it is impossible to tell how deep the problem could run. Essentially, any client/server implementation will pass along information to, and retrieve information from, all nodes on your network. Once an intruder gains access to one aspect of your network, he or she will have access to all areas, using the client/server system as a passageway in which to travel. And, of course, the intruder will not try to gain access to the most secure part of your client/server system, but will go for the weakest area, the PC or workstation.

A standard PC is not very secure without the addition of third-party security products. This is because the PC's user has access to the operating system commands that can directly affect any data files without restriction. For this reason, application data should not be stored on this platform unless a separate security product is installed.

In the client/server environment the PC should only be used to work on select application records for a brief period of time. These records should not be stored on the hard disk of the PC. If you do allow the user to store data on the PC, then you must perform a data sensitivity, or risk analysis, to determine the level of exposure in order to install the proper third party security product. Many of these products are on the market. The best ones would not only authenticate the user's activity but would also encrypt sensitive data to the hard disk and to the network.

Usually, within a client/server environment the processors that share the work must communicate their intentions to each other over a network. In order to effectively communicate, a set of rules called a protocol is established. These protocols vary from situation to situation, but, regardless of the specific protocol, the communication process actually sends the data across the network in the form of packets that are constructed according to the protocol rules. This means that all the data for any activity is available for reading with the proper equipment. Data, such as user names and passwords, memos, or files, can be obtained and read as plain text.

The typical corporate workstation is easily broken into and exploited. Many software tools are capable of capturing keystrokes and passwords typed into a system. This means that a malicious user can grab password information when it is sent from the workstation to a server or as it is keyed on the workstation. Unfortunately, these programs are often designed to be difficult to detect.

With workstations being very vulnerable to hacker attacks, the whole client/server model becomes a large security risk. Once a workstation is controlled by an intruder, the intruder can access all of the resources provided by the client/server system. This is why a three-tiered client/server model with an application server can provide greater innate security. In simple terms: when less client/server software is operating on the workstations, less information is available to a hacker who breaks into the network.

Making the Connection

Storing Information: Network filesystems and databases are fundamentally based on client/server architecture.

Network Monitoring Systems/Logging and Analysis: Network monitoring systems and centralized log analysis systems use client/server technology to gather and process system diagnostic information.

Best Practices

In certain situations, it's not efficient for the client to talk directly to the server. A middleman can get between the client and the server to improve performance or provide other benefits such as security. This middle layer is often called *middleware*. The entire system is referred to as a *three-tier architecture*.

The next chapter of this book was originally about middleware, but it had to be cut for space reasons. Never fear—we've placed it on our companion web site, you can find it at www.sagesecure.com/nsi.

Final Thoughts

A number of client/server network systems have become rather popular over the years. These are the systems that most people use daily: the Web, email, file transfers, and remote login. Many of these systems were designed with features, convenience, and speed-to-market in mind. As a result, some pretty serious security issues need to be considered when using these technologies. In the next chapter we'll explore the protocols behind the most widely-used network applications, along with their security issues.

Chapter 30
Accessing Information: Internet Services

The most commonly used systems
for accessing data over the Internet.

Technology Overview

For all intents and purposes, the Internet would not be as big as it is today without the Web, email, FTP, and newsgroups. These four applications are responsible for the vast majority of traffic moving across the Internet. It's impossible to discuss network security without talking a bit about the fundamental four Internet applications.

What people think: The Internet is the Web/email/AOL.

What we think: The Net is much more than meets the eye.

The Web

Ah, the Web. We wouldn't have much of a book to write if the Web hadn't come along and brought networked computing to the forefront of public consciousness. But back in 1993, when the Web was created, nobody could foresee the massive changes it would instigate. As a result, security was not a big consideration. The Web was designed as an academic resource for sharing information, used by a closed community of people with little interest in hacking into each other's systems.

Web browsing is based on a really simple system called the *HyperText Transfer Protocol* (HTTP). Commands from a web client (the browser) to a web server are called "requests." The server sends back a "response," based on the nature of the request. All of the commands are sent as plain text. You can actually log into a web server using a command-line tool such as Telnet (described in Chapter 31, "Remote Access") and type HTTP commands directly to the web server.

Technically, any type of data can be sent back as a response. The only limitation is the web client's ability to understand the data. All web clients can process text documents formatted using the *HyperText Markup Language* (HTML). Most can also display images. Newer web clients can process documents created using the *eXtensible Markup Language* (XML).

Security Considerations

Web clients and servers have become increasingly more complicated during the past decade. Initially, HTML was a simple language that allowed academics to publish very basic documents. As the Web became commercialized, online publishers wanted to have greater control over page layout. web clients started extending HTML to provide more formatting control. Then came dynamic scripting—the ability for an HTML page to directly control certain aspects of the web client. Web clients also supported "plug-in" components, allowing third-party developers to extend the types of responses that the web client could understand. One commonly used plug-in is "Flash," a system for creating interactive animations.

At the same time, web servers were increasing in complexity. The second generation of web servers supported the ability to execute arbitrary programs and send the resulting output back to the web client. This meant that the Web could become a gateway to traditional computing applications. The technology was called the *Common Gateway Interface* (CGI), and it enabled HTTP to ultimately support e-commerce and many other types of complex applications.

These additional features had little to do with the original purpose of HTTP. The actual way in which the Web was being used far exceeded the protocol designer's expectations. Although this was positive in the sense that it meant the protocol was successful, it had dreadful implications for security.

The first major problem was that the web clients and servers had become very complex and full of bugs. These bugs led to numerous security vulnerabilities. Web pages could be designed in such a way as to give the creator full access to any machine that viewed the page in a client. Likewise, malicious requests could be used to obtain full access to machines running web server software. Although many such problems have been found and fixed, the client and server software continues to become more complex. Each time a new feature is added, the possibility that vulnerabilities are also introduced arises.

Another problem is that some of the newer features purposefully provide dangerous functionality. ActiveX, for example, allows a web server to directly control Microsoft web browsers, which under Windows really means the entire PC. Microsoft has placed a number of security measures in place, but it is relatively easy to fool people into disabling these security measures. For example, a hacker can set up a greeting card web site. Viewing cards on the site requires a special ActiveX control to be accepted. Many legitimate greeting card sites operate this way, so accepting the ActiveX control seems like a normal thing to do. But once the control has been accepted, it has full command of the user's PC and can then use any number of methods to install a trojan for permanent access.

One of the most serious problems is that all of the communications between web clients and servers occur in easily visible text messages. This makes information security impossible under HTTP. For this reason, a modified version of the protocol was created called HTTPS.

The HTTPS protocol is identical to HTTP, except that it uses a system called the *Secure Sockets Layer* (SSL) to communicate across the network. SSL is a standardized layer built on top of *Transmission Control Protocol/Internet Protocol* (TCP/IP) that provides encryption and authentication between a client and a server.

A number of potential security problems are also present on the server side. Many web sites generate pages dynamically by using some form of CGI program. Vulnerabilities in the CGI program can be used to gain access to the server machine. These are in addition to any vulnerabilities that might exist in the web server software itself.

Best Practices

The best solution is to constantly keep abreast of security news and ensure that all software is running the latest updates.

Disabling extended web client features is a good active approach to protection. The features can then be reenabled on a site-by-site basis. Although this will break the functionality of some web sites, the additional security benefit far outweighs the inconvenience. Furthermore, only a handful of major sites require these features to operate, and it is easy to add them into a "special case" list. Most modern web browsers offer this type of fine-grained administrative control.

On the server side, it's incredibly important to ensure that CGI programs are secure. Any vendor providing a CGI program for dynamic content management should have security as a high priority. It should be constantly seeking out potential security problems and should have a rapid response if flaws are detected within its system.

Email

The other major Internet application is email. Businesses can no longer function effectively and certain tasks are nearly impossible to do without email. The ability to add attachments and digital signatures makes email more effective than traditional mail for most document exchange needs.

Three basic components to email exist: the email transport system, the email repository, and the email client. The *Simple Mail Transport Protocol* (SMTP) is used by every Internet email transport system and is responsible for getting mail from one email server to another. Destination mail servers run additional software to allow email clients to connect. Many different types of email clients are in use. Some people access email through a program such as Outlook or Eudora, others access it through a web browser (Yahoo!, Hotmail, AOL mail, or something customized). Email clients use either the *Post Office Protocol* (POP) or the *Internet Message Access Protocol* (IMAP) to connect to the mail repository server and receive mail.

Simple Mail Transport Protocol (SMTP): When you send an email message, your email client connects to an SMTP server. This server takes in the details of the message (To: address, headers, message body). It then uses the *Domain Name System* (DNS) to look up the *Mail Exchange* (MX) record for each destination address. The MX record is the IP address of the mail server that handles email for the given domain name. The server finally makes a connection to the destination email server and sends the message across using the SMTP language. The system really is simple—you can actually log into an SMTP server with a command-line tool such as Telnet and directly interact with the SMTP server. For example, this is what a simple session with a mail server looks like. The lines in bold are lines that the user or a mail client would enter:

```
220 mail.example.com ESMTP Sendmail 8.11.6/8 .11.6; Mon, 10 Mar
2003 06:26:10 -0500
HELO myhost.example.com
250 mail.example.com Hello [149.2.55.31], pleased to meet you
MAIL FROM: spam@spamdomain.com
250 2.1.0 test@example.com . . . Sender ok
RCPT TO: info@sagesecure.com
250 2.1.5 info@sagesecure.com . . . Recipient ok
DATA
354 Enter mail, end with "." on a line by itself
From: "John Doe" <johndoe@aol.com>
To: info@sagesecure.com
Subject: Incredible offer
---------

For a limited time you too can own the book that revolutionized my
business and increased my wealth! Network Security Illustrated
showed me how to make money without leaving home, get an extended
line of credit up to $10,000, and meet the love of my life! Now you
can find happiness too! Just buy this book from us today! I saved
millions of dollars by reading Network Security Illustrated. Can
you afford not to buy this book? Get copies for all your friends
and become POPULAR! Act today!

.

250 2.0.0 h2ABRSG13453 Message accepted for delivery
RSET
```

```
250 2.0.0 Reset state
QUIT
221 2.0.0 mail.example.com closing connection
```

Post Office Protocol (POP): This is the simplest email repository server. This system does two things: it authenticates the user through a simple password exchange and then transfers mail from the mail server to the client. It doesn't do much more. Most web mail systems, such as Yahoo! and Hotmail, use POP to deliver mail to the web mailbox.

Internet Message Access Protocol (IMAP): This is a more complex system, which keeps mail messages on the server. Mail is stored on a central server, and temporarily downloaded locally for viewing. IMAP is good when your users need to access mail from multiple machines. It also is great if your users generally access mail from a desktop client system such as Outlook, but also want remote access to their mailbox through a web mail interface. By using IMAP for both, there's no need to synchronize the mail directories.

Security Considerations

Forgeries: If you look closely at our SMTP example above, you'll notice that the "from" address given in the MAIL FROM: line is different from that given in the message body. The recipient will see *From: "John Doe" <johndoe@aol.com>* in the message. Upon closer examination of the headers the MAIL FROM: address will also appear. The mail server actually doesn't care about either address. Verifying the "from" address is difficult and can lead to good messages getting dropped. As a result, neither of those addresses necessarily corresponds with the sender's actual name or email address.

The truth is that SMTP makes it incredibly easy to forge emails. Tracing a forged email is also incredibly difficult and can only be done by government agencies or through a highly organized effort. The only way to truly guarantee identity is via a digital signature.

SMTP Relaying: When neither the source or destination email address is local, the mail server is said to be operating as a relay. It's just relaying mail from one mail server to another. This is a bad thing.

In order to receive mail from other mail servers, an SMTP server needs to accept connections from everywhere. This means anyone can connect to any SMTP server on the Internet. A spammer can connect to any SMTP server that allows relaying and use it to send out millions of messages to any address on the Net.

In the Clear: Email is sent across the Internet in plain text. SMTP does not directly support encryption. Anyone listening can grab the entire contents of an unencrypted email message. If privacy is desired, encryption is the only effective solution.

Authentication: Neither POP nor IMAP natively support encrypted authentication at the protocol level. This means that all authentication information travels between the mail client and the server in plain text. Newer POP and IMAP servers have worked around this problem by incorporating SSL. If the mail client also supports SSL, the entire client/server connection can be properly authenticated and encrypted. A version of POP called APOP also exists that uses a different technique for securing POP authentication. SSL is more standardized and is often the preferable route if both client and server support is present.

Synchronization: POP has synchronization problems if you access email from multiple machines.

Performance: IMAP can be server intensive for complex operations such as searching and requires much more storage space on the server. You need to enforce mailbox limits on systems with many users.

Account Security: If an account is compromised, email messages can be downloaded or viewed and then deleted. Messages can be easily forged in a much more convincing manner.

Mail Clients: Viruses and trojans. Bad files (virus emails) get transferred to local machines and by the time the virus scanner checks the file, it might be too late. Client vulnerabilities arise due to support for nifty additional features.

Best Practices

Properly Configure to Prevent Relaying: The proper solution is to only allow incoming messages that have an approved destination address, and to only allow outgoing messages from within your internal network. Spammers can connect to your mail server and spam your own users, but they can't use your mail server to spam other networks.

Backup Mail: Backup sent and received mail independently of the mail client. Any message that is sent or received within a certain timeframe should be retrievable, regardless of whether the sender or recipient deleted the message. This can help if accounts have been compromised or in "he said, she said" disputes.

Deletion Policy: On the other hand, you don't want mail sitting around forever. Old email messages might prove problematic in a court of law. You certainly don't want to have to explain what somebody was thinking two years ago. As a result, most companies have a policy of deleting email messages that are over a certain age. That said, someone could always print out a message or download it to a floppy. But it's tougher to prove that the message was not tampered with if other records have been destroyed.

Encrypt and Digitally Sign Everything Important: This is the only real way to guarantee the authenticity and confidentiality of a message. There are many effective systems for encrypting email. *Pretty Good Privacy* (PGP) and its alter ego, the *GNU Privacy* Guard (GPG) are two popular freeware systems that are both very similar and compatible. Most mail programs offer some degree of integration with PGP and GPG. See the chapter on "Hiding Information" for more details on PGP/GPG.

FTP and TFTP

The *file transfer protocol* (FTP) is a well-known client/server application that enables a remote file transfer over networks. FTP is relatively easy to set up, requires little maintenance, and provides a straightforward interface for the user. Many different versions of FTP clients and servers have been written for every computing platform and are readily available. In fact, most server class operating systems include FTP by default.

TFTP, or the *trivial file transfer protocol* (TFTP) is a stripped-down version of FTP. It uses a client/server model as well, but is designed to work in a scaled-down fashion. TFTP is generally used by servers to boot diskless workstations, by routers to receive firmware updates, and by X-terminals to transfer small files.

Security Considerations

There is no place for an FTP server in an otherwise secure environment. Because of the way FTP is used today, FTP servers have requirements that make them generally incompatible with firewalls. Essentially, a well-designed firewall will constantly get in the way of an FTP client/server transaction.

FTP was designed with some inherent security, but it is largely based on permissions, not encryption. An FTP server can be set to allow guest or anonymous access to its files. Beyond that, most FTP servers get their permissions from the server's user access list. In Microsoft Windows, FTP server applications do not always integrate with the operating system's access controls. In this circumstance, a separate database of users and groups must be created explicitly for the purpose of FTP. This isn't necessarily a bad thing, as it can be used to protect the real user accounts from being compromised through an FTP insecurity.

TFTP has no security built into it at all. There are no permissions structures and no user base. TFTP was designed for temporary use in local environments, where security is not as important as convenience. For example, TFTP is frequently used to upload new firmware to a router. The firmware is first downloaded from the Internet and stored on a local workstation. A TFTP client is then installed on the local workstation and pointed at the IP address of the router. The router runs a TFTP server on its operating system. A connection is established and the new firmware is uploaded.

Best Practices

FTP servers should never exist on a private network, as they require many additional ports on a firewall to be opened, posing an increased threat to security. Any truly secure LAN segment should have as little exposure to the Internet as possible, which leaves no room for FTP. If an attacker is able to exploit the FTP server, access can easily be gained to the rest of the network. If a need to serve files over the Internet does exist, a *virtual private network* (VPN, Chapter 20) is the safest alternative. With a VPN in place, remote users can transfer files to or from an FTP server over an encrypted tunnel. An alternative to FTP over a VPN is placing the FTP server in a *demilitarized zone* (DMZ, Chapter 18). But remember: A DMZ should never contain any files that are critical to an organization.

News

Before the Web, email was the Internet's biggest application. But email is just a person-to-person communication system. What about person to group systems? Mailing lists were one solution, but there wasn't any formal or searchable system for organizing the lists. The Internet needed a hierarchical, distributed variation on the bulletin board system concept. The solution was a system called "Usenet," "the newsgroups," or "Netnews." Usenet is a major component of the Internet that few people know about.

Usenet newsgroups are organized hierarchically, similar to the system used with domain names, but backwards. For example, a Usenet newsgroup discussing the Java programming language might be named comp.lang.java. "Comp" is the top level. "lang" is the second level and "java" is the third. Some of the major top-level groups are

```
alt: alternative discussion, pretty much anything goes, such as
alt.paranormal.spells.hexes.magic, alt.animals.llama,
alt.religion.Zoroastrianism
comp: anything relating to computers
rec: recreational activities
sci: scientific disciplines
soc: social issues
misc: like alt, a little less random
```

The difference between Usenet newsgroup names and domain names is that a central authority does not control the naming system. Usenet is a truly distributed network. Anybody can create a new newserver, and therefore can create a new group with any name they wish to use. But the only way to get others to "carry" the group to their newservers is to have the group name officially adopted. This is a somewhat democratic process that takes quite a while and requires a pretty intense coordinated effort.

The first step in creating a new newsgroup is to figure out which top-level/sub-level group it belongs under. Some *Tag Library Descriptors* (TLDs) are easier to start groups in than others. Alt.* is relatively easy to create a group under. Comp.* is almost impossible unless you just invented a new area in computer science. The right subgroup is not usually obvious. For example, where would you place a news-group discussing the Segway personal transporter? Does it go under rec.bicycles.segway, rec.segway, alt.segway, or misc.transport.segway? One loyal Segway user tried to start a discussion about it in alt.scooter. His post was deemed "spam" by scooter aficionados who couldn't stand the Segway being mentioned in the same context as their lovable scooters. The discussion rapidly degenerated into a description of sodomy acts that would be performed on the poster. Apparently, sexual fantasies are within the charter of the scooter group, but Segways are totally off limits. By the way, this is par for the course with Usenet.

Most users will connect to only one news server—the one their *Internet Service Provider* (ISP) or corporation provides. ISPs will generally only carry the officially accepted newsgroups. But having a publicly accepted newsgroup isn't necessarily important. It's still possible to connect to other news servers. A number of compa-nies choose to run their own publicly accessible news server. They don't care about the group getting carried—they expect users to connect directly. These newsgroups are often used for support forums and other types of community chat. This can be less costly to set up than a web-based or email-based forum system, but the down-side is that relatively few users know how to access newsgroups.

Unlike email, Usenet uses a single system for the end-user client, the trans-portation system, and the delivery system. This system is called the *Network News Transport Protocol* (NNTP). When a newsreader connects to a newserver, they talk using NNTP. When two newservers exchange group data, they also use NNTP.

Security Considerations

Bugs: Usenet has been around for a long time and the protocol is pretty secure. The newservers and clients are another story. *InetNetNews* (INN), the standard UNIX news server, has had a number of bad vulnerabilities in the past. Microsoft's newserver hasn't done so well either. Most of the bad problems are probably gone from each system, but new ones could be introduced if new features are added.

Bandwidth: The biggest issue with Usenet is the amount of data. Running a full Usenet feed can easily exceed the capacity of a T1. Few companies will do this. On the other hand, selectively obtaining a few useful groups could be valuable, depend-ing on the needs of the company.

Server Resources: The massive bandwidth needs can drain server resources. If you are running a publicly available news server, a flood of postings may bog down the server.

Porn and Spam: Many Usenet groups are not moderated—anybody can post any-thing to the group. As a result, these groups are often filled with both porn and spam.

Not only does it use up bandwidth, but the presence of porn might violate company policy.

Illegal Warez: The piracy community loves Usenet. Supporting full Usenet access means employees can use company resources to download illegal software. It would be useful to simply not carry certain groups, such as alt.binaries (which has lots of porn and illegal software and accounts for 60–80 percent of the databulk on Usenet).

Web Gateways: Many newsgroups can be accessed via web-based gateways. If you're providing unrestricted Web access, then you're also allowing newsgroups in via the Web.

Best Practices

It makes little sense to carry your own news feed if you're looking to provide complete access. It's more cost effective to subscribe to a feed from a commercial provider. Get a couple of accounts and let your users access the remote newserver.

If you want to provide a small selection of Usenet groups to your organization, then an internal newserver makes more sense. Make sure the server is configured to only allow connections from within the network, otherwise spammers and freeloaders will use your server for their own purposes.

If you wish to provide publicly accessible newsgroups for support and community purposes, you should make sure the groups are moderated. Otherwise you'll have the same problems as with any publicly accessible news service—spam, piracy, and freeloaders. You can also require users to log in with a name and password. This is a little more of a management burden, but will completely prevent unwanted users from getting access.

Final Thoughts

The Internet is so much more than just the World Wide Web and email. Most people think the Internet *is* the Web and email *is* email. It certainly is not common knowledge that the Web and email are based on different protocols and are just a few of many well-established Internet services. Peer-to-Peer, which is discussed in chapter 41, is the first new Internet service that is gaining mainstream popularity and recognition. Unlike the technologies discussed here, Peer-to-Peer is not based on a single standardized protocol. In time, the global acceptance of services like Peer-to-Peer will change the balance of power among Internet protocols.

Chapter 31
Accessing Information: Remote Access

How to securely get at applications
and data on remote systems.

Technology Overview

Ever wonder why it is not trivial to access files that are stored on your office work-station or server from home? Most businesses today struggle with remote access solutions. The benefits of remote access are readily apparent, but the available solutions are often difficult to implement, tough to use, and very insecure.

Secure is the last thing remote access should be called. The very nature of remotely accessing data stored on your *local area network* (LAN) is insecure in principal. What you get with remote access is convenience, but the price of convenience is costly. It's the same reason why a loaf of bread is significantly more expensive at a 24-hour grocery store. Convenience almost always comes at a premium.

Remote access suggests you are going to take something that is stored privately on a LAN and be able to view it while on the road, at home, or anywhere besides the office.

Many people need remote access to a LAN to get at data they need to do perform their job. The goal is to have the same resources available when in the office, at home or on the road. Another function of remote access is directed at the *Information Technology* (IT) department. IT directors or network administrators often want to control multiple networks in the *Wide Area Network* (WAN) from one central location. In this scenario the administrators are not particularly interested in stored data, but wish to make system changes without having to travel to the site.

Remote access is made even more complicated by the variety of platforms that are in existence. Microsoft's Windows and UNIX, for example, are two different platforms that can provide many of the same services. However, both platforms rely upon different interfaces to manage their services. Different remote access tools are often needed for each platform. In the realm of remote access, the main difference is command-line interface management (UNIX) versus GUI management (Windows).

Command-Line Remote Access

Telnet is a remote access protocol that has been in existence for quite some time. Its main purpose is to give remote command-line access to a system connected to a network. Once a telnet connection is established, a login prompt appears, permitting the user to enter the system. It is then possible to log in to that computer and execute commands. Once a successful telnet login takes place, a user's control over the remote system is restricted only by permissions set on the server. In other words, if the user logs in remotely with telnet as the administrator, they will have administrator access privileges, but from an offsite location! For this reason it is imperative to issue administrative commands cautiously when using telnet. Most often, one "telnets" to a remote UNIX or Linux-based system. The commands used once the user has logged in are those supported by the remote operating system not telnet-specific commands. Telnet is a tool that simply acts as a gateway from one system to another. However, telnet is only one of several remote access systems that can provide this functionality.

Another tool that serves a similar purpose to telnet while providing inherent security is known as *Secure Shell* (SSH). It is a remote login system designed to give command-line access to a computer's operating system. SSH works in the same manner as telnet, only it uses encryption to secure the communications between local and remote endpoints. Unlike telnet, SSH encrypts every packet of data it sends to and receives from the host machine, including the initial username and password. As a result, any packets of data that may get intercepted will be indecipherable.

Graphical Remote Access

Although both SSH and telnet provide command-line remote access to a system, some users are looking for a remote *graphical user interface* (GUI) instead. GUI-based remote access increased in popularity dramatically when Microsoft Windows began to infiltrate the business world. Unlike UNIX, Windows was not designed with command-line remote access control in mind. As a result, some specialized applications need to be used in order to fully control a Windows server from a remote location.

Unix is quite the opposite of Windows in this way. UNIX was not designed with a user-friendly graphical interface in mind—it used to require the use of a command line or shell. However, over time, a complex but workable graphic interface was created for the UNIX environment. This application, known as X or X-windows is merely an optional add on for any UNIX environment. Unlike modern user-oriented operated systems, X windows is actually a client/server application. Its purpose is to give a UNIX user the flexibility of multiple windows and a desktop. This GUI environment haves made UNIX more accessible to new users.

The fact that the X has a client and server component to it makes it complex and extremely flexible. As a result, it is often used to provide graphical remote access interfaces to remote UNIX machines. The X Window System, more simply "X" or "X11," is the standard graphical engine for the UNIX and Linux operating systems. It provides an identical windowing environment for similar platforms, which gives the many different flavors of UNIX some common ground.

The X Window System operates independently from the operating system and hardware of a computer. This is what allows for its extreme level of compatibility and flexibility within the UNIX platform. As a result, X Windows has grown in popularity tremendously over the years. It is estimated that the worldwide community of users of the X Window System currently exceeds 30 million.

Another tool that can provide a remote graphical picture of a target system is known as *Virtual Network Computing* (VNC). It is, in essence, a remote display system that allows a user to view a computing desktop environment from anywhere on the Internet, and from a wide variety of machine architectures.

VNC works by sending a series of "photographs" of a source machine over a network to a target machine. It is a client/server application and it requires its own user name and password when it is set up. Then end result is a user can bring up a "window" and inside the window will be a live feed of the remote desktop of another system. Through this VNC window, the remote desktop can be controlled with the local mouse and keyboard. The effect can be astonishing to someone standing over the monitor of the remote machine. When VNC is operating, the mouse on the controlled desktop will show the remote user's movements. It can sometimes appear as if the system is haunted.

VNC is very small and platform independent and it doesn't require any additional hardware or software. This makes it easy to install on any system. For these reasons it has become the remote administration "tool of choice" for hackers.

The VNC system allows for desktop sharing and multi-user collaboration. It also remembers exactly how your windows were set up when you last logged in. If you connect to a remote machine from work and then connect again from home, your mouse will be in the exact spot you left it.

VNC and X Windows have been spawned from the world of open-source software. Both are very robust remote access methods and are completely free. Some vendors, however, have focused on creating remote access tools that are designed to work on the commercial side of the fence. These tools have been designed from the ground up to make Microsoft's Windows remotely accessible.

One popular commercial remote access tool is PC Anywhere. As its name clearly states, it will allow a remote user to connect to any Windows-based PC, whether it be a client or a server. The system being accessed remotely needs to be running a host version of PC Anywhere and has to be configured to receive outside connections. The host allows access based on a list of criteria, such as IP address and user names.

After installing the client software, the user can begin the remote login procedure. The user enters the IP address of the host machine, along with the necessary login information. If the connection is accepted, PC Anywhere loads a live "picture image" of the remote PC desktop. The user has total control of the host desktop from his or her remote machine.

PC Anywhere is very different from VNC as it is not platform independent. VNC is designed to send whole desktop images of any desktop from a server to a client. PC Anywhere is designed to integrate into the remote server's operating system and take advantage of this integration.

From a bandwidth perspective, the remote access tool is essentially sending a streaming video feed of the desktop to the remote user, which requires a relatively high bandwidth connection. The user will literally be moving the mouse pointer on the desktop of the host machine. If the remote user decides to load an application such as MS Word on the host machine, he or she could write a document from the remote location. To someone sitting in front of the monitor of the host machine it would appear as if a ghost has overrun the system. He or she would be gravely mistaken.

The lack of an efficient command-line interface has potentially tremendous drawbacks here. The high-bandwidth-only operation means that it can be very difficult to use PC Anywhere from a remote location over a modem. Unfortunately, this is one of the primary methods of remote access on Windows-based systems. Even though VNC also produces a desktop image for the remote user, in UNIX environments command-line access can always suffice as remote control.

How Remote Access Protocols Work

Telnet is undoubtedly the least sophisticated of all the remote access protocols. On one level, it works no differently from logging into a machine locally (it's even possible to telnet to your own machine). The host machine (often a server) activates the telnet service. The telnet service reads the access files that are native to the machine. The files contain information detailing the users that have access to the system, their passwords, and what level of access they can be granted. Telnet just extrapolates that information over a networked environment.

To work remotely over a network, telnet needs to operate over a TCP/IP port (23). A user on a remote machine only needs to run a telnet client to gain access to a remote machine running a telnet server. The remote user executes the telnet client and enters the IP address information of the remote host. Upon contacting the remote host, if a telnet server is running, the remote user will be prompted for a user name and a password. If the information entered matches that in the remote machine's host file, the remote user is then granted access to the server. Both telnet client and server applications are extremely small and extremely available on almost every computing platform known to humankind.

In principal, SSH works the same way telnet does. The major difference is it incorporates heavy security during all transmissions between the remote user and the server. The security is implemented in the form of encryption. Several types of encryption algorithms can be used with SSH depending on the situation. All forms of encryption serve the same purpose in the end: to ensure that no one eavesdropping can make heads or tails out of what he or she finds.

One method of authentication that SSH can support is RSA-based authentication. The scheme is based on public-key cryptography as defined in Chapter 26 of this book. With RSA and SSH, the server knows the public key and only the user knows the private key. A certain file on the host lists the public keys that are permitted to log onto the system remotely. At the time the remote user is logging in, the SSH program tells the server which key pair it would like to use for authentication. The server checks if this key is permitted to gain access to the system. If the particular key is on the list, the server sends the user's SSH client a challenge question, in the form of a random number, encrypted by the user's public key. The challenge can only be decrypted using the proper private key. The user's client then decrypts the challenge using the private key, proving that he or she knows the private key without disclosing it to the server.

SSH implements the RSA authentication protocol automatically. The user creates his or her RSA key pair by running a small application called ssh-keygen. To become more familiar with this Unix program you can reference the main pages of your local Unix system. In short, the private key pair is stored in a local directory called *identity* and the corresponding public key pair is stored in another local directory called *identity.pub*.

If other authentication methods fail, SSH prompts the user for a password. The password is sent to the remote host for checking; however, since all communications are encrypted someone listening in on the network cannot see the password. SSH on its own is the most secure version of remote access one can achieve from many points of view. However it is more limited in the scope of features it provides than some of its less secure contemporaries.

VNC offers the graphical remote access features that SSH lacks, but these features come at the expense of decreased security. VNC uses a challenge-response password scheme to make the initial connection: the server sends a random series of bytes, which are encrypted using the password typed in, and then returned to the server, which checks them against the right answer. After that, the data is unencrypted and could, in theory, be watched by other malicious users, though it's a bit harder to snoop a VNC session than, say, a telnet, rlogin, or X session. Since VNC runs over a simple single TCP/IP socket, it is easy to add support for *Secure Sockets Layer* (SSL) or some other encryption scheme if this is important to you. VNC can also be used with SSH, described later in "Best Practices."

Security Considerations

Any organization that is considering providing remote access to its network should question its ability to live with weak defenses. The bottom line is that remote access means punching a hole in your network structure and your firewall to allow users to get at information. The hole that remote access requires is a huge hole that malicious users can exploit. Each particular method of remote access has its own particular quirks and security holes, as a closer look will reveal.

Telnet is at the top of the heap for lack of security. Hackers can use a variety of telnet specific attacks to gain access to and control of a network. Telnet's underlying communications protocol is also inherently insecure. All communications between the local and remote machines are sent as simple, unencrypted text. This means that when a user logs in remotely to a telnet server, the username and password are easily captured by an eavesdropper. Once access is granted, any other data sent over the telnet connection is in plain text form as well. Passwords or confidential information sent across the connection can be intercepted with little effort. These transmissions can be read like an email when captured by a malicious user.

If a telnet-like service is required for remote access into a network, then SSH is a much better alternative. SSH provides the same terminal level access as telnet but encrypts all transmissions between the remote user and the access point.

Security gets much more difficult if you need a graphical remote access program. If the remote system is UNIX-based, don't directly use X Windows. It's massively insecure and very difficult to use correctly. You're much better off using VNC with SSH to gain remote access to your X desktop. Make sure any X-enabled machine is protected by a good firewall that is capable of blocking inbound X requests.

VNC is not secure by default either. Communications are unencrypted and the authentication system is easy to crack. The ability to share desktops allows hackers to see what you're doing—while you're still online! VNC should only be used with SSH when connecting over the Internet. This will slow down performance a bit when compared to using VNC on a LAN without SSH, but the security benefits more than make up for the loss in speed.

Most other remote access solutions are completely proprietary and are difficult, if not impossible, to use securely. If you must use an application such as PC-Anywhere, your best bet is to ensure that all connections are encrypted. This can be done using SSH, as explained in the "Best Practices" section.

Making the Connection

Connecting Networks: Many types of remote access require a large amount of bandwidth and as a result are not suitable over slow network connections.

Hardening Networks: Remote access and firewalls historically do not work well together. In general, remote access is best brought through a firewall via secure tunneling.

Hiding Information: Secure remote access requires sending all access commands over an encrypted channel.

Best Practices

Most remote access systems can be integrated with SSH for additional security. For example, SSH integrates with X Windows and VNC. X and VNC on their own are not secure.

SSH enables a user to redirect remote TCP/IP ports so that all traffic is strongly encrypted. This process is called tunneling. X or VNC data can be sent through an SSH tunnel, making it impossible for eavesdroppers to gather information from intercepted communications.

SSH can also compress the encrypted data. This provides a major advantage to remote users that are operating over a traditional modem connection. The only way to get reasonable performance out of VNC over a modem connection is to integrate the process with SSH.

Final Thoughts

Remote access should be used with the utmost caution if security is a priority. Even if SSH is being used to log into a machine remotely, security can be compromised. One of the biggest caveats to remote access is the integrity of the host machine. In general, remote access it requested from a machine that may be compromised. Even if encryption guarantees the transmission is secure, a keystroke logger on the host machine may be monitoring your username and password.

Chapter 32
Accessing Information: Peer-to-Peer Networking

Connecting individual, unrelated computers
directly as if they were on the same network.

Technology Overview

Peer-to-peer (P2P) computing is the sharing of computer resources and services by direct exchange between systems. These resources and services include the exchange of information, processing cycles, cache storage, and disk storage for files. Peer-to-peer computing takes advantage of existing desktop computing power and networking connectivity, allowing economical clients to leverage their collective power to benefit the entire enterprise.

The peer-to-peer method of computing has been around for decades. In recent years however, a certain aspect of P2P computing has reached the mainstream. The reason that P2P has come on so strong recently is simple. Currently a bevy of inexpensive computing power, storage, and bandwidth is available. All this cheap technology is also in the hands of the masses. One company in particular deserves credit for popularizing P2P: Napster. However, modern P2P is a very recent implementation of a very old concept. As many as 30 years ago, companies were working on architectures that would now be labeled P2P.

In a peer-to-peer architecture, computers that have traditionally been used solely as clients communicate directly among themselves and can act as both clients and servers. P2P nodes, as they can be called, will assume whatever role is most efficient for the network at any given time. This is possible because the P2P architecture allows its nodes to make role change decisions on the fly.

One result of this mechanized computing efficiency is that the load on traditional servers can be relieved. Now, instead of doing it all, traditional servers can perform more specialized functions such as email, Web, and domain name services, and do so with more efficiency. At the same time, peer-to-peer computing can reduce the need for IT organizations to grow parts of its infrastructure in order to support certain services, such as backup storage.

Backup is seldom needed in a peer-to-peer environment. Multiple versions of all files can be stored on multiple P2P nodes. If any given node breaks down, there will always be copies of the files located elsewhere. Another name for this polled resource environment is distributed computing. Many people feel that P2P and distributed computing will eventually replace all standard client/server environments. Some even think it's the next step in computing evolution. A closer look reveals that in real working environments it has tremendous benefits, and huge drawbacks.

What people think: I want my MP3's.

What we think: P2P networking has a potentially bright future, but the days of the great mp3-free-for-all are coming to an end. Security and legal risks will eventually outweigh the benefits of illegal filesharing.

P2P In Practice

In the business world, peer-to-peer computing can be integrated in many ways across many types of systems. Business applications for P2P computing come in several different flavors. In general, P2P offers a new and powerful method for businesses to use their combined hardware and software to its fullest potential.

Collaboration is one area that peer-to-peer computing can greatly enhance. A collaboration environment that is properly implemented with P2P will allow users an unprecedented level of real-time shared resources. Specifically, P2P collaboration tools give teams of users access to the latest data available. This can be a real advantage in organizations that have multiple office locations.

Collaboration increases productivity by decreasing the time for multiple reviews by project participants and allows teams in different geographic areas to work together. As with file sharing, it can decrease network traffic by eliminating email and decreases server storage needs by storing the project locally.

Peer-to-peer computing can help businesses deliver services and capabilities more efficiently across diverse geographic boundaries. In essence, edge services, acting as a network caching mechanism, move data closer to the point at which it is actually consumed. For example, a company with sites in multiple continents needs to provide the same standard training across multiple continents using the Web.

Instead of streaming the database for the training session on one central server located at the main site, the company can store the video on local clients, which act essentially as local database servers. This speeds up the session because the streaming happens over the *local area network* (LAN) instead of the *Wide Area Network* (WAN). It also utilizes existing storage space, thereby saving money by eliminating the need for local storage on servers.

Peer-to-peer computing can help businesses with large-scale computer processing needs. Using a network of computers, P2P technology can use idle processor power and disk space, allowing businesses to distribute large computational jobs across multiple computers. In addition, results can be shared directly between participating peers.

The combined power of previously untapped computational resources can easily surpass the normal available power of an enterprise system without distributed computing. The results are faster completion times and lower costs because the technology takes advantage of power available on client systems.

Peer-to-peer computing also allows computing networks to dynamically work together using intelligent agents. Agents reside on peer computers and communicate various kinds of information back and forth. Agents may also initiate tasks on behalf of other peer systems. For instance, intelligent agents can be used to prioritize tasks on a network, change traffic flow, search for files locally, or determine anomalous behavior, such as a virus, and stop it before it effects the network.

How P2P Works

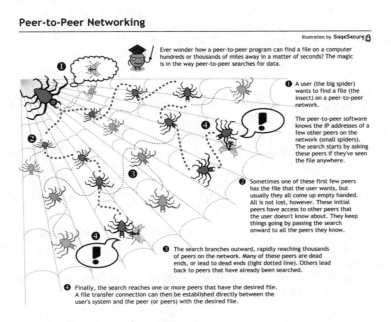

Figure 32-1

Security Considerations

The People Problem: There will always be malicious users who are intent on gaining super-user access to corporate networks. And no matter what security protocols are put in place, a skillful attacker, given enough time, will find a way around them. All that the security buffs need to do is to keep ahead of the hackers by creating bigger and better protocols. And after that is accomplished form a basketball team, petition to join the NBA, and beat the Lakers this year in the finals. Either task would be just as easy.

People are learning as they go with technology and this is well illustrated with the peer-to-peer computing model. The tools of the P2P revolution are becoming more advanced and more specific as time passes. Company files are made more available by being published to the world directly from a user's PC. Databases, spreadsheets, even entire applications, are becoming enabled with P2P features and critical information is flowing out from every PC. P2P systems typically provide mechanisms that include searching for specific content or documents, discovering other peers running the software, and implementing any number of other application level tools. Some of these tools include collaborative editing, *instant messaging* (IM), or remote wireless mobility support. It is easy to see why security is such a crucial factor in P2P networks.

Defending against the threats of an off-the-cuff P2P deployment, and managing or reducing the risks of loss of information or availability of systems is very difficult. It requires foresight, planning, and careful selection of the P2P infrastructure upon which your P2P enabled applications and services will be built. Many specific security challenges must be met and the following list is a good place to start:

External Threats: P2P networking allows your network to be open to various forms of attack, break-in, espionage, and malicious mischief. P2P doesn't bring any novel threats to the network, simply familiar threats such as worms and virus attacks.

P2P networks can also allow an employee to download and use copy-righted material in a way that violates intellectual property laws, and to share files in a manner that violates an organization's security policies. Applications such as Napster, Kazaa, Grokster, and others have been popular with music-loving Internet users for several years. As a result, many users take advantage of their employers' high-speed connections to download files at work. This presents numerous problems for the corporate network such as using expensive bandwidth and being subject to a virus attack via an infected file download.

Unfortunately, P2P networking circumvents enterprise security by providing decentralized security administration, decentralized shared data storage, and a way to circumvent critical perimeter defenses such as firewalls and *Network Address Translation* (NAT) devices. If users can install and configure their own P2P clients, all the network managers' server-based security schemes are out the window.

Theft: Companies can lose millions of dollars worth of property such as source code due to disguising files using P2P technologies. Using these disguising, or wrapping, tools people can hide one kind of file inside another. For example, an employee can make thousands of lines of intellectual property source code look like a typical MP3 file.

Now, an accomplice looking for the disguised file can download it from any location using a P2P client. Even if the company has a sophisticated file monitoring system that is watching all transactions in and out of the network, the transaction will seem perfectly normal.

Bandwidth Clogging and File Sharing: A major problem with P2P file-sharing programs is that they result in heavy traffic, which clogs an organization's networks. mp3 files, though highly compressed, are still quite large. Those who share them over P2P networks tend to do so in large quantity. We call these filesharing addicts "peerheads." Much like chain smoking, P2P is a hard habit to break and once addicted, downloading lots of files is the way peerheads get their fix. Although one 5MB file may be small, a hundred 5MB files add up and ultimately slow down a network.

Encryption Cracking: When using P2P for distributed computing many security related applications are available, the most relevant of which is code breaking. A prominent example of this occurred in 1999 as a combined effort between two organizations, Distributed.Net along with the Electronic Frontier Foundation. Together they launched a brute-force attack on the 56-bit *Data Encryption Standard* (DES) encryption algorithm and broke it in less then 24 hours. To accomplish this, the distributed computing model tested 245 billion keys per second. At the time, DES was the strongest encryption algorithm that the U.S. government allowed for export.

Trojans, Viruses, and Sabotage: A user could quite possibly download and install a malicious P2P application that could inflict serious damage. For example, a piece of code that looks like a popular IM or file-sharing program could also include a backdoor to allow access to the user's computer. An attacker would then be able to do serious damage or to obtain more information then he or she should have. In fact, a seasoned attacker can use these methods to gain entire control over a corporate network.

P2P software users can easily configure their applications to expose confidential information for personal gain. P2P file-sharing applications can result in a loss of control over what data is shared outside the organization.

P2P applications get around most security architectures in the same way that a Trojan horse does. The P2P application is installed on a trusted device that is allowed to communicate through the corporate firewall with other P2P users. Once the connection is made from the trusted device to the external Internet, attackers can gain remote access to the trusted device for the purpose of stealing confidential corporate data, launching a *Denial of Service* (DoS) attack, or simply gaining control of network resources.

Backdoor Access: P2P applications expose data on a user's computer to thousands of people on the Internet. This being the case, P2P applications were not intended for use on corporate networks. Of course, P2P applications are used on corporate networks every day, often without the permission of the *Management Information Systems* (MIS) department. The result is a serious exposure of the network's internals to the outside world. For example, if a user starts his P2P client and then clicks into the corporate Intranet to check his email, an attacker could use this as a backdoor to gain access to the corporate LAN. P2P clients open a big door to a network that cannot be blocked by conventional means. It is the equivalent to a large savings bank having a tunnel in the back where anyone can enter and exit with all the cash.

Confidentiality: P2P clients make it possible for a hacker to find out what operating system the peer computer has and connect to folders that are hidden shares, thus gaining access to folders and information that is confidential.

Authentication: Also, the issue of authentication and authorization is a problem. When using P2P you have to be able to determine whether the peers accessing information are who they really say they are and that they access only authorized information.

Making the Connection

Ensuring Availability: Peer-to-peer networking has the potential to be incredibly valuable as an enabling technology for distributed storage, making data highly available across a global network.

Hardening Networks: It's very difficult to restrict P2P access when hardening a network without engaging in a draconian network access policy.

Best Practices

At the moment, P2P technologies have no place in the corporate network. They bring much in the way of trouble, and little in the way of added value. Best practices for peer-to-peer don't really exist because from a security point of view it's a "worst practice." Take that, Napster.

Final Thoughts

One thing is clear, for all the benefits the P2P model brings to computing, it brings as many drawbacks. The real problem is not a pros vs. cons issue, but a perception issue. P2P brings about such change in the way computers are used that security tactics need to evolve fast enough to keep up with the pace. Reality shows, however, that security methods can hardly keep up with the more traditional client server model. So how will organizations change their whole mindset and keep ahead of the security curve?

Awareness through education is the only answer. P2P offers a great revolution in computing—faster, more reliable, and more efficient access to data at a lower cost. Now, if businesses can implement these systems strategically, with security in mind from the beginning, everyone will come out ahead. It is vital that users become confident in the ability of the security measures being utilized to protect them in order for P2P technology to reach its full potential.

**Chapter 32
Accessing
Information:
Peer-to-Peer
Networking**

XI
Ensuring Availability

Summary

For many organizations, a loss of service is just as devastating as a loss of information. The odds of service interruption or information loss decrease significantly if no single points of failure exist. Some interesting and powerful technologies exist that can keep mission-critical services available even in the face of a catastrophic disaster.

Key Points

- Recovering from a network catastrophe takes time.
- Sooner or later, a major system failure will happen.
- A truly redundant information system contains no single points of failure.
- Maintaining high availability becomes more difficult and more expensive as systems increase in size.

Connecting the Chapters

Ensuring availability is a challenge for any organization, large or small. There is much more to ensuring availability than making sure data is backed up. The following chapters cover a variety of methods to help keep data ready and waiting at all times:

- **Chapter 33, "RAID,"** introduces technique for using multiple synchronized hard drives to increase performance and/or reliability.
- **Chapter 34, "Clustering,"** describes a technology that allows a system to remain operational if its hardware or software fails by replicating its running processes on a simultaneous system in real time.
- **Chapter 35, "Backup Systems,"** actually perform two functions, copying critical data to another location and enabling the swift and painless recovery of the backed up data.

Introduction to Ensuring Availability

Nothing is more stunning than the dawning realization of a catastrophic failure. You know, when your first thought is a defiant, "It can't be that bad, can it?" followed by a doubtful, "Am I awake?" leading to a soft, awestruck, "Oh my." Newbies to the disaster process usually collapse in a blubbering heap somewhere around this point. Veterans break out the grain alcohol, wire cutters, and acetylene torch. With any luck, only the first item will be necessary. Of course, it never hurts to be prepared.

Massive resources are often devoted toward systems that mitigate or prevent failure. Live through one five-alarm disaster and you'll quickly understand why. If you haven't ever had such "experience", you might actually believe your backup plan is enough (we all have extensive backup plans, right?).

The real problem is that recovering from a major catastrophe takes *time*. Hardware failures take time to repair. Software failures take time to troubleshoot and work around. Some problems, such as software bugs and viruses, can be dormant in the backups as well. You can't merely restore—you need to fix the problem so it doesn't happen again. And all the while, Jim-the-exceedingly-helpful-guy-from-sales is calling every two minutes to inform you that he can't get his PowerPoint slides and that you should check to see if the server is down.

Putting Off the Inevitable

Sooner or later, a major system failure *will* happen. Usually, this means an interruption in service and possibly damaged data. We can't prevent failure, but we can make failure irrelevant. (We also want to earn millions selling a "Make Failure Irrelevant" motivational poster that depicts a single ant stumbling and getting crushed while the rest of the colony heedlessly marches onward. Or maybe we'll just settle with using ants as a theme for this chapter and hope lots of people buy the book.) But we digress.

How do we give our puny ant-like network the resilience of a massive colony? More importantly, can we make our network steal food from picnics and bring it to us when we're hungry?

Let's start by looking at ways to increase system and service reliability (we'll get to the food bit; don't worry). Two basic techniques are available: improving quality and adding redundancy.

Improving the quality of equipment is conceptually easy, but gets very costly at a relatively early stage. It's also impossible to totally prevent failure through only quality improvements—too many components can fail, and too many things can cause failure.

A well-implemented redundancy solution can prevent failure in all but the most absurdly catastrophic situations (meteor impacts, nuclear wars, biblical plagues, and so on). Don't write off quality completely: the last thing you want is a network filled with "redundantly broken" equipment. What you save up front with cheap gear will be lost when you waste time replacing broken bits every week. The trick is finding the right balance between cost and quality—the quality sweet spot.

The Anatomy of Redundancy

A truly redundant information system ensures that at least two ways to get information are present, regardless of what fails. In such a system, there are no *single points of failure*. The industry term used to describe this is "highly available."[1]

In theory, this appears to be as simple as having two or more of every system. In practice, drawing the boundaries of the system turns out to be rather difficult. If you're not sure what one system looks like, it's hard to know when you have two of them. Therefore, the difficulty of implementing redundancy depends on the nature and scope of the system in question.

The basic problem is that every system can be seen as a collection of smaller systems (until you break things down to the subatomic level), at which point quantum physics takes over and things get really hairy. For example, a network can be thought of as a system composed of many individual computers. Each computer is a system of hard drives, processors, memory, and other components (which are systems themselves). Likewise, you can group together a bunch of network systems and create a single, larger, internetwork system. The public Internet is the most well-known example.

Specific parts of a network can also be grouped together to create smaller "subsystems" within the larger network system. For example, a cluster of servers running database software is seen as a database system.

The upshot is that any device or network can be viewed as one big system, as many small systems, or as something in between. When we define a specific system we're really drawing a set of boundaries that are convenient to our needs. Much like the borders of a country, these boundaries can be totally artificial, or might coincide with natural, functional, or technical boundaries. For example, it's easy to think of a stand-alone PC as a system because it's physically self-contained. This is a natural

The Quality Sweet Spot

Every product line has a "sweet spot"; a point where you get the most bang-for-the-buck. To find it, get a product list from a reputable manufacturer and sort it by price. Go about 1/3 of the way down from their "top-of-the-line" model. This is the sweet spot. Revel in it. You should notice that you don't save much by going further down the list, you merely lose useful features/quality. You should also notice that the price curve is much steeper as you look up to the top.

---

[1]Ironically, IT professionals often find themselves more available than their systems. One could even suggest that there's an inverse relationship between the two types of availability. After all, someone who's constantly fixing systems has little time for a social life.

technical boundary, but in many cases a PC is useless without a human operator. The secretary's PC isn't going to create a memo on its own (yet). Functionally, the secretary is part of the system, too.

When planning redundancy it is important to align a system's boundaries with the natural, technical, and functional boundaries. Most of the problems in high availability appear wherever the boundaries diverge. For example, a system's hardware might be flawlessly redundant, but what happens if the software fails or an operator error occurs? The software and operator are outside of the technical boundaries of the hardware, yet are functionally critical to the proper operation of the system. In order to have true fault tolerance, the system boundary must also align with these functional considerations.

Size Matters

As if defining a system isn't hard enough, there are also cost and administrative factors. The cost of making a system highly available increases exponentially with the

---

Strangely Common Disasters

You're in the top floor of a tall office building. Are you worried about water damage to your systems? Probably not, but you should be.

Years ago, companies that had servers kept them in specially constructed rooms designed to protect sensitive equipment from fire, humidity, and water damage. These rooms had raised floors to minimize water damage, halon fire suppression to protect the electronics, and humidity- and temperature-controlled air circulation.

Today, many small and even mid-sized companies run their entire computer network from a walk-in closet or unused office. If a water pipe bursts or the building sprinkler system goes off, the servers could get soaked—either directly or from seepage from the floors above. The failure of a building's air conditioning unit during a heat wave could also lead to equipment damage from heat or humidity.

The best solution is to have a secondary location with just enough equipment to get critical business functions running. This should include spare servers that are ready to restore data from a tape backup and some basic networking equipment. Laptops make great temporary workstations—they don't need power all the time, they're easy to store, and people can work anywhere, including on the floor if necessary. Even a small company can afford to keep an extra server and hub at somebody's house, which is probably enough to keep things running.

---

size of the system. For example, adding redundancy to a single computer is much easier than a making a redundant database cluster. Redundancy also has a management overhead—broken parts need to be fixed or else the redundancy is lost. Too many redundant systems can create complexity, which leads to human failures (and these often do far more damage).

The largest practical system one can make redundant is the entire network. For example, major banks have their entire networks replicated in multiple countries across the globe in case disasters were to make a large portion of the world inaccessible (meteors, nukes, biowar, locusts, frogs, and so on). Creating network-level redundancy like this is extremely challenging, and often only necessary for very large organizations with offices around the world. Then again, some degree of network redundancy can prove invaluable during a mid-sized disaster, as explained in the sidebar.

Not all systems are designed with redundancy in mind. Firewalls are a good example. The rules of network routing prevent you from having two active firewalls operating side-by-side. Only one can be operational at any given time. You must switch off the failed firewall and activate the spare. Which means you need to know when a failure happens. You can't expect the failed firewall to turn on its spare—perhaps it failed by bursting into flames. Therefore, an external system, possibly within the spare firewall, is needed to detect failure and enact the transition. What this shows is that a highly available firewall solution requires more than two identical firewalls. It requires defining a larger system that includes additional hardware and software.

The key to successfully achieving high availability is choosing the right system boundaries. Pick too broad of a system and you'll be dealing with incredibly expensive issues such as global redundancy. Pick too small of a system and the complexity of managing the result will melt your brain. Straddle a functional or technical boundary and you might not be able to eliminate certain failure points. If you get it right you can plan that long vacation to a remote resort knowing that civilization will collapse before your network fails.

Final Thoughts

The following chapters will discuss some technologies that can be used to make your systems highly available. The simplest and most commonly used is the *Redundant Array of Inexpensive Disks* (RAID). The term RAID actually describes a group of similar technologies that spread data over multiple disk drives. Clustering is another technology that combines the memory, processing, and storage resources of multiple, specialized computers to improve performance and minimize the impact of an individual system failure. Distributed computing is similar to clustering, except that the computers involved are often unspecialized and spread out across multiple networks. High availability solutions will often use a combination of these technologies.

Oh, we almost forgot about the stealing food from picnics thing. We've been diligently tracking the research done with robot insects and distributed computing. We've come up with some innovative designs and algorithms. When combined with existing open-source software, it's relatively trivial to build a robotic insect army. Last summer we feasted like kings on stolen apples and sandwiches. Unfortunately, our publisher is very strict about our page count, and we simply can't fit the extra section into this book. But we think we'll be able to squeeze it into the second edition. So be sure to tell our publisher that you want a second edition (and that they should pay us handsomely for it) if you want to learn how to create your own RoboSwarm®.

Chapter 33
Ensuring Availability: RAID

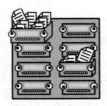

A technique for using multiple synchronized
hard drives to increase performance and/or reliability.

Technology Overview

Organizations and individuals store vast amounts of their data on hard drives today.
Hard drives offer large amounts of storage space and the ability to seek and retrieve
data very quickly. However, computers are imperfect machines made by imperfect
beings and warranties do not last forever. The result is a simple technological fact:
Hard drives fail. Eventually, you will be confronted with a dead hard drive; therefore,
it is important to be prepared. When using the right combination of technologies and
techniques, a dead hard drive is barely a blip on the radar screen.

What if you could store all the information that normally is saved on one hard
drive to two or more hard drives at the exact same time? What if when one of these
hard drives breaks, data could be instantly read from any of the other drives? It's
highly unlikely that two hard drives will fail simultaneously. Now, imagine that you
could employ this type of storage in real time. This very real technology that eliminates downtime due to drive failure is known to most of us as RAID.

How RAID Works

RAID is short for *Redundant Array of Inexpensive Disks*, or Independent Disks as some like to think. While many different types of RAID exist, the principle function is the same. RAID takes two or more hard drives and allows them to act together as a team. Normally, information or data is taken from the motherboard to one drive through its *drive* controller. With RAID, that same data is channeled through a *RAID* controller and then spun on to multiple disks simultaneously. The RAID controller is a device that enables multiple drives to remain in sync with one another. Once synchronized through RAID, a hard drive can share data in one of several ways, depending on how it is told to operate. The following are some examples:

Mirroring: The hardware requirements for mirroring are two hard drives (or any multiple of two) connected to each other with a RAID controller. A software interface writes data to the two drives simultaneously. If one drive fails, the other immediately takes over. The total size of storage space available is equal to half the number of total drives. The performance for writing is the same as writing to a single drive. Intelligent RAID controllers can get a read performance improvement by reading from the master and mirror drives in parallel, but most mirror controllers are not that intelligent.

Striping: This is defined as spreading information out over two or more drives. The main result and benefit is parallel performance for reading and writing. The RAID array essentially harnesses the resources of all the hard drives simultaneously, resulting in faster read and write access times. Striping alone does not provide redundancy —it simply enhances performance.

Parity: The main idea with parity RAID is to maintain a mathematical hash of the entire dataset, which can be used to rebuild any missing or corrupt data from the RAID array. A single parity drive can be used to recover any one failed drive in the RAID array. Two parity drives can rebuild two failed drives, and so on. Calculating parity can slow down the performance of writing information to the array.

Stripe + Parity (RAID 5): This concept takes a stripe array and adds one or more parity drives. A minimum of three drives is required. The striping gives a performance benefit; the parity ensures reliability. If one drive completely fails, the system is unaffected. The failed drive can be restored using a parity drive. The total storage area available is roughly the combined size of all the disks used minus the number of parity drives. While read performance is usually good, writing performance can suffer greatly in certain situations.

Stripe + Mirror (RAID 10, RAID 01, and so on): This is another way to combine performance with reliability. Here a number of drives are striped together, providing performance. An identical number of drives are used as a mirror, providing reliability. This gives great reading and writing performance. The total storage available is half the number of disks used.

RAID Building Blocks

Mirroring, Striping and Parity are the three concepts behind most forms of RAID. The more complex types of RAID systems are essentially combinations of these three basic ideas.

Mirroring is the simplest way to achieve high reliability. Striping provides high performance. Parity is a complex, but space efficient technique for ensuring data availability. The right choice or combination depends on the needs of the system being protected.

Legend
- ⊙ - Summary
- ⊚ - Reliability
- ⊖ - Storage Space Usage
- ⊞ - Reading Performance
- ⊘ - Writing Performance

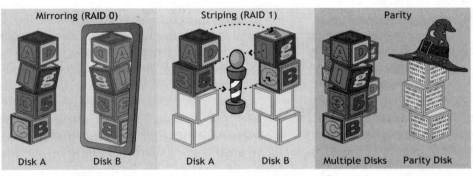

Mirroring (RAID 0)

Disk A   Disk B

Striping (RAID 1)

Disk A   Disk B

Parity

Multiple Disks   Parity Disk

⊙ : All disks have identical data (blocks).
⊚ : great
⊖ : poor (space of just one disk)
⊞ : normal to good
⊘ : normal

⊙ : Data is spread across all disks.
⊚ : none
⊖ : great (total space across all disks)
⊞ : good to great
⊘ : good to great

⊙ : Each parity can fix one failed disk.
⊚ : excellent
⊖ : good
⊞ : bad
⊘ : normal to great

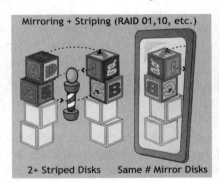

Mirroring + Striping (RAID 01,10, etc.)

2+ Striped Disks   Same # Mirror Disks

⊙ : All disks have identical data.
⊚ : great
⊖ : so-so (half the total space)
⊞ : good to great
⊘ : good to great

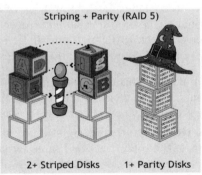

Striping + Parity (RAID 5)

2+ Striped Disks   1+ Parity Disks

⊙ : Data is spread across all disks with parity.
⊚ : great
⊖ : good
⊞ : bad
⊘ : good to great

Illustration by SageSecure

■ **Figure 33-1**

Security Considerations

RAID does not protect against software failures or hardware failures. This includes, but is not limited to, the following: Controller cards, motherboards, ram chips, viruses, NIC cards, faulty cable, hackers, fires, lightning, and earthquakes. The irony is many of these items listed may end up causing downtime long before hard drive failure ever does. Vegas anyone?

RAID makes for a good addition to nearly any environment where reliability is a concern. It should not be a stopping point when setting a precedent for data redundancy and availability. Above all else, backups are still necessary and should be considered as critical on a network with or without RAID employed.

Making the Connection

Storing Information: RAID arrays are made up of hard drives. They are often used to store large amounts of critical information that need to always be accessible. It is very common for RAID arrays to store databases, for example.

Best Practices

The different RAID types are best suited for specific environments. The first step to take when considering the purchase of a RAID array for a network is to prioritize performance, cost, and reliability. After you have established those needs, the advantages of each RAID method will be easier to analyze. The following information might be useful if and when you face this decision:

Mirror RAID: This is the peanut butter and jelly of RAID. It's ideal for simple environments such as workstations or less-critical servers. Mirror RAID hardware is relatively inexpensive and the simplest array only needs two disks.

Conversely, mirroring is not the best use of hard drive space, as you only get 50 percent of the total space as usable for storage. Lastly, it is not very fast when compared with other types of RAID. Seek times and write times are similar to those of normal hard drives that are not in an array.

Striping RAID: The most noticeable increases are in data seek times and overall drive performance. That's right, striping an array actually provides a speed boost to server performance, which is always a welcome feature. Striping is also the most efficient in use of drive space, providing nearly 100 percent of the total drive space for storage.

Striping + Parity RAID: This combination is a good balance between reliability, performance, and space. The drive array can survive as many failures as the number of parity drives. Usually, one or two parity drives are enough, as multiple simultaneous drive failures usually indicate a physical catastrophe that RAID will not cure. If a single drive fails, only that drive needs to be replaced and the replacement can be rapidly rebuilt from the parity data. The performance on reading will be somewhat better than mirrored reads, but writing performance may suffer due to the parity calculations. This is the best solution for environments that have large amounts of reading and high reliability requirements yet do not require frequent writing. Web servers might use this technique.

Striping + Mirror RAID: This is another balance between reliability, performance, and space. Here performance is boosted at the expense of available space. Only half of the total drive space is available for storage. In the event of a drive failure, the entire striped mirror needs to be rebuilt—a time-consuming process. This is the best solution for write intensive applications such as databases.

RAID Hardware: Choosing the right hardware is also important. Some vendors will sell RAID packages where it is not necessary to be concerned with the specific hardware that makes up the array. Other vendors will sell RAID kits or RAID compatible components. In this situation hard drives can be chosen by brand name. The particular brand of RAID controller card is also important.

A decision of hard drive architecture (*Small Computer System Interface* [SCSI] or *Integrated Drive Electronics* [IDE]) will be important. Many high-end RAID solutions use SCSI because it's been traditionally faster, more extensible, and more powerful. But SCSI drives are more expensive and produce greater amounts of noise and heat than their IDE counterparts. As a result, an air-conditioned server room is advised that is separate from the rest of the office. Adequate internal fans are also a must if you are constructing the server in-house.

Finally, a decision can be made to include a *hot pluggable backplane*, which allows any one of the drives to be removed on the fly and replaced if necessary. In other words, you can actually replace a damaged hard drive in a live environment, while the RAID array is still fully operating. This moves towards guaranteeing absolute availability of data, since the system never needs to be shutdown. This is a nice failsafe to be able to rely on in any type of environment.

While RAID will provide a good deal of redundancy, it only offers *drive* redundancy. You can have 15 SCSI hard drives in a RAID array and they will all be rendered useless if the power supply shorts out or the motherboard gets fried. Global redundancy is always a priority over local redundancy. On this note the following types of items should make any redundancy checklist:

- *Network Interface Card* (NIC) card redundancy
- Motherboard purchase date
- Dual power supplies
- *Uninterruptible Power Supply* (UPS) battery backup system
- Faulty network cables.

RAID Strategy: How much insurance is enough insurance? The more drives, the less chance of down time, but how many is enough? Consider total space requirements and total budget when making this decision. Next, consider speed requirements. Will you be serving an enormous volume of users? How much data does each user request in megabytes per hour? Is the server storing small bits of numbers or huge image files? These considerations will help in deciding how much to invest in the speed of the system. If larger, more expensive SCSI drives are purchased they will always provide the greatest possible speed and throughput. On the other hand, IDE drives are much cheaper and not *that* much slower.

- Mirror RAID: great for two drives, home PCs, and small office environments.
- With three or more drives, you need a good controller. Ask your controller manufacturer which RAID level they recommend for your performance specifications.
- Avoid software RAID because hardware RAID is not too expensive and gives you more flexibility and reliability.
- Buy a bunch of spare drives when you build the system. Hard drives are constantly changing. While most RAID systems let you use different sizes and types of drives, it's always easier to use the exact same type/size of drive.
- Never purchase the latest, greatest drives. Get drives that have been around for a year or more. Check the Internet for reviews and determine if any systematic problems with a specific model you are considering have occurred. Some motherboards have built-in RAID controllers.

Final Thoughts

WARNING: RAID levels (0,1,5, and so on) are not standardized and manufacturers uses the numbering system to fit their marketing needs, not the customer's technological needs. Because this is true, do not go to a manufacturer and insist on RAID 5. What they call RAID 5 may be some other RAID configuration you were not intending to use. Instead, propose to them your specific performance and space requirements and let them propose a configuration and call it whatever they please. Ultimately, make a purchase from the manufacturer that gives the best price for performance ratio and fits your redundancy needs.

Chapter 34
Ensuring Availability: Clustering

A technology that allows a system to remain operational if its hardware or software fails by replicating its running processes on a simultaneous system in real time.

Technology Overview

All the cool companies get high on clusters: high performance, high availability, and high costs. Sometimes, it seems like companies put in clusters because everyone else is doing clusters. Clusters are powerful, and they can do wonders for organizations that really need them. But they are very complicated, and can do more damage than good if they're not designed and installed correctly. In a poorly configured cluster, a single computer failure can create total data and service loss. In this chapter, we'll teach you to say no to clusters (unless you live in California and have a prescription from a specialist).

Cluster technology allows a number of computers to act as a single server by sharing resources. Exactly what resources are shared depends on the purpose of the cluster. For high performance, clustered computers can share memory and processing power. For high service availability, clustered computers can mimic one another by replicating data and application code to each machine. RAID (Chapter 32) can work across a cluster to combine the storage subsystem into a giant network drive. The more you want the cluster to do, the more complicated the configuration and management becomes.

In reality, a cluster is simply a group of independent computers that are combined to work as a system, a redundant system. Once the clustered group is properly set up, it exists as one single device on a network. After all the talk about RAID in the previous chapter, it would suffice to say that RAID is to hard drives as clustering is to servers. In other words, apply the concept of creating a redundant array of hard discs to a broader level. A redundant array of inexpensive computers you ask? Yes, except for the inexpensive part.

A comparison of clustering to RAID, while providing a friendly analogy, is not very accurate. Major differences exist between both concepts. For example, clustering does not have the hands-off internal management system prevalent in most modern RAID setups. A clustered group is managed by a single system, but that system requires the use of carefully arranged software and hardware.

Clusters can be a cheap way to get supercomputer-scale power. It also can be used as a high availability tool, useful for keeping critical services running 24x7. In some cases it's even used as an alternative backup system. Statistics show that it's cheaper and easier than ever to build a cluster. With all these benefits, it's easy to make a case for clustering in almost any organization.

How Clustering Works

The pieces that unite to create a complete cluster are known as nodes and resources. The role of each node is to provide a gateway to the resources. If one node or gateway goes down, another node takes over to provide continued access to data. In cluster speak this concept is widely referred to as fail-over.

A node is usually nothing more complex than a typical file/web/database server. However, the nodes are often connected using specialized hardware. This hardware is designed to handle the massive amount of data that needs to be rapidly moved from node to node. It also provides additional system status information that helps load balancing and failure management. In recent years the hardware required to set up a clustered system has dropped in price dramatically.

A number of different ways to share resources among cluster nodes exist. Of these, two models are most widely used: the *Shared Nothing model* and the *Shared Device model*.

Shared Device Model: With this type of cluster, applications can access any hardware that is connected to any node in the cluster. For this to properly occur, the data stored within the cluster needs to be synchronized. The synchronization process that occurs is handled by a service run on the cluster known as a *distributed lock manager* (DLM). Essentially, the DLM handles the coordination of the hardware being accessed by the cluster. If more than one node within the cluster needs to access the same piece of hardware simultaneously, DLM will resolve the conflict. Although the shared device model offers the advantage of applications being able to access all the hardware, it comes at a cost. DLM's effort to resolve software/hardware conflicts can put a significant strain on the whole cluster, resulting in reduced performance.

Shared Nothing Model: This is the common choice for most businesses today. It is cheap to employ and mainly software driven. Chances are if you have observed a working cluster in action, this is the type you have seen. With this type of setup, only one node has access to a shared resource at any given time. This eliminates the need for DLM, reducing the overhead of the cluster and increasing performance. If one of the nodes in the cluster does fail, the surviving node(s) can take over the resources of the failed node, ensuring their continuous availability to users.

Cluster Implementation: The most important point about implementation is that it takes time. Even if it is outsourced, cluster implementation can be time consuming because the tweaking and testing is a long process. Internal *Information Technology* (IT) staff in an organization should only take on cluster implementation if they have relevant past experience setting up the specific type of cluster being rolled out on the network.

The biggest challenge in setting up a cluster is getting the nodes to communicate—a process called "bonding" in cluster-speak. This is commonly done using two or more Ethernet channels to provide the greatest amount of bandwidth possible between the cluster nodes. Frequently, a specialized piece of hardware is purchased to make this possible.

Security Considerations

High availability is defined as a system that is always there for an organization despite any situation or circumstance. A cluster is considered highly available because it has many backup and fail-over systems built into its design. Being highly available enhances overall security from one perspective because data can be retrieved by an organization in the event of a system failure. However, the complexity of clusters also makes a network more insecure because it adds potential holes in the infrastructure. It also acts as a golden palace; a place that hackers, once they probe a network, will inevitably want to gain access to and control. It is imperative to follow good security measures before and after a cluster is implemented on a network. Here are some ground rules:

- Store all servers in physically secure locations. Clusters are implemented to provide extra reliability for critical data—respect that by supporting the data redundancy with physical security.

- Ensure that all typical security measures are in place and functioning across the network. This includes firewalls, *virtual private networks* (VPNs), Virus Scanners, and Sniffers.

- Secure all network infrastructure services such as *Domain Name Server* (DNS) and *Dynamic Host Configuration Protocol* (DHCP). If these are compromised it becomes trivial for a hacker to take control of a cluster.

- Be extremely cautious with cluster administrator privileges. Once administrator privileges are granted to a user, that user can make changes that could easily break or corrupt the cluster!
- The cluster administrator should be certain that connections from remote applications come from trusted, secured computers. A hacker can often leverage a workstation's connection to a cluster to compromise the cluster itself.

Many of these security tips assume that the network in question has been well designed and locked down. For more information on network design and hardening refer to the sections referenced below.

Making the Connection

Connecting Networks: Clusters are often connected by specialized hardware that eliminates single points of failure in the chain. The particular ways in which clusters are connected is critical to successful redundancy and fail-over.

Hardening Networks: Clusters are put into place on networks to make critical information highly available. It is obvious that this data needs to be kept from prying eyes. Before a cluster in implemented on a network, the network must be properly secured.

Storing Information: Clusters are made up of computers. Often, clusters are used to store large amounts of critical information that needs to always be accessible. Many different types of storage hardware can be used to build a cluster.

Best Practices

Clusters are usually implemented in organizations for one of three reasons: parallel processing, load balancing, or fault tolerance. Of course, more than one of these tasks can be handled by clustering, even the same cluster, in any organization.

Parallel processing is a concept applied to applications that require extraordinary amounts of computing power. Computers in a cluster can be organized such that the individual processing power of each node is harnessed to make one larger "super-computer." When clusters are used for parallel processing, a very powerful computer is simulated by the cluster for a fraction of the cost of a real supercomputer. Clusters are often seen used like this in large research facilities, educational institutions, or any organization that would require a large amount of processing power.

Load balancing is a very different need from parallel processing. In load balancing, the nodes of a cluster are set up to have network traffic and data distributed over them equally. This is often seen in organizations that use applications where the performance needs are not consistent and vary from mild to extreme over time. Web servers may typically behave this way. If one web server suddenly becomes overrun with requests for web pages, it may spill the requests over to another node in a cluster.

With fault tolerance, the cluster is prepared to handle the failure of any piece of hardware or software involved in the cluster. At least one node in a cluster is handling resources at a given time. If the node fails, another node in the cluster takes over the responsibility of those resources automatically. The benefit to fault tolerance is zero downtime. This is made possible by the mirroring of the critical hardware and software components of the cluster from node to node.

Not all cluster systems are created equally, and not all cluster systems are created for commercial use. As computer hardware has become faster and cheaper in recent years, clusters have become more viable. Clusters are no longer only for the rich and famous and as a result, projects like *Beowulf* have taken off. To quote the FAQ from www.beowulf.org, "Beowulf is a high-performance massively parallel computer built primarily out of commodity hardware components, running a free-software operating system like Linux or FreeBSD."

Projects such as Beowulf have encouraged people to build clusters from whatever computers they can get their hands on. These clusters are computing's version of Frankenstein. They aren't pretty and occasionally express their anguish by going on a rampage throughout rural German towns. If you decide to go this route, read Shelley's book. It might offer some insight on the love-hate relationship you'll be having with your pieced-together beast.

For practical business purposes, always build a cluster from equipment that has been properly designed for clustering. Projects like Beowulf make for a fine weekend hobby, but what they save in money they will cost in time. As it is, cluster-ready equipment is complicated and time-consuming to set up. When configuring a new cluster, one must begin by selecting all the components that will be part of the system. Each component will need to be treated individually until the completed cluster is set up and running. Here are some things to consider:

- Time is money, and getting clustering right takes time, a lot of time.
- Pay attention to specific applications needs. Compatibility is important with clustering on all levels, hardware and software.
- Clustering is not always the best solution, find other firms in similar situations to yours and determine what solutions they used to solve the problem.
- Has the vendor selling the cluster kit sold them to other companies who have been happy with the results? Find out and get referrals.
- Clustering sometimes simply doesn't work! This may sound extreme, but it can be true from time to time.
- It is only very recently that clustering has become accessible to the common man, as a result, it is not extremely tested at the mid-size business level.

Bring in outsiders to help you set up the cluster. Be sure they have specific and repeated experience setting up the type you have purchased.

Final Thoughts

Good long term-planning is an absolute requirement before implementing a cluster. A network should be proven reliable and secure long before a cluster is added. That said, clustering could provide a tremendous amount of services to an organization. Clusters are easily expandable, can be obtained at a relatively low cost, and are very highly reliable. If your business has data or applications that require enhanced performance, limitless uptimes, and good security, it may be time to get high on clusters.

Chapter 35
Ensuring Availability: Backup Systems

Backup systems actually perform two functions,
copying critical data to another location and enabling
the swift and painless recovery of the backed up data.

Technology Overview

Security is as much about getting back on your feet after a disaster as it is about preventing a disaster. It is ironic, actually, because recovery is rarely associated with security. Think of a backup system being the direct equivalent to a good homeowner's insurance policy. One can never really tell how good insurance is until he or she needs to use it in an emergency. Once a home has been robbed, however, insurance is actually put to the test. How quickly does your agent answer you phone calls? How fast is an appraiser sent out to your home to assess the situation? Most importantly, how quickly do you get made whole again? An insurance policy may look great on paper, like not having expensive monthly payments, but when the time comes to get back on your feet if the carrier's response is not quick and straightforward you will find yourself in an unfortunate situation.

Modern data backup systems, much like insurance policies, cannot be truly evaluated until restoration is needed. Of course, no business wants to be in a position that requires restoration. To complicate matters even more, many choices are available to an organization that needs backup solutions. It is a forgone conclusion that testing all of the available backup systems will not be an option. Instead, achieving a firm grasp on the different backup systems available is the best a company can hope to accomplish. With the proper knowledge, the appropriate backup system to match any security philosophy can be obtained and tested.

> **What people think:** We have installed our *Redundant Array of Inexpensive Disks* (RAID) and cluster systems, so we no longer need traditional methods of backup.
>
> **What we think:** Security is as much about recovery as it is about prevention and the only chance for clean recovery is good solid backup.

How Backup Works

Many types of backup systems are available and they all work slightly differently from one another. Specifics aside, all backup systems serve the same purpose: to make a duplicate of critical information. Once a duplicate is successfully made, a secondary goal of backup systems is to move the data to an alternate location. An alternate location is defined by any location other than the primary source of the data being backed up. The reason for taking backups off site, in any form, is in the event of on-site damage such as fire, flooding, or any other act of God.

Many different attempts at simplifying the backup process have been made. Modern backup systems come with software that provides a user-friendly management interface. The software is installed into a machine, automatically detects the available drives and interfaces with them. At this point, backups can be created and controlled. A backup procedure is generally referred to as a *job*.

Backup software is capable of setting up multiple jobs and executing them at multiple times of the day, week, month, or year. Jobs should be set to run at hours where the systems being backed up are not in use or are in limited use. The backup process will take a toll on system resources—the larger the backup, the greater the toll. Each job can be customized to specifically backup any files or directories on any computer or computers across a network.

The management interface provides total control over all aspects of backup on the network. Some backup software can even write data to drives on remote machines. To use media most effectively, the backup software will allow two basic styles of writing: *complete* or *append*. Complete will perform a total overwrite of the media placed in the drive to store the backup. Append will only write the changes that have been made to the files being backed up since the last backup. If a good backup plan is put in place, appending media can save time and storage space.

Choosing a storage device for backup will most likely occur at the time backup software is purchased. Most storage devices are packaged together with backup software. One common choice for backup media is tape drives. Tape drives and tape media have been around for much of computing history. They make an excellent source of backup media for several reasons. They can hold huge amounts of data on a small amount of physical space for very little money. A blank magnetic tape that is smaller than the palm of a hand can hold over 50 gigabytes of data. Find another medium that can hold that much information and still retail at less than twenty-five dollars.

However, tape media have their drawbacks. Tapes are fragile. They can be easily damaged or erased. Passing any form of magnetic field near a tape, for example, will delete or damage the contents. Tapes are also extremely slow. Modern tape drives will boast faster and faster recording rates, but compared to other forms of storage media tapes move like a tortoise. That said, the slow speed of tapes does not negate their effectiveness as a backup medium. There's rarely an immediate need for a completed backup, so a long backup process is acceptable for most networks. In many cases, the backup process can run in the background without affecting system performance whenever necessary. Even so, backups are often run during off hours to minimize any potential impact on usage.

An alternative to tape backup is CD-ROM backup. A typical blank CD-ROM can hold approximately 700 megabytes of data. A typical blank DVD-ROM can hold upwards of 13 gigabytes of data. While larger tapes beat out even DVD-ROMs in the storage area, CDs and DVDs are much more reliable. They are difficult to damage and are not susceptible to erasure upon exposure to magnetic fields.

Security Considerations

Ask yourself what happens when information is duplicated. The answer, of course, is that two or more copies of that information exist. When more copies of something is available it becomes much easier to get at and much more difficult to secure.

Backing up critical data systems essentially halves security. Security is even further reduced when the backed up data is moved off site to an insecure location, which is extremely common. It is always the most valuable data that is specifically backed up and relocated off site. It is this same valuable data that many companies want to keep from prying eyes. Backup and security are inherently at odds with one another.

The problem is most organizations spend so much time establishing a backup strategy that security never enters into the equation. Understandably so, as establishing an effective backup and restoration strategy can be an exhaustive process. Unfortunately, backup does not take place in a vacuum and securing the duplicated information is just as essential as securing the source.

A trusted and secure location should house backed up data. This could be a company with a great reputation that handles outsourced backup. Or it could be a fireproof safe that holds tapes carried off site by a trusted employee. Whatever the scenario, data that leaves the main network, in any form, should be secured and accounted for at all times.

Making the Connection

Outsourcing: Backup is frequently outsourced. Typically, large *Internet Service Providers* (ISPs) are capable of performing this type of service.

Storing Information: All different types of media can be used to back up primary computer systems. Though tape media and tape drives are commonly used, more options are available as storage becomes cheaper and larger.

Best Practices

Take small steps with backup. A good backup system should be able to grow with the needs of any organization. No one quick solution and no one fantastic answer to any backup problem is available. The best approach to take is to focus on the central backup needs and slowly build out from the middle. Prioritize the demands of the organization. Should cost come before speed? Should off-site removal be automated or handled manually? How much data really needs to be backed up? If all our systems were destroyed, how quickly would we need to be functioning again?

The process of restoring data is much more important than the process of backing it up. Ironically, most organizations spend their time concerned about backing up their data. Backing up data is a nice idea, but what if the data does not restore itself properly? Restoring data can be a tricky process. Sometimes it does not work and it leaves the company that spent all that time and money on backups in quite a bind.

To avoid restoration problems it is important to run restoration testing. Perform a full backup and then restore the data to another location. If the restoration fails, look closely at the possible causes. If tape media is being used, check the tape drive for dust or dirt. Run a tape-cleaning device through the machine. Tape drives are sensitive. Also check the files being backed up to see if they have been corrupted, either before or after the backup process. Sometimes files become corrupted during a long backup procedure while being transferred to a drive. In many cases, older files that have not been updated can be corrupted as well. Corrupted files may backup normally, but often will not restore at all.

Databases have special backup needs, as they do not commonly store flat data. Most backup software vendors will sell modules that can plug into the backup management interface. A database module can be licensed for an extra fee. Once it is set up, it should be thoroughly tested, and restoration should be tested again as well. Backing up and restoring databases successfully can be more difficult than traditional flat files.

Final Thoughts

The need for backup is common knowledge in today's fast paced world of information technology. Everybody has such a good understanding of backup that they forget to think about the real security issues. Data protection and recovery issues are complex and change from organization to organization. Consequently, strategic thought about backup is often brushed over or taken for granted. Don't put backup on the backburner when developing a security strategy—give it the time and thought it deserves.

XII
Detecting Intrusions

Summary

No matter how good your defenses, eventually a hacker will break through. How will you know when this happens? How will you catch the villain red-handed? Intrusion detection technologies can help spot hackers during and after the fact. Some of the tools can even identify places a hacker might attack before anything bad happens.

Key Points

- Properly deploying an *intrusion detection system* (IDS) is a massive undertaking that can only succeed if the organization has a compatible security philosophy and policies.
- Regardless of the marketing, intrusion detection systems are tools for experienced network administrators—not solutions that automatically solve problems on their own.
- Some hackers are people, but most are actually computer programs.
- Most intrusion detection systems are designed to catch people hackers, but end up being optimized to catch program hackers.
- As used in practice, intrusion detection systems are glorified virus scanners (and we all know how effective those are).

Connecting the Chapters

Detecting intrusions can only be accomplished with persistence and consistency. The technologies covered here can help to verify the integrity of systems and data and detect if an intrusion has taken place.

- **Chapter 36, "File Integrity,"** looks at tools can detect unauthorized modifications to critical system files and data.
- **Chapter 37, "Virus and Trojans,"** covers malicious applications intended to give third parties some form of control over remote computer systems.
- **Chapter 38, "Network Scanners,"** describes programs that examine critical network systems services for configuration errors and vulnerabilities.
- **Chapter 39, "Network Sniffers,"** captures network traffic for the purpose of analysis and intrusion detection.
- **Chapter 40, "Logging and Analysis,"** explores gathering and analyzing diagnostic status information from network devices and software.

Introduction to Detecting Intrusions

An organization's security philosophy will dictate its stance against hackers and cyber crime. If the philosophy dictates aggressive defense against malicious users, an *intrusion detection system* (IDS) will be a necessary tool.

Properly and effectively using an intrusion detection system can be costly and time consuming. It often requires bringing on specialized staff or consultants. The right staff will have the expertise to set the traps, hide in the trees, and snare the enemy when he or she walks over the net.

An intrusion detection system is made up of several parts. Each of the parts needs to integrate seamlessly into the physical network on which it resides. In addition, the design philosophy of the system as a whole needs to match the security philosophy of the organization.

Intrusion Detection Is an Art and a Science

Certain prerequisites are necessary to make an intrusion detection system worthwhile, including the following:

- Network administrators with intrusion detection experience.
- Network administrators with the spare time to monitor a network.
- A security philosophy that includes the willingness to integrate business processes with logging technologies.
- A network infrastructure that has been well designed and hardened.

Without these elements, the intrusion detection system will be an expensive system that won't produce results. IDSs are powerful, but complicated. They can be used to catch hackers, but they need to be wielded by an experienced network administrator. However, when they are purchased by an organization that mistakenly thinks the system will create results automatically, they fail miserably.

Despite the way vendors often market intrusion detection systems, they are not automated solutions. Virus scanners are an example of an automated solution that is used to detect, prevent, and delete viruses. Virus scanners are able to accomplish this with some efficiency because they are programmed to chase other programs. All of the movements that viruses may make are predetermined patterns. The virus scanner is constantly updated to detect and delete the latest viruses. Without the updates, it would be helpless against the newest viruses whose actions cannot be predicted.

A real hacker may use some programs as tools to help get at what he or she wants, but ultimately the high-level hacking will be performed manually. This is why catching a hacker is an art as well as a science. Intrusion detection tools can be used to catch an intruder in action if they're focused properly on the right things. Knowing where the intruder might be at a given time (how to focus) takes expertise.

What's a Hacker?

There are three basic levels of "hacking." At the highest level are the hackers capable of discovering security vulnerabilities in software. These hackers are often students, academics, or security professionals. They are motivated by curiosity, the desire for recognition, or a direct mandate to detect flaws in a system's security.

Next are the programmers, called crackers, who take a discovered vulnerability and turn it into a practical program known as an exploit ('sploit). The program is capable of attacking a system and gaining access by taking advantage of the vulnerability. Sometimes the high-level hackers will create harmless exploits as a "proof of concept." Usually the exploits created by crackers are more malicious and install backdoors (trojans) on the compromised system.

At the lowest level you have script kiddies, people most often characterized as teenage pseudo-hackers. They simply download exploits made by crackers and launch them against large swaths of the Internet. When they find an exposed machine, they leave their mark, often disabling a Web site or computer network with a *Denial of Service* (DoS) attack. Real hackers —the Net's elite, look down upon script kiddies because they lack real skill. Script kiddies take shortcuts; they use prewritten programs to take advantage of documented loopholes in computer systems. With this approach to hacking, little planning is involved and almost no engineering is required.

*Worm*s are exploits that have been combined with programs that automatically search for new susceptible target machines. Many of these find-and-exploit programs are designed to identify a target, gain access, install a copy of the worm software, and use the infected machine to search for new victims.

The vast majority of hack attempts on the Internet come from worms and other automated programs. Script kiddies are constantly launching new worms, and old worms can take a while to die out. As

Full Disclosure

The details of a newly discovered vulnerability are often released to the public through various security discussion lists (the "bugtraq" list being the most popular). "Full disclosure"" happens if enough information is given to create a workable exploit, or if an exploit is released as proof of concept.

It is considered bad form to fully disclose details of a vulnerability without first notifying the software vendor and giving them adequate opportunity to create and release a patch. Most hackers will wait one to two weeks, although some will give more or less time, depending on the vendor and the application.

Once the patch is out, the details also come out. It's never long before a workable exploit is released and launched against unpatched systems. This is why it's critical to ensure that security patches are applied as soon as they're released.

a result, the Internet is full of digital slimy wigglers that are endlessly scouring the net looking for susceptible computers to exploit.

The truth is, the average hacker is actually a computer or a teenager who's acting like one. The highly skilled human hackers are far more capable, but are generally not interested in pointless hacking. These are the hackers of legend, but the odds of one attacking your network are slim.

Needless Complexity

Is the goal of an intrusion detection system to spot the vast majority of hacking attempts (the computers) or the small number of professional attempts (the hackers)? Stopping automated hacking attempts would require an infrastructure similar to a virus scanner. A device would filter all incoming/outgoing traffic, matching against known automated hacking signatures. The signature database would have to be frequently updated. There'd be little need for a network administrator's involvement. Properly done, a system like this would stop the vast majority of all hacking attempts on the average network.

What about those hackers? Well, security companies have successfully convinced many major corporations that the handfuls of skilled hackers are really the problem. They have designed their commercial intrusion detection systems as massive intelligence gathering tools that are capable of tracking many professional hackers through a network (when used correctly). Ultimately, this means having a network administrator who knows how to deal with the high-end tools bundled with an IDS. Without the right network administrator talent, the IDS becomes a glorified virus scanning device.

For most organizations' real needs, the current crop of intrusion detection systems are needlessly complex. The functionality for the uber-administrator with tons of experience and time should be offered separately from the core, low administration virus scanner-like approach to detecting script kiddie activity. In the broader market, intrusion detection systems still maintain their original positioning. They are known to be high-level tools that can stop hacker intrusions when used properly. But this could all change in the coming years. IDSs could formally travel down the road of virus scanning software.

The Vicious Cycle

Currently, every corporation has a virus scanner and every few months a new virus is released that the scanner does not know about and can not protect against. Virus scanners are reactive programs: They can only chase new viruses once they are created. They will always be playing catch up. If intrusion detection goes this way, it will become another cat-and-mouse commodity. Security professionals want to see IDS's deployed correctly, but who's got the time and skill to monitor all that information?

Final Thoughts

Intrusion detection systems are a good thing. They need to be approached with respect and caution. Stopping hackers in their tracks is not easy and it is only getting more difficult. It is, after all, the hackers who hold all the cards. They can attack what they want, when they want, with little or no rationale. Creating an impenetrable defense will be impossible. But creating a practical defense that is consistent with your organization's security philosophy is a realistic and attainable goal.

Chapter 36
Detecting Intrusions: File Integrity

File integrity tools can detect unauthorized
modifications to critical system files and data.

Technology Overview

You suspect that a hacker got into your network. But how do you know if any data
was damaged or changed? Even worse: you have no idea that a hacker got into your
network, but accounting just called and found a really strange discrepancy in the
books.

What people think: If a hacker gets into our network and does
damage, it'll be obvious.

What we think: It's easy to tell if you've been hacked if the hacker
changes your company Web page. But what if he breaks in and subtly
change a few files? What if the programs you use to check up on your
system have been replaced by hacked versions that hide the hacker's
activities? How will you know?

File Integrity tools help determine if critical system and data files have been tampered with or altered. When something looks unusual the integrity checker will send out some type of alert (an email, a message to a pager, and so on). Some integrity systems will automatically replace the tampered file with a version that's known to be safe. This process can help detect and recover from intrusions. It can also help with general system problems such as corruption due to hard drive failure. Many situations that might result in the destruction of data can be identified, remedied, and possibly prevented with file integrity tools.

Most of the major file integrity tools will first make sure that critical system files haven't been altered. Under Windows, this means the registry, startup files (autoexec.bat), and many of the files that live in the Windows, or WinNT directory, especially the major system libraries (*Dynamic Link Libraries* [dlls]). It might also include major Microsoft programs such as Outlook, Word, and Excel. Under UNIX systems, the core files are the system configuration files, the boot files (kernel), the standard system programs (/bin, /sbin), standard library files (/lib), and some critical user applications (/usr/bin, /usr/lib, /usr/sbin).

Past the basics, the rest is often up to the administrator to configure. Checking all files for changes is not an option as some files change too frequently while others are simply not important. The administrator will usually have to approve all changes to files that are being monitored. This can be a timesink, so finding the right balance between effective monitoring and critical monitoring is important.

How File Integrity Works

The fundamental concept behind file integrity is the ability to compare a file to a known good version of that file. Imagine you had a copy of every critical system file burned to a CD. You know that it's impossible to alter the CD (the CD drive on your server can only read), so you can safely assume that none of the files on the CD have been modified. The files on the CD are referred to as known good. There might be a serious problem if a file on the hard drive is different from the corresponding known good file on the CD.

If it were possible to maintain a CD drive with an updated copy of a systems' critical files, checking file integrity would be easy. The problem is that many critical files on a system occasionally change, but need to be monitored anyway. These files cannot be burned once, and considered "known good" forever, so a more flexible system is needed.

One of the better solutions is to take a snapshot of critical system files and place the snapshot in a protected location. While more flexible than the CD solution, this requires the use of hard disk space. To improve efficiency, most file integrity systems do not make a complete copy of each file. Instead, a mathematical process called a hash function is used to take a fingerprint/signature of each file. These fingerprints are securely stored in a protected location. If any portion of a critical file changes, the fingerprint of the altered file won't match the fingerprint of the original file.

File Integrity

Illustration by SageSecure

Certain files are critically responsible for the integrity of a computer system. It's important to ensure that these files are never altered or stolen. A file integrity tool creates a unique "fingerprint" for each critical file on the system. If the file changes even slightly, the fingerprint will be noticeably different.

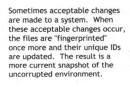

Sometimes acceptable changes are made to a system. When these acceptable changes occur, the files are "fingerprinted" once more and their unique IDs are updated. The result is a more current snapshot of the uncorrupted environment.

If files are altered or deleted by a third party, a file integrity system will alert the system administrator. The integrity database will help pinpoint the exact files changed, which will enable a speedy recovery from backup.

Figure 36-1

An automated process continuously takes fingerprints of all the critical files and compares them to the known good fingerprints in secure storage. An alarm is triggered whenever a mismatch occurs. This often results in an email or pager message being being sent out to the system administrator(s).

Security Considerations

Sneaking Past the Guard: Integrity checkers do not constantly check all of the files—it would hurt the performance of the system. Therefore, most checkers use a schedule. With a little bit of monitoring, a hacker can figure out the schedule and do all the dirty work in between scans. With any luck, all the damage can be done and cleaned up before the checker makes its rounds a second time.

Blocking the Alarm: Most file integrity systems send out alerts via email. This is an obvious solution, but what if the email doesn't get out? A smart hacker may launch a *Denial of Service* (DoS) attack against the network's mail server. If the integrity checker can't reach the mail server, the message will stay on the hacked system. The resulting delay can buy the hacker enough time to delete the outgoing message and prevent the alert from ever being sent. The hacker could also use a number of tricks to cause the mail server to send all further alerts to an invalid system.

Compromising the Comparison Process: The program that actually does the file comparisons is vulnerable to being directly attacked. If a hacker can obtain access to the program files, he or she may be able to replace it with a non-functional version. The hacker can also alter the database of fingerprints, deleting or updating entries to reflect the new changes. Most current systems are vulnerable to this type of attack if the hacker gets full access to the system.

Altering Files without Changing the Fingerprint: Technically, it should be impossible to alter a file and still get the same fingerprint. In reality, it is possible, although very difficult, to accomplish. Of course, if somebody creates and distributes a point-and-click program that can automate this process, it's no longer all that difficult!

Memory Hacking: A number of critical system programs are loaded into memory when the computer first boots or when the program is first run. The core portion of the operating system (called the kernel) is one such system program. Other programs are placed into memory for efficiency purposes. Few, if any, file integrity systems can check the integrity of software that resides in memory. Hackers can either alter the version of these memory resident programs, or create a new version in memory that replaces the original. When the system or a user runs the program, the hacked version in memory gets executed instead of the clean version on the disk drive. This will continue as long as the system doesn't reboot and the memory doesn't get flushed. The result is a complete evasion of the integrity checking system. The most advanced hackers directly manipulate the kernel of the machine, giving themselves ultimate control over the system.

Database Data: Some information is highly valuable, yet changes often. Databases tend to hold information that is constantly being added, deleted, and edited by applications and users. How can you tell which modifications are legitimate and which are not? In most cases, the task is simply impossible. File integrity tools won't work. Instead, the database and related applications need to rely on their own security and integrity systems. Some interesting systems have been created to establish patterns of normal database use and observe anything out of the ordinary. Massive changes to table structures, large inserts/deletes, or modifications to password tables can trigger alarms in these systems.

Monitoring Too Much: Configuring the system to look at too many files will cause unnecessary alerts. For example, user data and temporary data should not be be monitored. It is important to flag only critical files for monitoring. While this may take a greater deal of time initially, it will save time in the long run.

Monitoring Too Little: If you don't monitor enough, you might miss something important. For example, some applications create their own user accounts and store configuration information in nonstandard locations. If you're just tracking system files, you'll miss any changes to application configuration files.

Ignoring Reports: On a busy network, things will change frequently. Rarely will these changes be the work of an intruder. After a while, the frequent alterations that result from routine changes can lull those watching into an apathetic state.

Poor Initial Design: If the systems being monitored are not designed well, or if the monitoring system is deployed inaccurately, there will be lots of spurious reports.

Making the Connection

Local Filesystems: File integrity systems check the local file system. If a security issue with the local file system exists, it creates a weakness that can be exploited to bypass the integrity checker.

Storage Media: As with local file systems, issues with the storage medium itself can be exploited by a hacker to work around an integrity checking system.

Internet Services: Alerts are often sent via email, which can be blocked or intercepted.

Intrusion Detection: Integrity checkers are often key components of any IDS.

Viruses and Trojans: Integrity checking systems and virus checkers are opposites of each other. The virus checker assumes that all files are good unless they match a "bad" virus signature. The integrity checker assumes files are bad unless they match a "good" signature.

Best Practices

Proper Configuration: It sounds so simple, but it's so rarely done. Spending the time planning out the files that need to be integrity checked and then activating the system *before* putting it online is critical. We've been guilty of putting off installing an integrity checker. By the time we got around to it, we couldn't guarantee that the system was in a known good configuration. So we had to reinstall the whole thing from scratch. Not fun. If high security is a must, think about this next suggestion.

Using Read-Only Media: A good strategy for systems that need to be very secure is to keep the integrity checking program, configuration files, and fingerprints on read-only media. In an ideal situation, you'd have a read-only CD drive in the machine and a CD-R in it with all of the critical integrity checking files and fingerprints. Whenever the fingerprints or configuration changes, an external CD burner is attached to the machine and the CD is updated. In such a situation, the hacker would have no opportunity to alter the the signature database as the CD is read-only.

This method doesn't completely stop the hacker—a number of ways to work around the integrity checking system are available, but it does mean that you can always trust the signatures on the CD. Therefore, in a clean controlled environment, files can always be compared from a clean source.

Period Reboot-Comparison: Every once in a while, you should shutdown critical machines and then load them using a custom boot disk. This boot disk should have a clean version of the integrity checking program and clean versions of all the critical signatures. The hacker can't control the environment on your boot disk—it stays locked away in a safe somewhere. Therefore, you're guaranteed to get a reliable status report on the cleanliness of your system.

Alternate Alerting System: It's a good idea to use an alternate system besides an email alert, since email service is easily disrupted. Hooking up a pager system is one option. Another is to invert the system: have an "OK" message sent out periodically. If too much time passes between "OK" messages an alert can be triggered.

Final Thoughts

File integrity is one of those ideas that sounds great, until it comes time for implementation. In practice, file integrity takes determination and patience to effectively implement. Furthermore, integrity is only good if the baseline file is clean. Can you trust most systems to have clean files? The best time to effectively implement file integrity tools is when a system is newly installed. Otherwise, relying on the integrity of the system files usually requires a file wipe and rebuilding the system from scratch.

Chapter 37
Detecting Intrusions: Viruses and Trojans

Viruses and trojan programs are malicious applications
intended to give third parties some form of
control over remote computer systems.

Technology Overview

Viruses are one of the most publicized aspects of the modern computing world.
Reporters love to write articles about them and people love to ooh and ah over the
concept. Even the least technologically savvy people on the planet seem to be puz-
zled and naturally curious about the fact that computers can get sick, too.

Of course, the rest of us know computers do not get sick; they simply inadvertently download small and pesky programs that have been written with malicious intent. These programs have historically been known as a trojan or a virus. Today, the
differences between these types of malicious programs have blurred. Most scanning
software on the market claims to address both problems with equal proficiency. The
fact remains that viruses and trojans, although lumped together, perform different
functions for the attackers who create them.

A virus is a small program that is designed to infect one or more of a computer's
files. As it spreads, it may cause a variety of serious or benign problems. Sometimes

the goal is to annoy and sometimes the goal is to destroy. Sometimes a virus will delete files or even format an entire hard drive. In other cases, a virus may display messages across the screen merely to make itself known. In almost all cases however, a virus will render an operating system unstable, often due to its exploitative code.

To meet the definition of a virus, a program must meet two criteria:

- It must execute itself. To accomplish this it will often place its own code in the path of execution of another program.
- It must replicate itself. For example, it may replace other executable files with a copy of the virus-infected file. Viruses are capable of infecting any machine on a network, whether categorized as a workstation or a server.

Once a system has been infected with a virus, the virus can execute its *payload*. A malicious virus might have a payload program that deletes critical files from a hard drive. Some viruses are highly destructive, some are just annoyances or spread harmlessly, and most fall somewhere in between.

Viruses present themselves in a variety of ways with a variety of purposes. Computer history has revealed five officially recognized categories of viruses. They are as follows:

File Infector Viruses: File infector viruses infect program files. These viruses normally infect executable code, such as .com and .exe files. They can infect other files when an infected program is run from a floppy, hard drive, or from the network. Many of these viruses are memory resident. After memory becomes infected, any non-infected executable that runs becomes infected.

(Master) Boot Sector Viruses: Boot sector viruses infect the system area of a disk—that is, the boot record on floppy disks and hard disks. These viruses are always memory resident in nature. Boot sector viruses are seldom seen today because they were designed to exploit *Disk Operating System* (DOS) systems. Once the boot record of an infected disk is accessed, the virus remains in memory, and all floppy disks that are not write protected will become infected when the floppy disk is accessed.

Multi-partite Viruses: Multi-partite (or polypartite as they are sometimes called) viruses infect both boot records and program files. As a result, it is not easy to repair the damage caused by these viruses. If the boot area is cleaned, but the files are not, the boot area will become infected again.

Macro Viruses: These types of viruses infect data files. The damage done by these viruses has had the most direct impact on corporate *Information Technology* (IT) departments in recent years. The sidebar describes this pest in greater detail.

Trojans serve a different purpose. Their goal is not to spread, or even to annoy. A typical trojan application is cleverly disguised as another program, so as to not be detected. Because it cannot replicate itself, a trojan needs to be invited in to a system, much like the original gift from Troy. Once invited, it contains a hidden surprise that will give the enemy (malicious cyber criminals) a tremendous battle advantage, thus the assignment of the not-so-clever name.

A trojan is often extremely difficult or even impossible to detect. They are designed to leave little or no digital paper trail. When a user inadvertently executes a trojan program, it buries itself into the operating system, residing undetected for potentially infinite periods of time. It can delete log files that the operating system may create, thus destroying any record of its functions. The purpose of a trojan is often to give

> ## The Etymology of a Macro Virus
>
> Modern software has been designed to integrate flawlessly with a user's everyday needs. Companies behind this, such as Microsoft, want to make the job of developing applications easy as well. They have provided development tools that work with their applications flawlessly. Examples of this include Visual Basic and its relationship with Microsoft Office. It allows users and developers to simplify common tasks with the click of a button. These little programs are ironically called *macros*. Visual Basic also makes MS Office development so easy, it has inadvertently given the average developer fantastic tools to create viruses. As a result, viruses created as Visual Basic macros are the most common seen in circulation today.

an external user total backdoor control of a system, without the user's permission or knowledge. Scared yet? Don't panic; keep reading to learn the best way to deal with these problematic devils.

A virus with a trojan as its payload creates a particularly dangerous combination. The virus can get into systems and replicate itself, leaving little trojans in its wake. Within a short while, a vast network of machines will be under the virus creator's control. These machines are called zombies; a large collection of zombies under centralized command is called a zombie-net. These zombie-nets are often used to launch untraceable, massive distributed Internet assaults such as a Denial of Service attack.

For every crime a law has been written, and for every criminal a cop is out there looking to bag the bad guy. Computer viruses scanner (anti-virus) applications are analogous to police, as they have always focused on identifying, catching, and eliminating the latest viruses in circulation. Most recently they have also set their sights on trojan applications, despite the fact that these programs are extremely difficult to identify.

What people think: Antivirus software will protect a network from all intrusive attacks. Once it is set up and running, a virus or trojan has no chance of being contracted. The reason I have never gotten a virus is because I have always run antivirus software diligently on my system.

What we think: In our extensive years of computing experience we can count the number of viruses we have come into contact with on our hands. Did we mention we have six fingers on each hand?

How Antivirus Software Works

Antivirus software is not a very complicated concept. The software is designed to scan files, folders, or entire drives on a system. This includes media drives such as a CDROM or floppy. Once directed at a drive location, the software recursively scans all the files within the selected location. During the scanning process, the software looks for viruses and trojans that it knows about. The known viruses and trojans are stored in a database, which is maintained by the developer of the scanning software.

The scanner's fundamental ability to do its job is based on the strength and accuracy of an ever-growing database of virus and trojan signatures. The companies that have successful antivirus software products have maintained their reputation through their ability to keep these large databases effectively up to date.

Once a virus scanner is installed, the first thing it should do is request a connection to its virus database so it can download the latest virus definitions. This update process is routine and automated. Once it updates itself, the scanning process begins. Most scanning software will also become resident on a system by default. This means it will load when the system boots up and run in the background, periodically checking files for viruses.

All virus scanning software allows the administrator to set its functions through some kind of properties dialogue. Most modern scanning software can be told which files to check, at what time, on which drive, and how often. In addition, the administrator can tell the software if it should run resident or not.

Security Considerations

Acquiring a virus or trojan on a computer system can be a big problem. The worst-case scenario is that a malicious remote user gains control of a network. Once this is accomplished, the malicious user can perform almost any task they wish. Here are some popular choices:

Spoofing: Spoofing occurs when a malicious users launches attacks on other networks and makes it look as if they are coming from any network but their own. Sometimes hackers will break into high-profile companies and "spoof" attacks from inside. This is why every now and then CNN reports that Ronald McDonald hacked into the pentagon.

Theft: Plain and simple: a user sees items on your network they like, then they steal them. For some institutions this can be a huge problem.

Vandalism: Some malicious users do pointless damage to systems by deleting files, changing configurations, or even shutting the systems down.

Unlike trojan horse programs, viruses are usually more of a nuisance than a long-term problem. Though they rarely result in an outside party having direct control of your system, they usually automate some of the evil processes, particularly vandalism. Another nuisance associated with typical viruses is their ability to spread. They spread over all sorts of paths, including disk-based media (floppy, CD-ROM), and networks, local and wide.

Making the Connection

Privacy: Spyware, or software installed on a system to keep track of what the system is doing, is often considered a trojan. Spyware is categorized this way because it is often delivered to a user who is downloading some other application from a web site. The spyware is bundled with the application and installed on the user's machine when he or she install the intended application. It gathers information about the user's activities and sends it off to an undisclosed remote location where dangerous chemicals are tested on *baby seals*!

Anonymizers: For many users, keeping anonymous on the Internet is of major importance. For those doing legitimate work, it may be for privacy reasons. For those using computers for illicit or illegal reasons, it is to keep from getting caught. People in the second category often use trojans as one of many tools to keep themselves anonymous. Once a trojan acquires control of a remote system, the malicious user can act through the "zombie" system and the source of his or her mischief will be identified as coming from the compromised system, as opposed to his or her own. When hackers launch major attacks at organizations, they often need many anonymous controlled systems in their arsenal. The process of controlling many unwitting users' computers and using them to launch a major cyber attack is known as droning.

Best Practices

Viruses and trojans comprise only one category of security threat to a computer system or network. Antivirus software is a decent tool used to guard against this threat, but nothing more. To think that running antivirus software gives your system some kind of immunity or absolute security is incorrect.

The best way to use antivirus software is as a part of a more robust, three-dimensional security philosophy. Any good security philosophy will have an established policy for dealing with viruses and malicious applications. That policy should be specific but not exist in lieu of other forms of security policies and practices.

Antivirus software has its share of caveats. For one thing, it needs to be running constantly to be effective. It also needs to constantly connect to the Internet to up-date itself, which inevitably opens your systems to more security holes. It can also be a tremendous resource hog and is quite expensive. One can take certain steps to reduce the amount of negatives associated with this software.

To reduce the total amount of resources used by antivirus software in workstation environments, consider disabling its auto-resident function. Despite what the instruction sheet may say, virus scanners can be as effective if used periodically. Running them in the background of your system will slow down everything! Instead, set the antivirus service to manual, and run the software once a day at your own discretion. This advice applies differently to antivirus server software that runs on a dedicated system. However, antivirus servers can often be safely told to check the whole network for viruses with less frequency.

Final Thoughts

Viruses are undoubtedly a legitimate threat to all computers and networks. The reality is that most of the huge database of viruses that scanners reference when searching systems will never be used. Many viruses are written, distributed, discovered, deleted and never heard from again. Furthermore, an individual is far more likely to download a file from the Internet with a trojan concealed in it than contract a virus from another user. For this reason, watching one's behavior is more effective than behaving wildly and scanning for viruses.

Chapter 38
Detecting Intrusions: Network Scanners

Programs that examine critical network systems
services for configuration errors and vulnerabilities.

Technology Overview

Every good network administrator has a set of network scanning tools in his or her
utility belt. These applications not only help find problems in existing networks, but
also are invaluable when configuring new equipment. A network scanner can quickly
determine if a machine is working properly and if the desired services are running. It
will also identify any other services that are running.

The most basic network-scanning tool is the "ping" command. When typed at a
prompt, this will send a special "echo" packet out to the target machine. Often, machines will respond by sending back the same type of packet—thus the term "echo."
Often, network administrators use this technique to determine if a machine has been
properly connected to the network. In the "security considerations" section we'll explain why this is a really bad thing.

A more advanced level of tool is called a port scanner. Network services that use
TCP/IP (Transmission Control Protocol/Internet Protocol, explained in Chapter 23)
accept incoming connections using a system called "ports." A port is simply an

additional piece of address information attached to network data—sort of like an apartment number. Most common Internet services use standard port numbers. Web servers (*Hypertext Transfer Protocol* [HTTP]), for example, listen for connections on TCP port 80. Mail servers listen on TCP port 25. Custom applications will use ports with higher numbers, such as 4009, or 63335.

A port scanner looks at each port within a specified range. The scanner notes each port that has a service actively listening. It then cross-references these ports against a database of applications to figure out the type of application that is running. Advanced port scanners will actually connect to the service to obtain additional information. For example, connecting to a web port will let the scanner figure out the type of Web server software that is running. "Nmap" is the name of one of the most popular port scanning tools.

A "fingerprinting scanner" attempts to determine the operating system of the target machine. It does this by using a number of obvious and obscure clues. Some services, such as telnet and some versions of *Secure Shell* (SSH), report the operating system and version upon connection. This is pretty obvious, but some systems are harder to figure out. Advanced techniques can also be used, such as looking for patterns in the way connections are handled. These patterns can be matched against a signature database. Current fingerprint systems can be eerily accurate with extremely little information. The Nmap network scanner has fingerprinting ability built into it.

A "vulnerability scanner" tests the target machine for susceptibility to exploits. Two approaches to this process exist. The first is to fingerprint the target machine. If the target appears to be running vulnerable software based on the fingerprint, an alert is generated. The second approach is to actually run programs that take advantage of all the known vulnerabilities. These programs usually don't harm the target machine; instead they simply note whether they were able to successfully gain access. This is a much more effective strategy, but certain types of exploits will damage the target system due to their nature. Therefore, a combination of both approaches is often the best choice for comprehensive and safe scanning. Popular vulnerability scanners include "Nessus," "SAINT," "SANTA," "SATAN," and so on.

Many of the best network scanning devices were designed to operate in the Unix environment. The following are some reasons for this:

- Historically, network scanners have evolved from hacker tools. The most valuable network servers (firewalls, *Domain Name Server* [DNS], email, and Web) generally run UNIX operating systems (Solaris, Linux, Free/Net/OpenBSD, and so on). The tools were designed to run on the victim computers.

- Hackers want their software to be as small and efficient as possible, making it easy to quickly download and install remotely. Therefore, most tools use a command line interface, which results in a smaller program that is faster and easier to develop than one with a full graphical interface. The command line interface of UNIX is ideal for this type of interaction and makes it easy to take the output of one program (a network scanner) and pass it to another (an automated exploit system).

- UNIX systems generally offer much better low-level access to network devices and most UNIX systems come with a large number of programming tools pre-installed.

This means that you'll need access to a UNIX machine in order to run many of the best network scanners. Some effort has been made to translate these programs to Windows, but the process is difficult and the final result is usually less effective than the original UNIX version.

How Network Scanners Work

Network scanners usually require a number of parameters to operate. The most important is the target machine or network. While some scanners focus on a single machine, most can be configured to scan an entire network segment.

Port scanners work by attempting to open connections to each port on the target computer. There are two types of ports: *Transmission Control Protocol* (TCP) and *User Datagram Protocol* (UDP). Most services run on TCP, but some use UDP. TCP is easier to scan because it requires an initial handshaking process regardless of the application. If an application is waiting on a TCP port, the scanner's connection will initiate the processes.

Scanning UDP is much more difficult. No connection is established; therefore it's not easy to see if an application is actually running on the port. It's possible to test the presence of a particular application by sending data that will generate a response from the application, but this approach requires extensive custom programming and is rarely used.

One trick that can be used for detecting both TCP and UDP services is the absence of a response. The manner in which a connection is refused can indicate whether a service is present or not.

Security Considerations

Ping Is a Bad Diagnostic Tool: The proper way to test if a server is working is to test the service in question. Pinging a web server means nothing. Many dead servers will still respond to pings. By creating a system for testing the actual service, you're guaranteed to get a more reliable status report. Furthermore, allowing a system to receive and generate ping messages makes the machine vulnerable to a number of *Denial of Service* (DoS) attacks that can be conducted using this tool.

Reality Check: Vulnerability scanning only can search for known vulnerabilities with well-understood exploits. Not every vulnerability has a simple exploit. For example, TCP connections are fundamentally vulnerable to being hijacked, but this usually requires a large amount of knowledge and is often very difficult to accomplish in practice. No well-known "quick and dirty" programs that can do this reliably against modern machines are available. Skilled hackers might be able to write a custom program to attack a specific connection.

Fingerprinting Can Be Wrong—Wildly Wrong: Fingerprinting tries to make guesses based on information that can be altered by many different factors. For example, TCP sequences can get completely skewed by *Network Address Translation* (NAT) devices. Transparent firewalls can create non-standard responses for machines behind the firewall.

Making the Connection

Connecting Networks: The topology of a network will have a large impact on the effectiveness of network scanning devices.

Hardening Networks: Properly hardening a network will make it difficult for network scanners to probe systems and network devices.

Best Practices

Learn How To Use Them: These are the same tools that hackers will use against your own network. Use them and understand how they operate. Anything that these tools show you, they'll also show a hacker.

Know What They Look Like: Skilled hackers will use stealth techniques to hide the fact that these tools are running on your network. Use a packet sniffer to examine the network traffic when one of these tools is running in stealth mode. Eventually, you'll be able to spot telltale signs of an unauthorized network scan.

Tuning Intrusion Detection Systems: Many of these *Intrusion Detection Systems* (IDSs) will look for network scans as an indication of a pending hack attempt. See if you can evade the IDS using network-scanning tools. If you can, then you need to make your IDS more sensitive. If you can't, then either you don't fully understand how the tools work or you've effectively tuned the IDS (assuming you're also not getting a million false alarms).

Final Thoughts

Network scanning tools are quite powerful. They do not replace the need for solid network engineering skills, however. Useful inferences can be made using programs like Nmap, but education is the key to great insight. Network scanners will provide a networking guru with much more than a networking newbie.

Chapter 39
Detecting Intrusions: Network Sniffers

A sniffing device captures network traffic for the
purpose of analysis and intrusion detection.

Technology Overview

You're sitting at your desk, surfing the Web for business-related sports scores when
you notice the network becoming sluggish. Then it dies altogether. A few minutes
later, it's up again. What just happened? Looking over at the network rack, you can't
help but admire the Christmas decorations. Wait—it's June! Those aren't decorations, those are red warning lights flashing next to the green happy lights. Clearly,
something has gone wrong, but what is it?

Nine times out of 10, the cause of this type of problem gets traced back to a single malfunctioning computer or a single software failure simultaneously plaguing a
number of systems. How can you identify the culprit? Looking at flashing lights isn't
going to tell you much. What you need to see is the actual data traffic moving around
your network. Network data capture and analysis tools, also known as sniffers, are
designed to help.

The most basic network sniffing tools simply capture all of the data on the network and store it to a file for later analysis. Most will also summarize the network traffic in a human readable format and can display the summary in real-time.

More advanced tools can break down the traffic by factors such as the type of application generating the data and the source or destination system. These tools might provide graphic visualization and may also be capable of identifying malicious traffic patterns. Some can even interpret the application data itself, extracting application specific information such as commands and login information. In the case of web traffic, this could include web addresses, usernames, and passwords.

Network sniffing devices can also be used to detect intruders. Unfamiliar network traffic can indicate the presence of intruder activity or an active backdoor on a system. For this reason, sniffing devices are a critical part of many intrusion detection systems.

This capability for data analysis sounds like it would be incredibly useful to hackers. It is. A major hacking goal is to control or establish a sniffing device on a network. This provides a continuous feed of critical information, such as captured passwords, confidential files and email messages, and so on.

Sniffing devices leave hackers in a quandary. Sniffers are great hacking tools, but are also great intrusion detection tools. If a sniffer is on the network, a network administrator has a good chance at detecting the hacker.

Always a step ahead of the cat, the mousy hackers created the anti-sniffer. This is a program that can detect the presence of sniffing devices on a network. If a sniffer exists, it's either part of an *Intrusion Detection System* (IDS) or something that was installed by a previous hacker. A hacker can use evasive techniques to invade the network without being detected if a sniffer is being used. If no IDS is present, the hacker can brashly overrun the network defenses, raping and pillaging systems along the way with impunity.

Anti-sniffers do provide benefits to the network administrator. An anti-sniffer can be used to detect the presence of unauthorized sniffing devices on the network. The presence of such a device would be a good indication that the network had been hacked. Of course, hackers might also design an antisniffer detector, which would prompt network administrators to develop an antisniffer detector-detector. The good news is that this sort of one-upmanship seems to have ceased at the anti-sniffer level.

How Sniffers Work

The basic function of a sniffing device is to capture all of the traffic that passes across a network segment. A sniffer is essentially a malfunctioning computer. Normally, a computer will only record packets that have the appropriate *Internet Protocol* (IP) address and Ethernet hardware address in the header fields. Other packets will be ignored by the network interface. A sniffing device changes the settings on the network card to record every packet that passes by, regardless of the destination address. In this situation, the network card is said to be in "promiscuous" mode.

Creating a device capable of recording network traffic is only half the battle; the other half is finding the right spot on the network to place the device. A sniffing device can only see the data that's present on its network segment. For the majority of locations on the average network topology, this may amount to little, if anything. The illustration shows how different placements and different network topologies affect the information available to the sniffing device.

**Chapter 39
Detecting
Intrusions:
Network
Sniffers**

Network Sniffer Deployment Options

Illustration by SageSecure

In order for an intrusion detection system to be successful, sniffing devices must be properly deployed throughout the network. Some locations are better than others, depending on the type of attack being detected. For the best results, an intrusion detection system should have a number of sniffers in different locations. Each sniffer should focus on the attacks that can be best detected from its vantage point.

❶ *Outside of the Network*

This is the best place to detect and analyze external threats such as: Denial of Service (DoS) attacks, spoofing attacks, attacks directed at firewalls/routers, and compromised systems in the upstream provider's network.

❷ *Within or After the Firewall*

Here you can spot attacks that have managed to get past the first line of network defense. Attacks that can be detected include: viruses, trojans, attacks at servers, and application specific attacks, such as web browser exploits.

❸ *Internal Routers and Switches*

The first two sniffer locations can detect external threats. But what about attacks from within? Sniffers placed at junction points where data travels between servers and workstations can detect the activities of hackers who have already gained access to the inner network.

❹ *Inside Server/Workstation LANs*

A sniffer positioned within a LAN segment can detect attacks between internal machines. This might catch the spreading of a virus, the actions of a rogue employee or the movements of an external hacker.

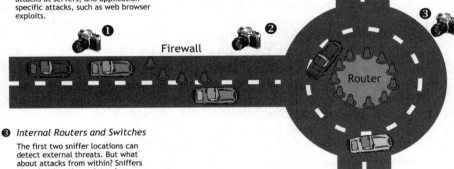

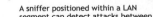

Figure 39-1

Network Taps

Much like the concept of a telephone wiretap, it's possible to tap into a network connection. A tap device allows a computer to listen in on a conversation between two machines without either machine detecting the tapping computer's presence. Network administrators can use taps to prevent hackers from detecting their intrusion detection systems. Hackers with physical access to the network might use a tap to hide the presence of an on-site sniffing device.

A number of companies sell tap devices. These are very professional and are intended for network administrators. It is also possible to create a network tap by hand. This involves strategically altering a network cable by disabling the "transmit" part of the cable. The result is that data goes in, but nothing ever comes out. Hackers generally don't use this technique because it gives up the ability to remotely control the device. It's more of a poor man's version of the commercial network tap devices. Handmade cables also don't work directly with switches, and rarely work on 100 Mbps networks.

In practice, the capabilities of a sniffing device depend on who installed it and for what purpose. If a network administrator is installing a device as part of an intrusion detection system, the topology can be altered to maximize the effectiveness of the sniffing device. Special switches can be used to allow a sniffer to monitor all of the traffic across a switched segment directly from the switch itself. Devices also can be placed between firewalls and routers, capturing all Internet traffic. Network taps, described in the sidebar, can be used to securely add sniffing devices to critical parts of the network in a manner that is undetectable or exploitable by hackers.

On the other hand, a sniffing device installed by a hacker is probably going to be in a less ideal location on the network. Hackers often convert a workstation or server into a sniffing device. The particular machine chosen may not have access to very much information, especially if it's connected to a switch. For this reason, hackers have devised numerous techniques to divert additional information to their device. The implications of this will be described in more detail below.

Security Considerations

Hubs: If you're using hubs, don't. Switches have dropped significantly in cost—now nearly the same price as hubs. The problem with hubs is that they make sniffing incredibly easy. Network traffic is broadcast to every device on the hub. A sniffer will have full access to all of the communications that pass across the hub. Switches create a direct connection between the source and destination machine. Other systems connected to the switch will only see directly relevant traffic.

Bad Switches: Cheap switches, or poorly configured switches, can be tricked into acting like hubs. Many switches that can be remotely configured have default passwords that are rarely changed. A hacker can log into the switch and set it to act like a hub. Hackers can also exploit vulnerabilities in switch designs to force hub-like behavior.

Traffic Diversion: Even with a properly functional switch, hackers can use many techniques to divert traffic to their sniffing device. Fundamental vulnerabilities in the *Address Resolution Protocol* (ARP) (see Chapter 6) allow hackers to trick other computers connected to the switch into talking directly to the sniffing device. The sniffing device then passes on the communication to the intended recipient. This is known as a man-in-the-middle attack. Other protocols also have weaknesses, allowing the hacker to trick machines into thinking that the sniffing device is actually the router.

Network taps: Hackers who gain physical access to a network may install a sniffer using a tap device. This sniffer will be very difficult to detect using network analysis tools. Only a physical sighting of the device will alert an administrator to its presence. This is yet another reason why physical security is a critical factor when securing a network.

Making the Connection

Connecting Networks: The topology of a network will have a large impact on the effectiveness of sniffing devices.

Hardening Networks: Sniffers can't do much with encrypted traffic. Properly hardening a network will safeguard much of your data against the possibility of sniffing-based intrusion.

Best Practices

Switch Configuration: Make sure the switches on your network can be remotely configured. Change the password on the switches to prevent hackers from altering the settings. Periodically check the configuration of the switches to ensure that everything is the way you left it.

Intrusion Detection: If your security philosophy calls for the use of a complete intrusion detection system, then one or more sniffing devices will certainly be part of the package. Upgrading critical switches to network tap switches will allow you to maximize the potential for information collection.

Network Diagnostics: Even if you don't use an intrusion detection system, it's incredibly useful to have a laptop sitting around that can act as a network diagnostic device. Installing sniffing software on a laptop is ideal because you can easily move the laptop around the facility, connecting it to different network segments and monitoring performance. A low-end network tap (or a homegrown solution) will help in sniffing switched environments.

Anti-Sniffers: It's also useful to periodically run sniffer-detection software on your network. Most modern intrusion detection systems incorporate this functionality. If you don't have an IDS, you can download one of a number of programs for detecting sniffing devices.

Final Thoughts

Sniffing a network can provide a network administrator insight into the type of traffic that routinely flows across a network. Sometimes,observing a day's worth of network traffic is similar to watching people go by while sitting at an outdoor café in Paris – there's a lot of interesting activity to watch while sipping from a delicious café au lait. Unfortunately, most of the time it's more like sitting in a dark room with Twinkies and a Coke, feeling isolated and lame. But hey, it builds character, right?

After several weeks of routine sniffing, Parisian style or not, patterns will certainly develop. Once an administrator becomes familiarized with these patterns, it will make it easy to pick out unusual traffic, which could indicate malicious intruder activity.

Chapter 40
Detecting Intrusions: Logging and Analysis

Gathering and analyzing diagnostic status
information from network devices and software.

Technology Overview

Logging could be the most boring concept ever. It's fundamentally wasteful—billions of bytes of data are put into digital filing cabinets, never to see the light of a monitor. You know that guy who keeps a spare copy of every receipt organized alphabetically in a file? That guy is a logger. His friends? Meet Bobby Paperclip, Johnny the Stapler, Sara Hole Punch, Frank File Cabinet, and Steve Super Glue. Who wants to be a logger?

Boring or not, logging is the most important concept in intrusion detection and recovery. Without logs, the only way to know about a problem is to observe it happening (or it's aftermath).

Logging can be used to:

- Make sure things are going smoothly, according to routine.
- Figure out what went wrong.

- Determine performance, effectiveness, and so on.
- Hold individuals accountable for actions.
- Build historical records that can be useful during audits.

Once you start logging, you'll begin to realize that logs are very valuable and useful in many situations. You're might even want to start logging right away. That's a great idea, but try not to get too friendly with the office supplies.

How Logs Work

A log is a record of an event or a snapshot of information. Most applications and operating systems are specifically designed to record certain events and system status. Log files are a collection of log entries.

The following are the four basic parts to the logging process:

Generation: An important event happens and is recorded. In order for an application to generate a log, it needs to be able to recognize and report on important events. Not every application has been written with logging in mind. Some applications simply don't provide any opportunity to gather runtime information. Other applications can report on nearly anything that happens.

Collection: During the collection phase, logs are gathered together. Different applications log information into different locations throughout a system. Many separate systems might be running, each generating their own logs. It's critical to securely gather and centralize these log files. If a system goes down, its historical logs will still be accessible from the secure archive of logs. Furthermore, log analysis is usually most effective when all of the logs are centralized.

Filtering: Logging generates a lot of information because it's hard to predict ahead of time what will be needed. Once a crisis happens, only the logs pertinent to the problem at hand are necessary. Filtering removes the logs that aren't of use to the analysis at hand.

Analyzing: Finally, this is where practical sense is made of the logging information. Many different types of analysis can be performed on logs. Determining the cause of a problem involves looking for entries that shouldn't be in the logs, or for things that should be there, but aren't. When using log files to gauge performance or effectiveness it means looking at only the normal data and ignoring the strange stuff that shouldn't be there.

Techniques for Secure Logging

Hackers are smart. They will attempt to cover their tracks whenever they invade a system. They know that logging systems will probably record their entry into the network. Good hackers go through the logging systems and ensure that records of

Part XII Detecting Intrusions **407**

**Chapter 40
Detecting
Intrusions:
Logging and
Analysis**

their activities are wiped. Hackers will also snoop through log files looking for information that could help them further compromise a network.

The way to avoid this problem is to increase the security of the logging process. There are two basic approaches to secure logging.

Batching: Log data accumulates locally and then batches of logs are sent to the remote repository at regular intervals. This is the easiest system to implement as most programs log information locally. A number of systems are available that can securely transfer log files to a remote repository on a periodic basis. The problem is that an intruder can examine and/or modify the logs before they get batched to the remote log system.

Secure channel: The better approach is to create a secure channel from the application directly to a secure log repository. Not every application supports this process. With some systems, a creative trick can be used to capture the log as it's written to the local file and pipe it across the network to the remote logging facility.

Analysis Techniques

Once logs are collected and centralized, they can be analyzed. A good analysis strategy uses a combination of automatic and manual analysis techniques. Automatic analysis can raise flags that should be checked when manually analyzing the logging data. A brief description of each follows:

Automatic Analysis: A number of commercial and freeware software programs exist that are capable of analyzing and correlating logs from many different sources. Most *Intrusion Detection Systems* (IDSs) are fundamentally automatic log analysis tools. The idea is that these software programs can scan through millions of lines of log files and spot trends and suspicious log entries that no human could find. These systems often match log files against databases of signatures or patterns that various hacking/virus/Trojan intrusions leave behind. Automatic systems are limited, however, as they can only do what they've been programmed to do. An experienced network administrator may be able to see things that an automated analysis tool was not programmed to recognize.

Manual Analysis: Going through log files by hand can be either incredibly tedious or more effective than the best automated system available. It all comes down to the skill of the analyst. With the right tools, sifting through millions of log entries by hand is relatively easy. Having a good command of powerful searching tools is a must. Being able to write custom scripts is also helpful. For example, the authors frequently examine network packet dumps by eye. Our experience allows us to pinpoint traffic anomalies within a few seconds. We have even been known to accurately diagnose network problems based on how fast the activity lights on a switch were blinking. Then again, we've also been known to nearly electrocute ourselves by attempting to plug RAM directly into a running computer.

Security Considerations

Valuable Information in Logs: Log files can contain lots of interesting bits of information. In some cases they intentionally record juicy stuff. In others it's purely unintentional. For example, web logs can end up gathering far more data than anyone would anticipate due to buggy browsers that give out too much information. Usernames, email accounts, passwords—all this and more may be sitting in "innocent" log files scattered throughout a system. Consequently, logfiles are major hacker targets.

Capturing Logs In Transit: When logs are centralized, they might travel across the net. A smart hacker will follow the path of the log packets to the true central logger. The hacker will then either capture the logging packets for later analysis or simply attempt to compromise the log server. Encryption and dedicated logging connections are solid techniques for protecting log files in transit.

Overconfidence: The analysis system says all is fine, so all is fine, right? Some of the more advanced products on the market are marketed as infallible. In reality, dedicated hackers can get by these systems. Even if the system is good, chances are it's been partially disabled to allow more "convenient" use of the network. A good human-based analysis on a regular basis will do wonders for network security and performance.

False Positives: The flip side of the coin is the ever-alarmed analysis system. The minute somebody in accounting opens up the wrong web site—WHAMMO! Sirens blaring, net admins flying out of their bat caves, total pandemonium, followed by excessive cursing about the highly sensitive analysis system. Eventually, somebody unplugs the damn thing by "accidentally" tripping on the power cord and makes employee of the month.

Making the Connection

Hardening Networks: Log files are used extensively during the process of tightening a network. When firewall rules are written to restrict more and more inbound traffic, overzealous network administrators can accidentally lock the front door. Log analysis can help determine what pathways in are blocked that should be opened and vice versa.

Best Practices

The biggest trick to successful logging is gathering the right amount of information. If you gather too little, the critical information you need might not be there. If you gather too much, you might never find the needle in the silo full of hay. The following are points worth considering:

Encryption: Ideally, all logging data should be encrypted. This prevents hackers from modifying or reading the log data.

Transit: Get your logs to one place securely. Make sure all of your transfer channels are encrypted or otherwise secure. Many high security systems send logs via one-way cables to the remote logging system.

Secure the Central Logging System: A central logging system is a prime target for hackers. Make sure they can't get onto it. One technique is to ensure that the system can't be remotely accessed. This might mean keeping the system off the network. Logs can be sent to the log system over a secure serial link that is only capable of writing whatever it receives. It would be impossible to control the log computer over the serial link. Another trick is to write the logs to write-once media such as a CD-R or DVD-ROM giving the hacker no chance of deleting or modifying logs. If you're not into rainforests and own a sizeable storage facility, you can also dump logs straight to a printer.

Consistency: Getting all of the logs into a standard, consistent format can be very helpful when analyzing information. Some of the more advanced analysis tools can cross-correlate different types of logging information to spot potential problems. For example, firewall logs can be cross-referenced against Web server logs.

The UNIX system logger, syslog, is one of the most widely used and understood logging systems. Most applications can be configured to log to syslog. Even Windows applications can log to syslog. This ensures that syslog is the only system writing the consolidated log files, which is good for consistency.

Baseline Data: It's incredibly useful to have some sort of baseline information on the systems being logged when everything is healthy. This acts as a control in the evil experiment known as hacking prevention. A number of analysis tools will actually detect problems by comparing information against such a baseline.

Divide and Conquer: Separate your logs into two groups: the things you expect to see, and everything else. First look through the "everything else" and examine the weird stuff that you can't explain. Then simplify the things you're expecting and look for what's missing. If you expect daily logs from a certain application, get worried if one of the days is not there.

Erasing Tracks: Put yourself in the hacker's shoes. If you were a hacker, what tracks might you leave, and how would you erase them? These are the things you should be securely logging. A secure log will not only show the hacker in action, but it will also show the hacker attempting to erase tracks. Here's a trick: Hackers will attempt to compile their attack software once they break into a machine. Log all attempts at running the compiler and capture the input. This will expose the hacker and his or her trojan programs.

Final Thoughts

Implementing and maintaining a system for logging is a lot like sliding down a long razor blade into a pool of concentrated malt vinegar. It's normally something you would rather not do. Once you get into the habit of logging though, it can be quite uh, never mind, bad analogy. Logging is worth the trouble, but it will take some time to get adjusted to storing, reading, and interpreting logs. It may make sense to start logging only one machine first to get comfortable with the process and then add machines to the central log over time.

Index

Note: Boldface numbers indicate illustrations.